Open-Channel Microfluidics (Second Edition)

Fundamentals and applications

Online at: https://doi.org/10.1088/978-0-7503-5507-0

Open-Channel Microfluidics (Second Edition)

Fundamentals and applications

Jean Berthier
Affiliate Professor, Department of Chemistry, University of Washington, Box 351700, Seattle, WA 98195, United States of America

Ashleigh B Theberge
Assistant Professor, Department of Chemistry, University of Washington, Box 351700, Seattle, WA 98195, United States of America

Erwin Berthier
Affiliate Assistant Professor, Department of Chemistry, University of Washington, Box 351700, Seattle, WA 98195, United States of America

With contribution from

Mark Weislogel

IOP Publishing, Bristol, UK

ISBN 978-0-7503-5507-0 (ebook)
ISBN 978-0-7503-5505-6 (print)
ISBN 978-0-7503-5508-7 (myPrint)
ISBN 978-0-7503-5506-3 (mobi)

DOI 10.1088/978-0-7503-5507-0

Version: 20240401

IOP ebooks

British Library Cataloguing-in-Publication Data: A catalogue record for this book is available from the British Library.

Published by IOP Publishing, wholly owned by The Institute of Physics, London

IOP Publishing, No.2 The Distillery, Glassfields, Avon Street, Bristol, BS2 0GR, UK

US Office: IOP Publishing, Inc., 190 North Independence Mall West, Suite 601, Philadelphia, PA 19106, USA

To Noam, Eden, Mila, Myron, Zoya, Thurston, and Alphonse

Contents

Preface

The rapid advancement of open microfluidics in recent years has prompted the need for updating our inaugural book, "Open-Channel Microfluidics," initially published in 2019. Additionally, we aim to expand the scope of our earlier publication, "Open Microfluidics," released in 2016, to encompass the examination of the dynamics associated with open capillary-driven microflows.

The second edition delves into capillary flow across a diverse array of geometrical channels and solid structures. We explore novel concepts pertaining to channels and guides tailored for microflows, drawing parallels between these designs and structures observed in nature. Additionally, a new chapter has been incorporated into the book, authored with assistance from Mark Weislogel, focusing on the applications of Open Microfluidics in spacecraft.

Acknowledgments

We are very pleased that this second edition includes a contribution by Dr Mark Weislogel (PhD, Principal Scientist, IRPI LLC, Wilsonville, Oregon, USA) who wrote the chapter 'Open-Capillary Fluidics Aboard Spacecraft,' highlighting the importance of open fluidics for spacecraft and space exploration.

The authors also thank the Chemistry Department of the University of Washington and the PhD students of the Theberge laboratory: Jodie Tokihiro, Anika MacManamen, Wan-Chen Tu, Jing Lee, Ulri Lee, Jian Wei Khor, Samuel B Berry, and Ashley Dostie for their experimental work, which is abundantly used to illustrate many developments presented in this book.

Author biographies

Jean Berthier

Jean Berthier
Affiliate Professor
jean.berthier@wanadoo.fr; berthj2@uw.edu

Jean Berthier is presently an Affiliate Professor at the University of Washington. He received a MS in Mathematics from the University of Grenoble, an engineering diploma from the Institut National Polytechnique in Grenoble and a PhD from the University Pierre et Marie Curie in Paris. After spending four years at Sandia and Los Alamos National laboratories focused on the reactive interaction between liquid and gases, he joined the CEA-Leti in Grenoble, France where he was involved in the development of microfluidic solutions for liquid–liquid extraction, bio-encapsulation of living cells, capillary solutions for portable point-of-care devices and development of "open micro-fluidics" theory and applications. In 2018 he joined the University of Washington in Seattle. He is the first author of the book "Microfluidics for Biotechnology" published by Artech House (second edition 2010), he is the author of the book "Microdrops and Digital Microfluidics" published by Elsevier (second edition 2012), and the first author of the books "The physics of micro-droplets" and "Open Microfluidics" published by Scrivener-Wiley Publishing in 2012 and 2016. He is the author of many publications in scientific journals, conferences articles and patents.

Ashleigh B Theberge

Ashleigh Theberge
Assistant Professor
abt1@uw.edu

Ashleigh Theberge is an Associate Professor of Chemistry at the University of Washington and Adjunct Associate Professor of Urology at the University of Washington School of Medicine. She received a BA in Chemistry from Williams College and a PhD in Chemistry from the University of Cambridge, UK with Wilhelm Huck in droplet-based microfluidics. During her graduate work, she was a Visiting Scientist with Andrew Griffiths at the Université de Strasbourg, France. She completed her postdoctoral fellowship in Biomedical Engineering, Toxicology, and Urology with David Beebe, William Ricke, and Wade Bushman at the University of Wisconsin-Madison. She joined the faculty at the University of Washington in 2016, where she leads the Bioanalytical Chemistry for Medicine and the Environment group with Erwin Berthier. This group studies chemical mechanisms underlying urologic disease, kidney disease, asthma, and infectious diseases and develops analytical chemistry tools to advance medicine, including open microfluidic systems for cell culture and analysis. A related emerging area of research of this group centers on developing new platforms for remote biospecimen collection and analyte preservation for diagnostics and clinical research.

Erwin Berthier

Erwin Berthier
Affiliate Associate Professor
berthier@uw.edu

Erwin Berthier is presently Affiliate Associate Professor at the University of Washington in Seattle. He is also cofounder and CTO of Tasso Inc. a Seattle-based company developing patient-centric, distributed, health technologies. He received a Diplome D'Ingenieur in Fluid Mechanics from ENSTA (Ecole Nationale Supérieure des Technologies Avancées) in Paris, a Masters of Electrical Engineering from the University of Canterbury (New Zealand), and a PhD in Biomedical Engineering from the University of Wisconsin in Madison. He is the author of over 70 publications in peer-reviewed scientific journals and 12 patents with technologies that have been licensed in current commercialized products. He is also a co-author of the book "Open Microfluidics" published by Scrivener-Wiley Publishing in 2016. Erwin currently works closely with Ashleigh Theberge as a co-PI in the Bioanalytical Chemistry for Medicine and the Environment group in the Department of Chemistry at the University of Washington. His current research interests focus on advancing the theory and applications of open microfluidics as well as distributed sensing technologies for healthcare applications, agriculture, and public health.

Foreword to the first edition

In 2012 I joined the University of Oxford as associate Professor in the department of Engineering Science. The aim of my research is to develop microfluidic technologies with a low barrier to entry from a biologist's perspective. I met Jean Berthier in 2018, when he was the external examiner for an Engineering Science Doctor of Philosophy thesis for a member of my group. For many years, I had been an advocate of his microfluidic books within my research group. In fact, there were several occasions when his books provided the required inspiration and direction for solving problems that my research group struggled with. Hence, his books have practically become reference texts for my lab.

This book, with co-authors A Theberge and E Berthier, represents a timely addition for anyone involved in microfluidics at a technical level. It boasts a pedagogy that will appeal to student, professor, and industrialist alike, incorporating a seamless transition from fundamental physics to innovative implementations. This encompassing text is necessary; open-channel microfluidics behave in nonintuitive ways due to interfacial forces dominating gravitational ones. In such systems, fluid interfaces, rather than solid materials, confine fluids and give them shape. This uniqueness provides both challenges and opportunities relative to classical microfluidics, and hence practitioners require a breadth of knowledge to identify and exploit such opportunities.

As a field, microfluidics continues to deliver many fascinating proofs of concept but, despite significant public and private investment, translation to industry has fallen short of product-level expectations set several decades ago. However, the wide range of basic operations demonstrated using open-channel microfluidics combined with its advantages over classical microfluidics will leave the reader with an exciting platform for the future. I believe this book will serve to educate the existing and next-generation microfluidics communities and thus provide the foundations upon which future open-channel microfluidic applications will emerge. These approaches could be the key to unlocking the promised potential, economic and societal, of manipulating tiny volumes of fluid.

Professor Edmond Walsh
CEO of iotaSciences
Associate Professor at University of Oxford: currently on leave

Foreword to the second edition

My journey in microfluidics commenced in Professor George Whiteside's group at Harvard, marked by the advent of soft lithography, which led to the explosion of polydimethylsiloxane (PDMS)-based microfluidic devices. My research endeavors have been centered on the integration of cells within microfluidic channels over a span of numerous years, notably through the creation of organ-on-a-chip (OoC) devices and microphysiological systems (MPSs) employing conventional 'closed microfluidic channels.'

In order to surmount a multitude of engineering challenges inherent in this pursuit, our research group has systematically embraced the 'open microfluidic' design paradigm. Today, my research focus is the development of high-throughput MPSs, wherein the potential of the open microfluidics approach is harnessed to its fullest extent.

The course of our group's pursuit to comprehend fluid behavior in open microfluidic systems has been significantly influenced by the seminal work of Jean Berthier and colleagues Kenneth Brakke and Erwin Berthier, 'Open Microfluidics,' which imparts invaluable insights into the fundamental principles governing these intricate systems. Within our group, the first edition stands as an indispensable resource and is deemed essential reading for all incoming lab members.

The forthcoming second edition of 'Open Microfluidics' not only builds upon the foundational concepts of open microfluidics but also explores a diverse array of applications spanning from biotechnology to space applications. It is poised to serve as an invaluable resource for researchers, students, and individuals intrigued by the dynamic intersection of microfluidics with various disciplines. The second edition offers a comprehensive guide to both the fundamental principles and advanced applications of open microfluidics, empowering its readers to actively contribute to the development of next-generation microfluidic systems; it thereby holds significant implications for human health and biomedical research.

Professor Noo Li Jeon
Department of Mechanical Engineering
Seoul National University

Introduction: open-channel microfluidics and open microfluidics

Open microfluidics is an emerging technique, a subset of microfluidics, which is useful in many domains, such as biology, biotechnology, space science, and thermics.

Although the term did not exist at the time, open microfluidics (OM) may be considered to have started with the Concus–Finn theory of capillary filaments [1, 2]. These investigations were done in the context of space studies. More recently, solder flows in open V-grooves were developed to produce microscopic conductors [3–5]. At the same time, the development of capillary filaments for space applications continued [6, 7]. Recently, the emergence of biotechnology has boosted microfluidics in general and OM in particular. The name 'open microfluidics' is currently used [8], even if the domain is sometimes denominated as 'open-surface microfluidics' or 'free-surface microfluidics' [9].

The definition of OM is still approximate. It varies in the literature, depending on the authors. One definition of OM could be: fluid flow at the microscale that has at least one free boundary (with air or vapor) other than the advancing meniscus. Figure I.1 summarizes the different subdomains of OM. 'Sessile droplet microfluidics' is placed inside a dotted circle because it does not strictly belong to OM but is sometimes referred to as such by some authors.

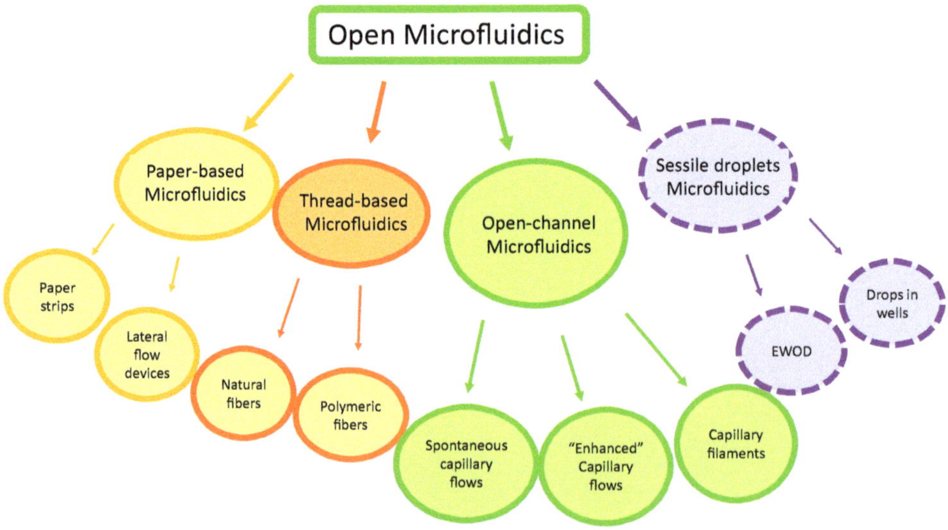

Figure I.1. The different domains of open microfluidics for biotechnology. Note that EWOD stands for ElectroWetting On Dielectric.

Let us first present the different domains listed in figure I.1.

I.1 Paper-based microfluidics

Paper has long been known to be prone to wicking due to the hydrophilic properties of cellulose fibers. Morgan and Wilson reported that, 2000 years ago, a very primitive form of chromatography was performed in Egypt using papyrus: bilirubin, a substance contained in urine, was used as a red dye on papyrus leaves [1]. Later, in the nineteenth century, the French physicist Gay–Lussac used sunflower paper to measure the acidity of liquids [2]. More recently, a formulation for the concept of paper-based detection systems was presented by Yagoda in 1937 [3]. Today, the most common portable diagnostic devices use paper as a capillary material.

Two slightly different types of paper-based devices are used for fluid testing: paper strips and lateral flow devices. Paper strips are slender pieces of paper fiber in which wicking is nearly unidirectional, in contrast to the two-dimensional flow present in lateral flow devices, in which hydrophobic walls are implanted to confine and guide the liquids (figure I.2).

Imbibition in a paper matrix can be considered to be open microfluidics, because the flow partly contacts the surrounding air. Wicking in a paper matrix is governed by the Lucas–Washburn–Rideal (LWR) law in conjunction with an experimentally adjusted coefficient that accounts for the porosity and morphology of the fibers in the matrix [5–11]. Once wetted, the diffusion of species in the paper is governed by Darcy's law [12].

Paper-based devices are fast evolving and their use for medical diagnostics is increasing. Many applications of micro paper-based analytical devices (μPADs) and labs-on-paper have appeared in recent years, from enzyme-linked immunosorbent assay (ELISA) tests [13–15] to recent DNA amplification [16]. On the other hand, cell culture devices are appearing, based on a technique called cells-in-gel-in-paper (CiGiP) [17].

On the one hand, paper-based systems have the advantages of being portable, autonomous, low-cost, and compatible with telemedicine. On the other hand, their sensitivity is still below that of conventional laboratory microsystems, even if

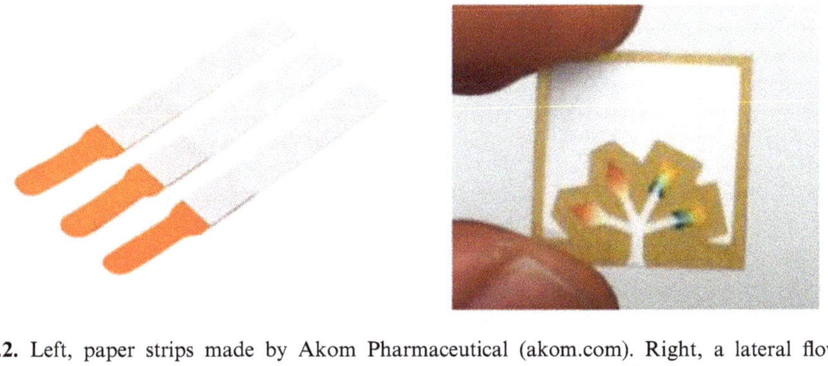

Figure I.2. Left, paper strips made by Akom Pharmaceutical (akom.com). Right, a lateral flow device, reprinted with permission from [4], Copyright (2008) American Chemical Society.

considerable progress has recently been made. However, avoiding the immobilization of biological targets in the paper matrix is a major challenge.

We do not present paper-based microfluidics in this book. There is an abundant literature and excellent reviews on this subject [18–21].

I.2 Thread-based microfluidics

Thread-based microfluidics uses the wicking of fiber bundles or yarns to guide the motion of liquids. The principle is similar to that of paper matrix wicking. However, paper-based substrates are assemblies of random fibers, while threads are constituted of more or less aligned or woven fibers (figures I.3(B) and (C)).

Threads may be natural, such as cotton, or synthetic, such as polymers or nylon. Natural fibers such as cotton or cellulose are usually very hydrophilic, while synthetic fibers need hydrophilic treatment, usually achieved by oxygen plasma exposition [24–26]. The effect of O_2 plasma treatment is to increase the surface energy of the solid structure, so that the surface becomes hydrophilic (figure I.4(A)).

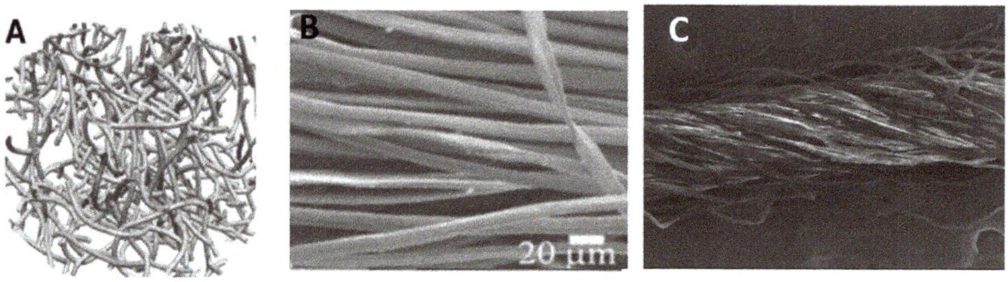

Figure I.3. Different types of fiber matrix: (A) a random matrix corresponding to paper; (B) aligned fibers found in yarns, reprinted from [22] with the permission of AIP Publishing; (C) a thread made of twisted fibers, reprinted from [23] with the permission of AIP Publishing.

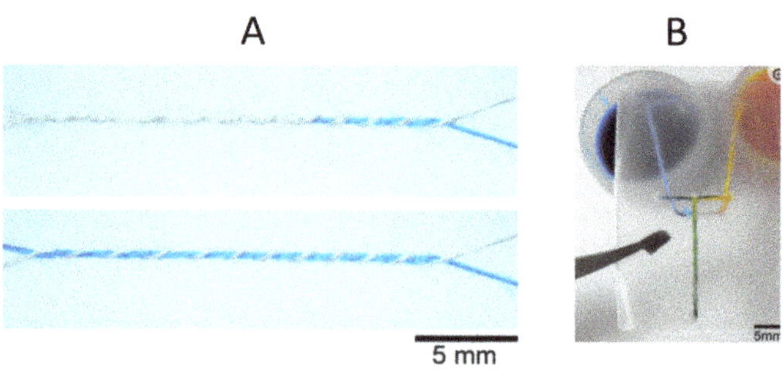

Figure I.4. (A) Twisted treated and untreated threads showing the hydrophilic and hydrophobic properties of the two threads, reprinted with permission from [27] Copyright (2010) American Chemical Society; (B) capillary flow in cotton yarns resulting in the mixing of the colored aqueous liquids, reprinted from [23] with the permission of AIP Publishing.

In a similar experimental approach to that taken for paper-based microfluidics, the dynamics of the capillary flow in threads is governed by the LWR law, except that the radius in the LWR law is replaced by an adapted characteristic dimension.

The spontaneous flow condition (SFC) for aligned fibers was derived in [28]. It shows that capillary flow may occur at a contact angle as large as 70° if the thread is composed of more than 30 densely packed fibers.

Capillary flow in threads is one-dimensional, but threads are flexible and they can be knotted together, which gives them three-dimensional properties. Thread-based microfluidics (TMB) has many applications in biology and biotechnology, for example passive blood–plasma separation using fibers coated with immunoglobulin [25] and potassium arrays based on ionophore extraction chemistry for the optical recognition of potassium [29]. In medicine, smart bandages have been developed to monitor the healing of skin wounds [30, 31] or recovery from surgery [31, 32]. Tissue engineering also makes wide use of TBM to achieve the seeding and immobilization of cells which are left to proliferate on the surface of the woven substrate [33].

This book does not treat the case of thread-based microfluidics. There is an abundant literature and excellent reviews on the subject [34–36].

I.3 Sessile droplet microfluidics

Some authors have referred to and published the details of systems using sessile droplets or droplets under the denomination of 'open microfluidics.' These configurations can be considered to be OM even if the liquid volume of the flow in contact with the surrounding air is very small.

There are many different types of sessile droplet microfluidics. Open electrowetting on dielectric (EWOD) is one of them. In such systems, droplets are moved from one electrode to another by an electric force applied asymmetrically to the interface of the droplet [37, 38]. Another example is the use of sessile droplets moved by magnetic fields acting on magnetic beads attached to biologic targets [39] (figure I.5).

Sessile droplet behavior has been widely investigated. There are many books and reviews available in the literature [38–40]. Hence, we do not consider sessile droplets in this book.

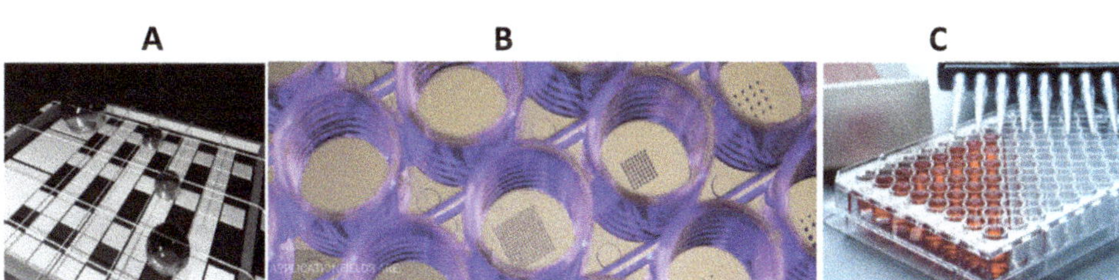

Figure I.5. (A) Open droplet electrowetting, courtesy of Yves Fouillet, CEA-Leti; (B) microdrops in wells from Microdrop Technology (https://www.microdrop.de/); (C) a typical microwell array.

I.4 Open-channel microfluidics

Open-channel microfluidics is the focus of this book. Open-channel microfluidics can be defined as the study and use of capillary flows guided by channel-like solid structures. To make this notion more precise, let us say that it corresponds to the general case of confined (closed) channels from which parts of the lateral, top, and bottom walls have been removed. The most frequent case is the removal of the top cover, but it can also take the form of the removal of the bottom, the removal of the channel sides, or the drilling of holes in the sides of confined channels (figure I.6). Contrary to the randomness of paper-based microfluidics and the semirandomness of thread-based microfluidics, the solid structures of open-channel microfluidics have a precisely defined geometry. And contrary to sessile droplet microfluidics, the liquid volumes are such that they can fill a whole channel.

Open-channel microfluidics is found in biology and biotechnology, space applications, electronics, and thermics.

I.5 Book contents

The book chapters are as follows:

(1) Chapter 1: capillarity basis, in which the principles of capillarity are recalled
(2) Chapter 2: the dynamics of capillary flow and the fundamental approach of Lucas, Washburn, and Bosanquet
(3) Chapter 3: spontaneous capillary flow in open channels, including the geometric conditions required for a spontaneous capillary flow
(4) Chapter 4: flow dynamics in open channels of uniform cross-section, presenting the adaptation of the Lucas–Washburn law to open channels
(5) Chapter 5: SCF in usual channel geometries such as open channels with rectangular, triangular, trapezoidal, and rounded cross-sections
(6) Chapter 6: capillary filaments in corners, flowing by themselves or preceding a bulk capillary flow
(7) Chapter 7: open-capillary flow in nonuniform channels, presenting the effects of cross-sectional enlargement and restriction
(8) Chapter 8: capillary flow in microporous media, such as paper pads and threads

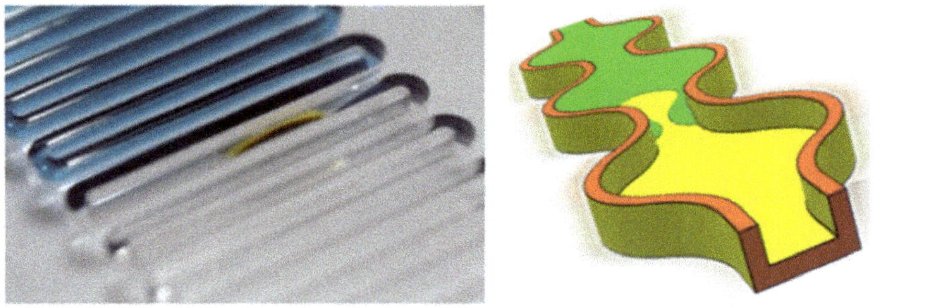

Figure I.6. Left: a view of capillary flow in an open winding channel (photo by Jing J Lee and Jean Berthier, University of Washington); right: a view of an open microflow in an open channel with constrictions and enlargements (Jean Berthier and Kenneth Brakke, Surface Evolver calculation).

(9) Chapter 9: open microflows found in nature, biomimetic open flows
(10) Chapter 10: two-phase open-channel microflows
(11) Chapter 11: applications to biosensing and biomedical domains
(12) Chapter 12: open-capillary fluidics aboard spacecraft. This chapter was written by Mark Weislogel
(13) Conclusions

References

[1] Morgan E D and Wilson I D 2004 An early description of paper chromatography *Chromatographia* **60** 135

[2] Okoroanyanwu U 2010 *Chemistry and Lithography* (Bellingham, WA: Wiley-SPIE Press)

[3] Yagoda H 1937 Applications of confined spot tests in Analytical Chemistry: preliminary paper *Ind. Eng. Chem. Anal. Ed.* **9** 79–82

[4] Martinez A W, Phillips S T, Carrilho E, Thomas S W, Sindi H and Whitesides G M 2008 Simple telemedicine for developing regions: camera phones and paper-based microfluidic devices for real-time, off-site diagnosis *Anal. Chem.* **80** 3699

[5] Bell J M and Cameron F K 1906 The flow of liquids through capillary spaces *J. Phys. Chem.* **10** 658

[6] Lucas R 1918 Ueber das Zeitgesetz des Kapillaren Aufstiegs von Flussigkeiten *Kolloid Z.* **23** 15

[7] Washburn E W 1921 The dynamics of capillary flow *Phys. Rev.* **17:** 273

[8] Rideal E K 1922 On the flow of liquids under capillary pressure *Philos. Mag. Ser.* **6** 1152–59

[9] Bosanquet C H 1923 On the flow of liquids into capillary tubes *Philos. Mag.* **45** 525

[10] Lutz B R, Trinh P, Ball C, Fu E and Yager P 2011 Two-dimensional paper networks: programmable fluidic disconnects for multi-step processes in shaped paper *Lab Chip.* **11** 4274–8

[11] Songok J and Toivakka M 2017 Modelling of capillary-driven flow for closed paper-based microfluidic channels *J. Micromech. Microeng.* **27** 065001

[12] Darcy H 1856 *Les Fontaines Publiques de la Ville de Dijon* (Paris: Dalmont)

[13] Cheng C-M, Martinez A W, Gong J, Mace C R, Phillips S T, Carrilho E, Mirica K A and Whitesides G M 2010 Paper-based ELISA *Angew. Chem.* **122** 4881–4

[14] Curry P S, Elkin B T, Campbell M, Nielsen K, Hutchins W, Ribble C and Kutz S J 2011 Filter-paper blood samples for ELISA detection of Brucella antibodies in caribou *J. Wildl. Dis.* **47** 12–20

[15] Martinez A W 2011 Microfluidic paper-based analytical devices: from POCKET to paper-based ELISA *Bioanalysis* **3** 2589–92

[16] Seok Y, Joung H-A, Byun J-Y, Jeon H-S, Shin S J, Kim S, Shin Y-B, Han H S and Kim M-G 2017 A paper-based device for performing loop-mediated isothermal amplification with real-time simultaneous detection of multiple DNA targets *Theranostics* **7** 2220

[17] Derda R, Laromaine A, Mammoto A, Tang S K Y, Mammoto T, Brock A, Ingber D and Whitesides G M 2009 Paper-supported 3D cell culture for tissue-based bioassays *Proc. Natl Acad. Sci. USA* **106** 18457–62

[18] Yager P, Edwards T, Fu E, Helton K, Nelson K, Tam M R and Weigl B H 2006 Microfluidic diagnostic technologies for global public health *Nature* **442** 412–8

[19] Martinez A W, Phillips S T, Whitesides G M and Carrilho E 2010 Diagnostics for the developing world: microfluidic paper-based analytical devices *Anal. Chem.* **82** 3–10

[20] Yetisen A K, Akram M W and Lowe C R 2013 Paper-based microfluidic point-of-care diagnostic devices *Lab Chip* **13** 2210–51

[21] Ahmed S, Bui M-P N and Abbas A 2016 Paper-based chemical and biological sensors: engineering aspects *Biosens. Bioelectron.* **77** 249–63
[22] Nilghaz A, Ballerini D R and Shen W 2013 Exploration of microfluidic devices based on multi-filament threads and textiles: a review *Biomicrofluidics* **7** 051501
[23] Ballerini D R, Li X and Shen W 2011 Flow control concepts for thread-based microfluidic devices *Biomicrofluidics* **5** 014105
[24] Adekunle K F 2015 Surface treatments of natural fibers, review, part 1 *Open J. Polym. Chem.* **5** 41–6
[25] Ferrero F and Periolatto M 2015 Wetting and wettability *Modification of Surface Energy and Wetting of Textile Fibers* ed M Aliofkhazraei (Rijeka: InTech) ch 6
[26] Mrsic I, Bauerle T, Ulitzsch S, Lorenz G, Rebner K, Kandelbauer A and Chassé T 2021 Oxygen plasma surface treatment of polymer films—pellethane 55DE and EPR-g-VTMS *Appl. Surf. Sci.* **536** 147782
[27] Li X, Tian J and Shen W 2010 Thread as a versatile material for low-cost microfluidic diagnostics *ACS Appl. Mater. Interfaces* **2** 1–6
[28] Berthier J, Brakke K A, Gosselin D, Berthier E and Navarro F 2017 Thread-based microfluidics: flow patterns in homogeneous and heterogeneous microfiber bundles *Med. Eng. Phys.* **48** 55–6
[29] Erenas M M, de Orbe-Payá I and Capitan-Vallvey L F 2016 Surface modified thread-based microfluidic analytical device for selective potassium analysis *Anal. Chem.* **88** 5331–7
[30] Sharp D 2009 The development of smart-bandage technologies *Thesis* Nottingham Trent University https://irep.ntu.ac.uk/id/eprint/356
[31] Smart bandages reveal healing (https://csiro.au/en/Research/MF/Areas/Chemicals-and-fibres/Advanced-fibres/Smart-clothing-and-textiles/Smart-bandages-reveal-healing/)
[32] Mostafalu P, Akbari M, Alberti K A, Xu Q, Khademhosseini A and Sonkusale S R 2016 A toolkit of thread-based microfluidics, sensors, and electronics for 3D tissue embedding for medical diagnostics *Microsyst. Nanoeng.* **2** 16039
[33] Nilghaz A, Ballerini D R, Guan L, Li L and Shen W 2016 Red blood cells transport mechanisms in polyester thread-based blood typing devices *Anal. Bioanal. Chem.* **408** 1365–71
[34] Young C D, Wu J R and Tsou T L 1998 High-strength, ultra-thin and fiber-reinforced pHEMA artificial skin *Biomaterials* **19** 1745–52
[35] Safavieh R, Mirzaei M, Qasaimeh M A and Juncker D 2009 Yarn-based microfluidics: from basic elements to complex circuits *13th Int. Conf. on Miniaturized Systems for Chemistry and Life Sciences (Jeju, Korea)*T S Kim *et al* (San Diego, CA: Chemical and Biological Microsystems Society) pp 1–5
[36] Reches M, Mirica K A, Dasgupta R, Dickey M D, Butte M J and Whitesides G M 2010 Thread as a matrix for biomedical assays *ACS Appl. Mater. Interfaces* **2** 1722–8
[37] Jain V, Devarasetty V and Patrikar R 2017 Effect of electrode geometry on droplet velocity in open EWOD based device for digital microfluidics applications *J. Electrost.* **87** 11–8
[38] Berthier J 2012 *Micro-Drops and Digital Microfluidics* (Amsterdam: Elsevier) 2nd edn
[39] Berthier J and Brakke K A 2012 *The Physics of Microdroplets* (New York: Wiley)
[40] de Gennes P-G, Brochard-Wyart F and Quéré D 2004 *Capillarity and Wetting Phenomena: Drops, Bubbles, Pearls, Waves* (New York: Springer)

IOP Publishing

Open-Channel Microfluidics (Second Edition)
Fundamentals and applications
Jean Berthier, Ashleigh B Theberge and Erwin Berthier

Chapter 1

The theoretical basis of capillarity

1.1 Introduction

In this chapter, the theoretical basis of capillarity is reviewed. Capillarity is based on two notions of the surface—or interfacial—tension of fluids and the surface energy—or excess surface energy—of solids. In this chapter, we present the notions of surface tension for different pairs of materials (liquid–air, liquid–liquid, solid–air, solid–liquid). The fundamental relations of capillarity and wetting are reviewed, such as Laplace pressure, Young's law, Zisman's approach, and the Wenzel and Cassie relations.

1.1.1 Surface tension

Two parameters determine the miscibility or immiscibility of two fluids: a density discontinuity and an intermolecular potential difference, as sketched in figure 1.1. This figure is a simplification proposed by Antonoff [1]. It does not include all the physics at the interface but is very instructive. In this figure, the molecules far from the interface have equilibrated interactions with their neighbors, while the molecules close to the interface have unbalanced interactions. Consider an interface between a liquid and a gas. In the bulk of the liquid, a molecule is in contact with n molecules (n varies between 4 and 12, depending on the liquid); at the interface, this number is divided by two if the interactions with the gas are neglected. There is a local dissymmetry in the interactions, which results in an excess of surface energy. At the macroscopic scale, a physical quantity called 'surface tension' has been introduced in order to take account of this molecular effect. Surface tension is expressed in N m^{-1} (sometimes, it is more practical to use mN m^{-1} as a unit for surface tension).

A consequence of this analysis is the fact that a fluid system will always act to minimize its surface area: the larger the surface area, the larger the number of molecules at the interface and the larger the cohesive energy imbalance [2, 3]. Hence, it is energetically costly to expand or create an interface. This is especially the case in open microfluidics, in which the free surface (the surface in contact with the

doi:10.1088/978-0-7503-5507-0ch1

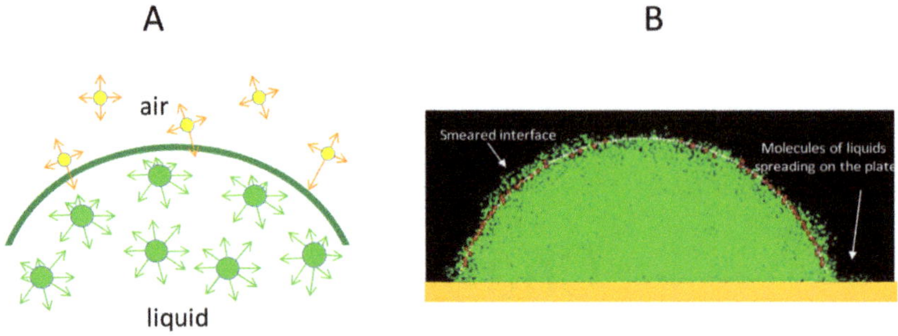

Figure 1.1. (A) A simplified sketch of the interface between a liquid and air; (B) a molecular dynamics simulation of a sessile droplet interface showing the fluctuating behavior of molecules at the interface (courtesy Terry Blake).

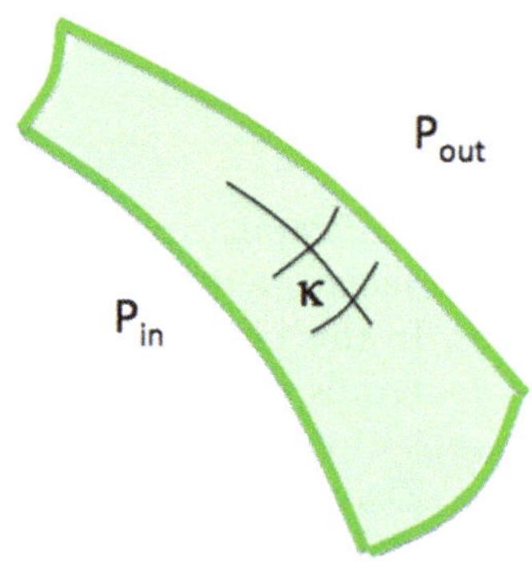

Figure 1.2. An illustration of Laplace's law for an interface.

surrounding air or gas) increases during the flow. Hence, flow restrictions should be expected when switching from a closed channel to an open channel.

The phenomenon depicted in figure 1.1 can be interpreted differently: because, in the vicinity of the interface, the molecules in the liquid phase are attracted inwards, a capillary pressure appears. Capillary pressure and surface tension force are two views of the same physical reality [4].

1.1.2 Laplace pressure

The pressure difference between the two sides of a liquid–gas or liquid–liquid interface—which is also the capillary pressure—is related to the curvature of the interface (figure 1.2). The Laplace law gives the relation between the pressure difference across an interface and the surface tension

$$\Delta P = 2\kappa\gamma, \tag{1.1}$$

where κ is the curvature of the surface, ΔP is the pressure difference between the two sides of the interface, and γ is the surface tension of the liquid (figure 1.2).

For a spherical droplet of radius R, the Laplace relation yields

$$\Delta P = \frac{2\gamma}{R}, \tag{1.2}$$

while for a cylindrical interface of radius R

$$\Delta P = \frac{\gamma}{R}. \tag{1.3}$$

Note that, in the case of a concave liquid interface, the capillary pressure is negative. Observe also that small droplets have an internal pressure that is much higher than that of large droplets. This property can be correlated to Antonoff's approach presented above: in the small droplets, the proportion of molecules close to the interface is higher.

The definition of surface tension appears to be straightforward, but it hides conceptual difficulties. A conceptual difficulty stems from the usual representation of the interfacial tension as a force tangent to the interface. If the capillary pressure is normal (perpendicular) to the interface, one wonders why the surface tension is represented as a force vector tangent to the interface, as, for example, in Young's law (see below)? In fact, a line must be present on the interface to use such a representation. The line can be a 'physical' triple line or any 'imaginary' line drawn on the surface. We recall that a triple line is a line delimiting the contact between a fluid and two other materials, for example, a solid and the surrounding air. The surface tension is assumed to be similar to pressure in the sense that the pressure P is a scalar, which results in a vector force $\vec{F} = P\,A\,\vec{n}$ when applied to an area A with the normal $\vec{n}$. Similarly, the surface tension γ is a scalar that results in the force $\vec{F} = \gamma\,l\,\vec{n}$ when applied to a line l with the normal $\vec{n}$ (the normal lies in the plane of the surface) [5]. A demonstration of the equivalence between surface tension force and capillary pressure was given by Finn [4, 6]. It is based on the integral relation

$$\int_S 2\kappa\,\vec{n}\,dA = \oint_C \vec{n} \times d\vec{l}, \tag{1.4}$$

where S is the liquid surface bounded by the contour C, κ is the curvature of the surface S, $\vec{n}$ is the unit normal to the curved surface S, and $d\vec{l}$ is the unit tangent to the contour C (figure 1.3). Here, C can be a triple line but also any imaginary line delimiting the surface S. Multiplying (1.4) by the surface tension γ_{LA} and using the Laplace relation (1.1), one finds that

$$\int_S \Delta P\,\vec{n}\,dA = \oint_C \gamma_{LA}\vec{n} \times d\vec{l} = \vec{F}_{LA}. \tag{1.5}$$

In the case of a triple line, a capillary force appears. The capillary force is the force exerted by the solid surface on the liquid. Hence,

$$\vec{F}_{cap} = -\vec{F}_{LA}. \tag{1.6}$$

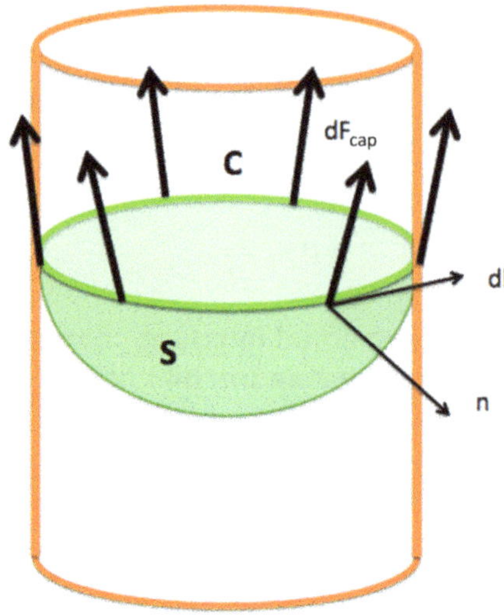

Figure 1.3. A sketch of the capillary force exerted by a solid cylinder on a liquid interface (the opposite of the forces exerted by the liquid on the solid surface).

1.1.3 Liquid–liquid surface tension

The same reasoning applies to the interface between two liquids, except that the resulting dissymmetry is less. For example, the surface tension between water and air is 72 mN m^{-1}, whereas it is only 20–50 mN m^{-1} between water and oils. Interfacial tension between two liquids may be zero: fluids with zero interfacial tension are said to be miscible. For example, there is no surface tension between fresh and salt water: salt molecules diffuse freely across a boundary between fresh and salt water.

One question has been intriguing researchers: if the surface tension of liquid 1 and air is known and if the surface tension of liquid 2 and air is known, is it possible to deduce the surface tension between liquid 1 and liquid 2? Antonoff was one of the first to propose a relation for the liquid–liquid surface tension [1, 5]:

$$\gamma_{L1L2} = |\, \gamma_{L1} - \gamma_{L2} \,|. \tag{1.7}$$

However, it appeared that Antonoff's rule was not always true. The only general relation between the three surface tensions is

$$\gamma_{max} < \gamma_{min} + \gamma_{med}, \tag{1.8}$$

where max, med, and min refer to the maximum, median, and minimum surface tensions amongst the three surface tensions. Relation (1.8) is a direct consequence of Neumann's construction (see below). Good and Girifalco proposed another relation [7, 8]

$$\gamma_{L1L2} = \gamma_{L1} + \gamma_{L2} - 2\Phi\sqrt{\gamma_{L1}\,\gamma_{L2}}. \tag{1.9}$$

While the exact value of the interaction parameter Φ depends on the two liquids, it turns out to be on the order of unity in most cases; in fact, it lies between 0.5 and 1.1 [8].

It was suggested that the contribution to the surface energy involves a polar component and a dispersive (van der Waals) component [9, 10]

$$\gamma = \gamma^d + \gamma^p, \tag{1.10}$$

where d and p respectively stand for the dispersive (van der Waals) and polar interactions. The interaction parameter ϕ is then

$$\Phi = \frac{\sqrt{\gamma_{L1}{}^d \gamma_{L2}{}^d} + \sqrt{\gamma_{L1}{}^p \gamma_{L2}{}^p}}{\sqrt{\gamma_{L1}\,\gamma_{L2}}}. \tag{1.11}$$

[In the case in which only dispersion forces are present, $\Phi = 1$ [11]. In this case

$$\gamma_{L1L2} = \gamma_{L1} + \gamma_{L2} - 2\sqrt{\gamma_{L1}\,\gamma_{L2}}. \tag{1.12}$$

Note that (1.12) can be cast in the form $\gamma_{L1L2} = \left(\sqrt{\gamma_{L1}} - \sqrt{\gamma_{L2}}\right)^2$.

As we remarked, the interaction parameter Φ depends on the liquids on both sides of the interface. Let us take the example of different organic liquids and water (table 1.1). The values of Φ —found by fitting (1.9)—all lie between 0.5 and 1.1.

Note that Antonoff's formula (1.7) is valid only for dispersive interactions (not polar interactions) corresponding to an interaction parameter $\phi \sim 1$. A comparison between the experimental values for the liquids of table 1.1 and the Good–Girifalco model (with the adjusted interaction parameter Φ) and Antonoff's rule is shown in figure 1.4.

1.1.4 Contact with solid surfaces: Young's law

In the case of solids, the surface energy is defined as the excess free energy per unit area of a particular interface—i.e. this is the energy required to create the surface [12]. We see here the particularity of the capillary actuation of liquids: the energy source for motion is embedded in the solid walls during fabrication. This is a fundamental difference from conventional microfluidics that uses external pumps or syringes to move the fluids.

Young's law defines the contact angle of a liquid with a solid surface. It is a 'boundary' condition that makes the link between the free interface and the solid wall. Cohesion forces dictate the behavior of the liquid–vapor interface expressed by the surface tension γ_{LA}, whereas adhesion forces impose the boundary condition

Table 1.1. The surface tensions of six organic liquids and their interaction parameters with water.

Organic liquid	$\gamma_{L1,A}$ (mN m^{-1})	$\gamma_{L1,\text{water}}$ (mN m^{-1})	Φ
1-Pentanol	26.0	4.4	1.07
1-Nonanol	28.0	8.5	1.00
Chloroform	26.7	32.8	0.75
Toluene	28.4	36.1	0.70
FC40	16.0	52.1	0.50
n-Dodecane	25.3	52.8	0.50

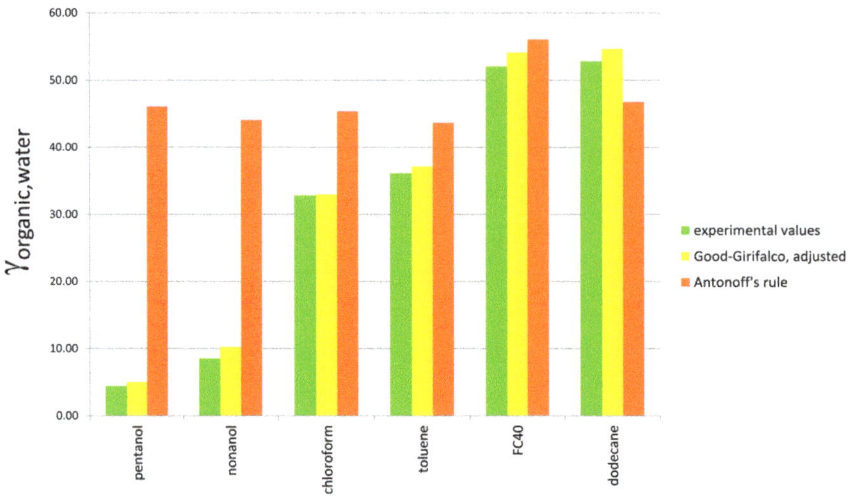

Figure 1.4. A comparison between experimental data, the adjusted Good–Girifalco model, and Antonoff's rule for the surface tensions between organic liquids and a water surface.

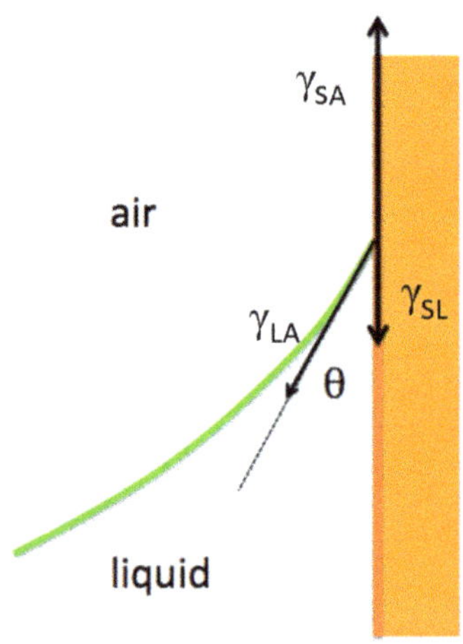

Figure 1.5. Contact at an interface with a solid wall.

through the contact angle θ. Both quantities are independent, although they usually show up in the same scenario.

Using the notations of figure 1.5, Young's law reads

$$\gamma_{LA} \cos\theta = \gamma_{SA} - \gamma_{SL}. \tag{1.13}$$

Consider the case of water in contact with a wall made of poly(methyl methacrylate) (PMMA, a common polymer used in biotechnology). The observed contact angle is close to 85°. The value of the left-hand side of (1.13) is close to zero and $\gamma_{\text{PMMA, air}} \approx \gamma_{\text{PMMA, water}}$. On the other hand, pentanol and toluene respectively have contact angles of 12.5° and 7.8° with PMMA, resulting in $\gamma_{\text{PMMA, air}} \approx \gamma_{\text{PMMA, pentanol}} + \gamma_{\text{pentanol, air}}$ and $\gamma_{\text{PMMA, air}} \approx \gamma_{\text{PMMA, toluene}} + \gamma_{\text{toluene, air}}$.

1.1.5 Neumann's construction

Young's law does not take into account the component of the capillary force perpendicular to the wall, because the wall is assumed to be rigid. Neumann's construction is an extension of Young's law to 'liquid' or 'deformable' walls. Consider three immiscible liquids, denoted by 1, 2, and 3. The equilibrium condition at the triple line is

$$\sum_{i=1,3} \vec{\gamma}_{i,\, i+1} = \vec{\gamma}_{1,\, 2} + \vec{\gamma}_{2,\, 3} + \vec{\gamma}_{3,\, 1} = 0. \tag{1.14}$$

This configuration is called Neumann's construction. In the usual case of two liquids and air, Neumann's relation is

$$\vec{\gamma}_{\mathrm{L1,\, L2}} + \vec{\gamma}_{\mathrm{L2,\, air}} + \vec{\gamma}_{\mathrm{L1,\, air}} = 0. \tag{1.15}$$

Even if the density of liquid 2 is somewhat larger than that of liquid 1, the droplet may 'float' on the surface, as shown in figure 1.6 [2].

Neumann's construction also applies to deformable materials, such as gels used to form flexible walls. Figure 1.7 shows an example of Neumann's construction for a cantilever [13]. The beam is curved by the presence of the droplet.

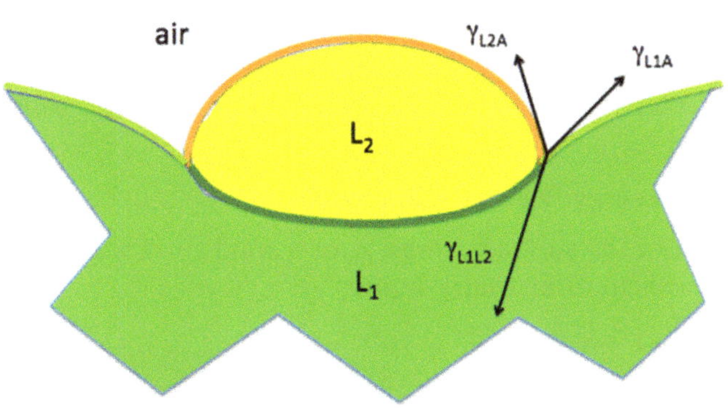

Figure 1.6. Neumann's construction for two fluids and air.

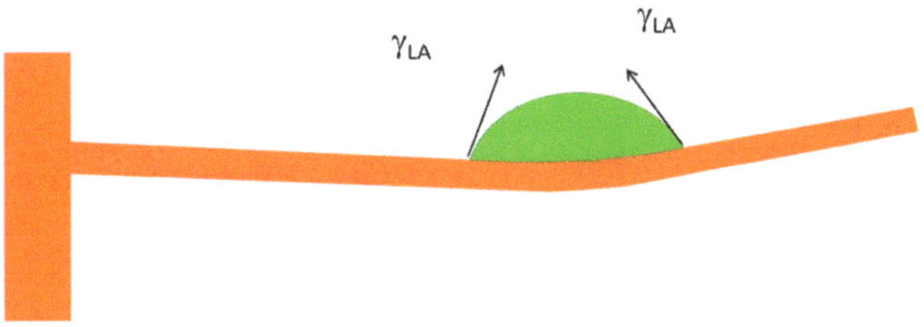

Figure 1.7. A cantilever deformed by the presence of a droplet [13].

1.1.6 The work of adhesion, the work of cohesion, and the Young–Dupré equation

By definition, the work of adhesion is the work necessary to separate two bodies with a common surface (figure 1.8). Using the notations in figure 1.8(A), the energy of the common surface S is

$$E_{12} = \gamma_{12}\, S. \tag{1.16}$$

After separation, the surface energy is

$$E = E_1 + E_2 = \gamma_1 S + \gamma_2\, S. \tag{1.17}$$

The work of adhesion is then

$$W_a = \gamma_1 + \gamma_2 - \gamma_{12}. \tag{1.18}$$

The work of cohesion is defined in the same way, except that the two surfaces are identical (figure 1.8(B)). Then

$$W_c = 2\gamma_1. \tag{1.19}$$

Let us now consider a solid surface and a liquid. The work of adhesion is

$$W_a = \gamma_S + \gamma_L - \gamma_{SL}. \tag{1.20}$$

Upon substituting Young's equation into this, we obtain the Young–Dupré equation

$$W_a = \gamma_L(1 + \cos\theta). \tag{1.21}$$

For a super-hydrophobic contact, $\theta = \pi$ and $\cos\theta = -1$; we deduce that $W_a = 0$: no work is needed to separate a super-hydrophobic liquid from a solid.

1.1.7 Solid surface energy: Zisman's approach

The determination of the surface energy γ_S of a solid surface is not straightforward. One has to use the help of auxiliary liquids.

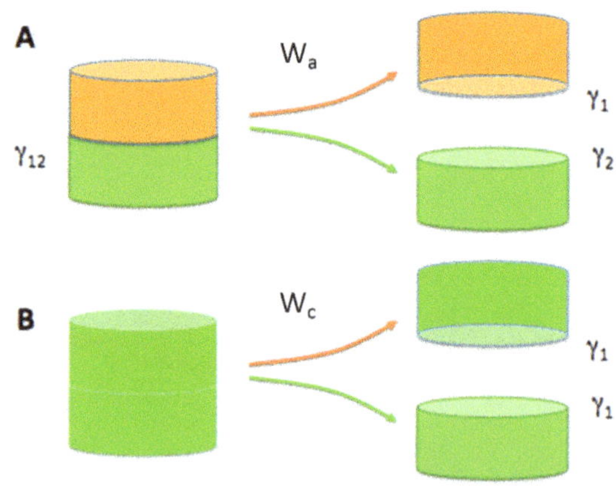

Figure 1.8. Schematic views of the work of adhesion and the work of cohesion.

1.1.7.1 Using Young's law

The first method is based on the use of Young's law. On the one hand, liquid surface tension with air γ_L can be measured by different methods, the most usual being the pending drop method [14, 15]. On the other hand, the liquid–solid contact angle θ can be determined by the visualization of a deposited droplet [16]. In Young's law, the determination of the solid–liquid surface tension γ_{SL} is needed to obtain the solid surface tension γ_{SA}:

$$\gamma_{SA} = \gamma_{SL} + \gamma_L \cos\theta. \tag{1.22}$$

For simplicity, the reference to air (A) can be dropped. The use of the semiempirical relation (1.9) proposed by Good and Girifalco enables the closure of the system. In the case of a solid surface, (1.9) becomes

$$\gamma_{SL} = \gamma_S + \gamma_L - 2\Phi\sqrt{\gamma_S\,\gamma_L}. \tag{1.23}$$

Upon substituting (1.23) into (1.22), we obtain the so-called Young–Dupré–Good–Girifalco (YDGG) relation [17]

$$1 + \cos\theta = 2\Phi\left(\frac{\gamma_S}{\gamma_L}\right)^{0.5}. \tag{1.24}$$

Hence

$$\gamma_S = \left(\frac{1+\cos\theta}{2\Phi}\right)^2 \gamma_L. \tag{1.25}$$

In fact, the unknown solid–liquid surface tension γ_{SL} of equation (1.22) has been replaced by the interaction parameter Φ. But, for low-energy solid surfaces, $\Phi = 1$, and we obtain [17]

$$\gamma_S = \frac{(1+\cos\theta)^2}{4}\gamma_L. \tag{1.26}$$

Relation (1.26) indicates that $1 + \cos\theta$ is a linear function of $1/\gamma_L{}^{1/2}$. An example for a PMMA solid wall (plasma oxygen treated) is given in figure 1.9.

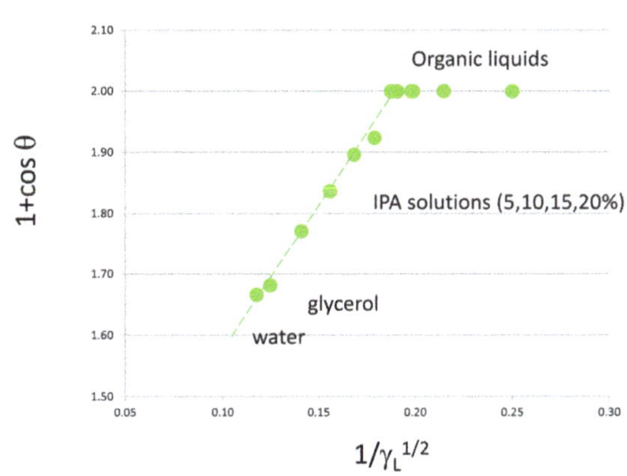

Figure 1.9. The YDGG relation for an oxygen-treated PMMA solid surface.

1.1.7.2 Using the Zisman plot

The second method is based on the Zisman plot [18]. For low-energy surfaces, Zisman experimentally observed that the cosine of the contact angle with the wall θ was approximately a linear function of the liquid surface tension

$$\cos\theta = 1 + b\,(\gamma_C - \gamma_L), \tag{1.27}$$

where γ_C is the critical surface tension, i.e. the surface tension below which the liquid spreads on the wall. In such cases, the critical surface tension is approximately the solid surface energy $\gamma_S \cong \gamma_C$. Figure 1.10 shows the Z-plot (or Zisman plot) for different solid walls and different test liquids.

Z-plots (Zisman plots) are very convenient for the experimenter: The critical tension γ_C can easily be obtained by taking a set of two or three liquids whose surface tensions are known and measuring their Young contact angles. The contact angles for the liquids of interest are then determined from their surface tension alone (with some restrictions, as we shall see in the following). The literature is abundant with discussions about the validity of the Z-approach [24, 25]. However, there is a theoretical background to the Z-approach. We refer the reader to the very instructive publication by Zhu and coworkers [26]. We give here the essentials of the approach, including the so-called Young–van der Waals equation.

Consider a low-energy wall and a series of homologous liquids. The Young–Dupré equation (1.21) can be cast in the form

$$1 + \cos\theta = \frac{2W_a}{W_c}. \tag{1.28}$$

[For simple molecular liquids and low-energy surfaces, van der Waals forces are dominant and we have

$$\begin{aligned} W_a &= k\,\alpha_S\,\alpha_L \\ W_C &= k\,\alpha_L^{\,2}, \end{aligned} \tag{1.29}$$

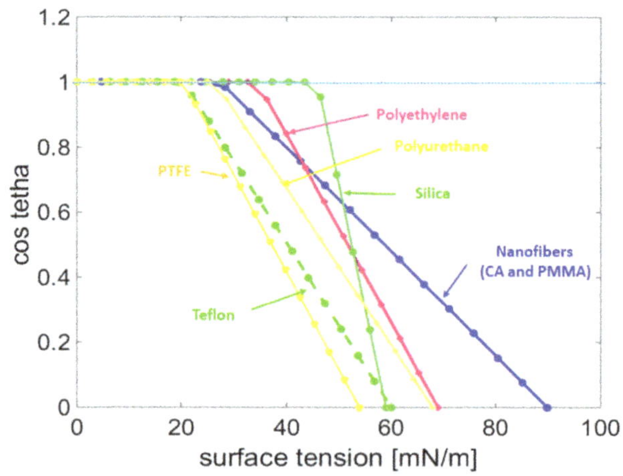

Figure 1.10. The Zisman plots for six solid substrates: nanofibers (cellulose acetate or PMMA) [19], polyethylene film [20], Teflon [21], silica [22], polytetrafluoroethylene (PTFE), and polyurethane [23].

where α_S and α_L are the dielectric polarizabilities of the solid and liquids and k is a constant characteristic of the homologous series of liquids. Substituting the works of adhesion and cohesion (1.29) into (1.28), and introducing the critical tension γ_C for $\theta = 0$, one obtains

$$\cos\theta = 2\frac{\alpha_S}{\alpha_L} - 1 = 2\sqrt{\frac{\gamma_C}{\gamma_L}} - 1. \tag{1.30}$$

This equation is called the Young–van der Waals (YvdW) equation. Note that this equation is similar to the Young–Dupré–Good–Girifalco equation (1.24) with $\Phi = 1$. Relation (1.30) is plotted in figure 1.11 for an untreated PMMA solid surface, and test liquids (water, glycerol, isopropyl alcohol (IPA) solutions, and different organic liquids).

Zisman's plot is shown to be a linearization of the Young–van der Waals (YvdW) equation. In the case of PMMA, this linearization is given by the equation

$$\cos\theta = 1 + \frac{(\gamma_C - \gamma_L)}{2.5\gamma_C}. \tag{1.31}$$

There are two essential conditions for the validity of the Young–van der Waals relation: first, the solid surface must be low-energy so that the interactions are of the van der Waals type (the London potential), and second, the test liquids should be sufficiently homologous (they should have the same constant k).

However, it seems that the Z-plots have a broader domain of validity. Figure 1.12 shows the Z-plot for a PMMA substrate—treated with oxygen plasma—obtained using six different oils, IPA solutions (aqueous solutions), glycerol, and water. These cases fall outside the conditions given by Zhu *et al*, but the Z-plot is still nearly linear.

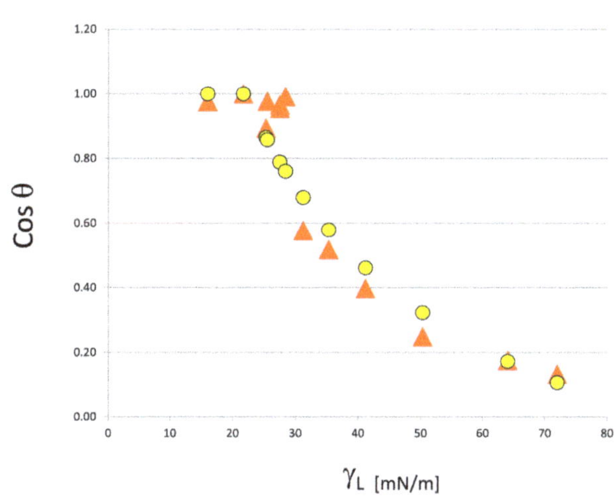

Figure 1.11. Plots of the cosine of the contact angle vs. the liquid surface tension (for PMMA): the orange triangles correspond to the measurements, and the yellow dots to the Young–van der Waals equation (1.30), for a critical tension $\gamma_C = 22$ mN m^{-1}.

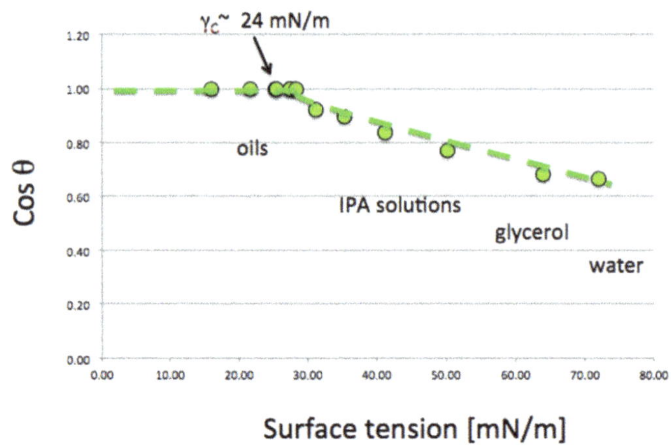

Figure 1.12. A Z-plot for a PMMA surface treated with oxygen plasma, showing a critical tension of approximately 24 mN m^{-1}.

Note that the use of Zisman or Young–van der Waals plots is of great use in open microfluidics. Using a Z-plot (1.27), the capillary force (per unit length) responsible for the capillary motion on PMMA can be written as

$$f_{\text{cap}} = \gamma_L \cos\theta = \gamma_L \min\left(1,\ 1 + \frac{(\gamma_C - \gamma_L)}{2.5\gamma_C}\right), \tag{1.32}$$

or, using the general YvdW equation (1.30),

$$f_{\text{cap}} = \gamma_L \cos\theta = \gamma_L \min\left(1,\ 2\sqrt{\frac{\gamma_C}{\gamma_L}} - 1\right), \tag{1.33}$$

Note that the minimum function is introduced in order to make the cutoff less than the critical tension. When we use (1.32) or (1.33), the measurement of the contact angle is no longer required. It is replaced by the critical tension γ_c. The capillary force (1.32) is shown in figure 1.13 for the different substrates of figure 1.10.

1.1.8 Wetting and pinning

1.1.8.1 Wetting

A drop of liquid deposited on a flat solid surface presents different morphologies depending on different surface tensions. The situation is called hydrophobic (or lyophobic for an arbitrary liquid) when the contact angle is larger than 90° and hydrophilic (lyophilic) when the contact angle is less than 90°. The complete wetting of a flat surface by a liquid corresponds to a contact angle of 0° (figure 1.14).

Liquid spreads on a substrate in a film (complete wetting) if the energy of the system is lowered by the presence of the liquid film [28]. The surface energy per unit surface of the dry solid surface is γ_{SA} and the surface energy of the wetted solid is

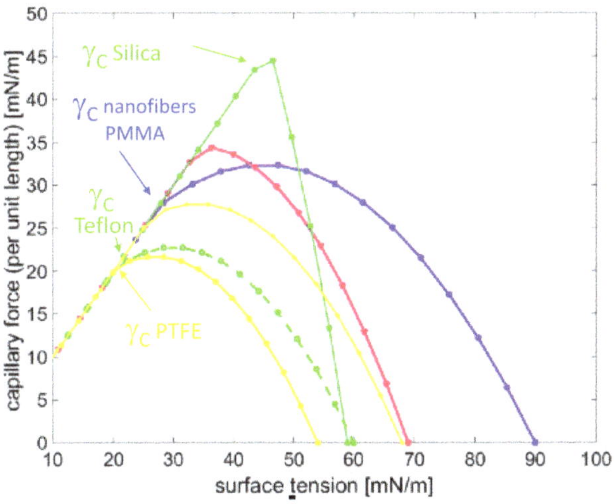

Figure 1.13. The capillary forces (Z functions) for the six different solid substrates: the surface tension of the liquid with air is plotted on the horizontal axis, and the capillary force per unit length is plotted along the vertical axis. The color coding used in figure 1.10 is reused here. The critical surface tension corresponds to the departures of the curves from the straight line $cos\theta = 1$ [27].

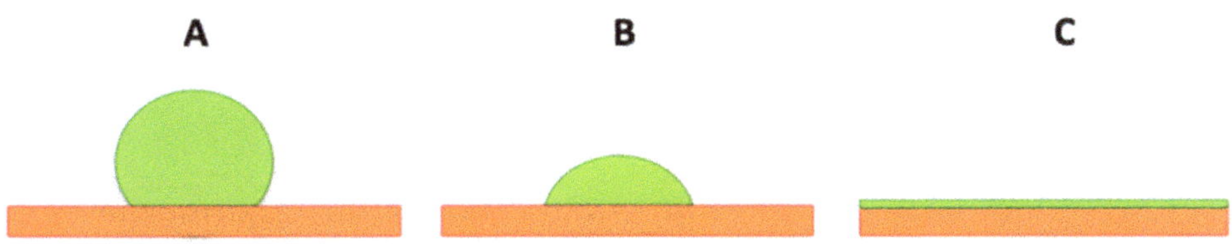

Figure 1.14. Different types of wetting: (A) the hydrophobic (lyophobic) case at a contact angle larger than 90°; (B) the hydrophilic (lyophilic) case at a contact angle less than 90°; (C) complete wetting at a contact angle of 0°.

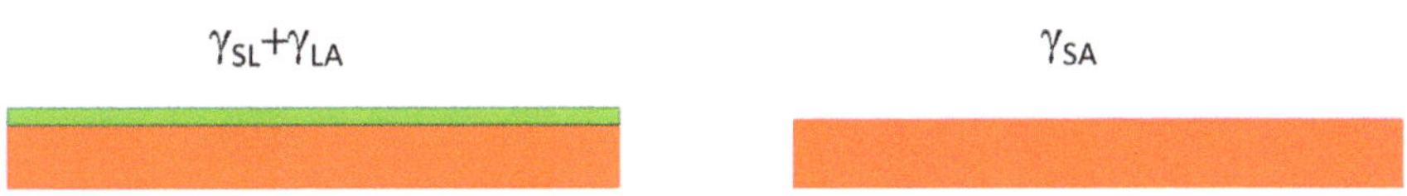

Figure 1.15. A comparison between the two surface energies.

$\gamma_{SL}+\gamma_{LA}$. The spreading parameter S determines the type of spreading (total or partial):

$$S = \gamma_{SA} - (\gamma_{SL} + \gamma_{LA}). \tag{1.34}$$

This definition (1.34) stems directly from Young's law. If $S > 0$, the liquid spreads on the solid surface; if $S < 0$, the liquid forms a droplet. If $S>0$, the surface energy is smaller when the liquid film covers the surface (figure 1.15).

1.1.8.2 Pinning and canthotaxis

Interface pinning—or anchoring—is the immobilization of the triple line during a wetting process. Interface pinning can occur at a sudden change of surface chemistry or geometry. These abrupt changes in surface properties may be unintended—such as surface defects—or deliberate, i.e. made in order to contain the fluid. Figure 1.16 shows examples of pinning [2, 29, 30].

Let us analyze the pinning at a wetting discontinuity, as sketched in figure 1.17. The wetting liquid is pinned at the discontinuity as long as its contact angle is less than θ_2, the Young contact angle with the substrate on the left of the figure. If the wetting pressure is sufficient to obtain a contact angle larger than θ_2, the pinning fails and the liquid wets the lyophobic surface. The interval $\{\theta_1,\theta_2\}$ is called 'canthotaxis' or the pinning interval [31, 32], i.e.

$$\theta_1 < \phi < \theta_2, \tag{1.35}$$

where ϕ is the interface angle with the solid surface. A similar configuration occurs on an edge (figure 1.18). The contact line on the angle stays fixed as long as the

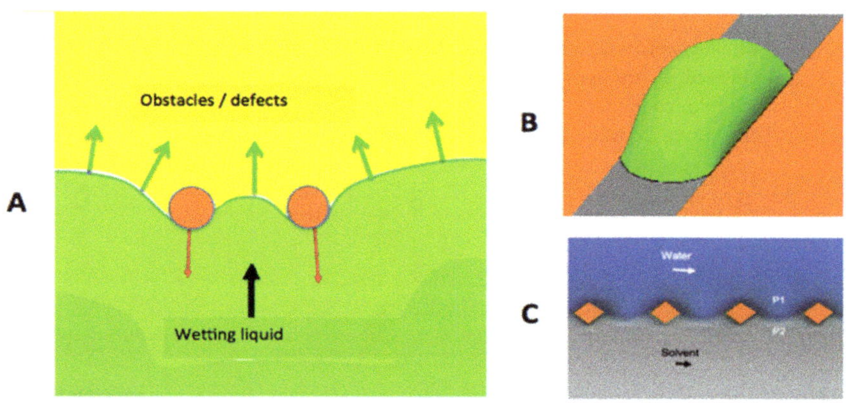

Figure 1.16. (A) A sketch of a wetting liquid stopped/pinned by obstacles or defects on a solid surface. The capillary forces acting on the triple line (green arrows) are balanced by the pinning resistance from the obstacles (red arrows) [2]. (B) A water droplet pinned by two hydrophilic–hydrophobic transitions [2]. (C) Cocurrent flows of water and solvent separated by the pinning of the interface on micropillars [30].

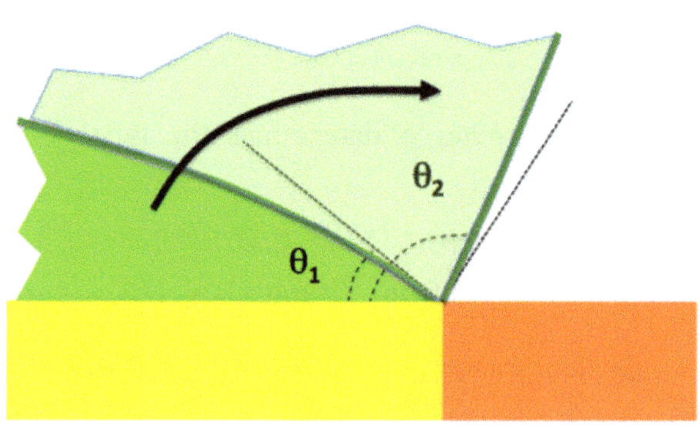

Figure 1.17. Pinning at a wetting discontinuity: the left-hand solid surface (yellow) is lyophilic and has a contact angle of θ_1, while the right-hand solid surface (orange) is lyophobic and has a contact angle of θ_2. The liquid (green) is pinned at the discontinuity as long as the contact angle is less than θ_2.

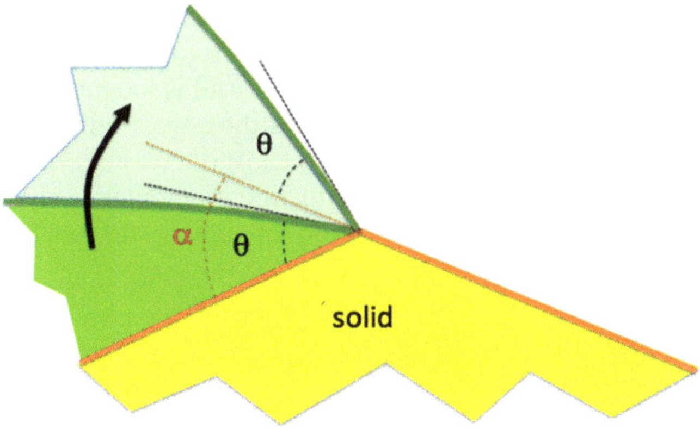

Figure 1.18. Pinning at an edge: the liquid (green) is pinned at the edge as long as the interface does not tilt over the contact angle $\alpha+\theta$.

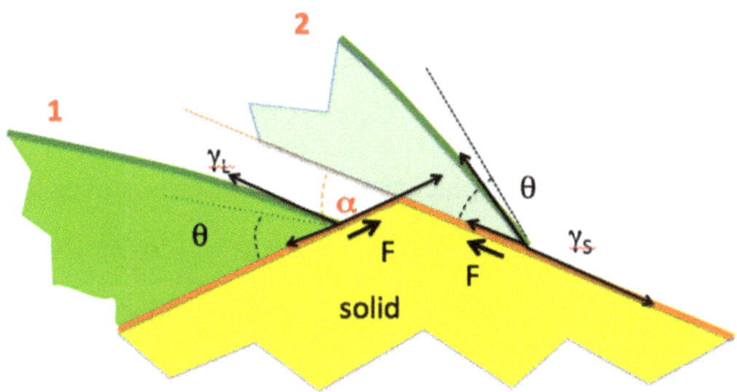

Figure 1.19. (1) If the interface is not located exactly on the edge and its contact angle is larger than θ, the resulting capillary force F is directed towards the edge; (2) if the interface is located beyond the edge and its contact angle is less than $\alpha+\theta$, the resulting capillary force F is directed back towards the edge.

contact angle is not forced over the limit $\alpha + \theta$, where α is the angle between the two planes. The interval $\{\theta,\theta+\alpha\}$ is the canthotaxis interval, i.e.

$$\theta < \phi < \alpha + \theta. \tag{1.36}$$

In the case in which the two planes have different chemical surfaces, with Young contact angles denoted by θ_1 and θ_2, the canthotaxis condition is

$$\theta_1 < \phi < \alpha + \theta_2. \tag{1.37}$$

If $\alpha = 0$, equation (1.37) reduces to (1.35). Simple reasoning based on the significance of the Young angle can explain canthotaxis (figure 1.19). Suppose that the interface makes an angle φ with the wall larger than the Young angle θ and is located slightly before the edge. The resultant of the forces is then directed towards the edge, because

$$\begin{aligned} F &= \gamma_S - \gamma_{SL} - \gamma_L \cos\phi = \gamma_S - \gamma_{SL} - \gamma_L \cos\theta + (\gamma_L \cos\theta - \gamma_L \cos\phi) \\ &= \gamma_L(\cos\theta - \cos\phi) > 0. \end{aligned} \tag{1.38}$$

Thus, the triple line moves towards the edge. Similarly, if the interface makes an angle ϕ with the wall smaller than the Young angle θ (or an angle $\alpha+\phi$ smaller than $\alpha+\theta$, depending on the chosen coordinate system) and is located slightly beyond the edge, then the resulting force is directed towards the edge, and the triple line recedes towards the edge. Hence the definition of the canthotaxis interval (1.37).

The pinning of liquids is useful in the production of capillary valves (see chapter 4) and is also currently used for the self-assembly of similar surfaces [33–35].

1.1.9 Wenzel's law

Solid surfaces are often imperfectly smooth and present roughness and asperities at the nano- or microscale. Wenzel demonstrated that Young's contact angle should be modified to take account of the roughness. A uniform roughness parameter r can be defined as the ratio between the real length between any two points of the surface and the flat distance between these two points, i.e.

$$r = \frac{l_{\text{real}}}{l_{\text{smooth}}}. \tag{1.39}$$

Given this definition, a linear surface element corresponding to the infinitesimal displacement dx is rdx (figure 1.20).

The work necessary for the displacement of the interface over a distance dx is

$$dW = \vec{F}\,\vec{dl} = F_x\,dx = (\gamma_{SL} - \gamma_S)\,rdx + \gamma_{LG}\cos\theta^*\,dx, \tag{1.40}$$

where θ^* is the apparent contact angle. In fact, if the liquid stays at equilibrium after the infinitesimal displacement dx, then

$$dW/dx = 0, \tag{1.41}$$

then,

$$\gamma_{LG}\cos\theta^* = (\gamma_{SL} - \gamma_S)\,r, \tag{1.42}$$

and, using Young's law, we obtain

$$\cos\theta^* = r\cos\theta. \tag{1.43}$$

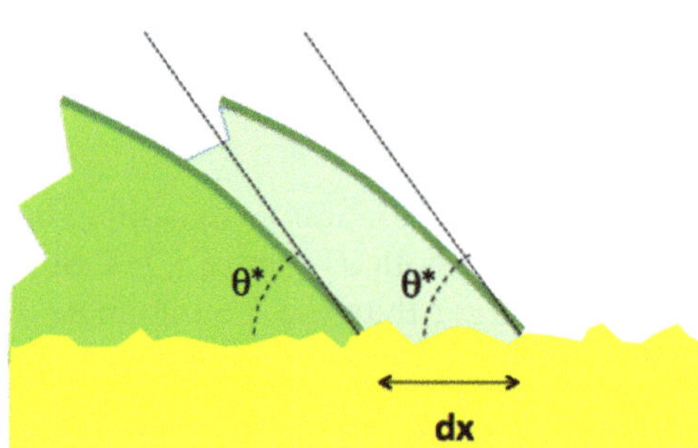

Figure 1.20. A sketch of an interface on a rough surface.

Relation (1.43) is called Wenzel's law [36]; it demonstrates that the contact angle is modified by the presence of roughness. As the parameter r is always larger than unity, a lyophilic rough surface has a contact angle with the liquid smaller than that of a smooth surface, and, conversely, a lyophobic surface has a contact angle with the liquid larger than that of a smooth surface. This property is nonintuitive, because it seems opposite to the notion of pinning on larger reliefs, as we have seen in the section above.

1.1.10 The Cassie–Baxter law

Now consider an inhomogeneous solid surface, such as that sketched in figure 1.21 [37]. For simplicity, it is assumed that at the nanoscale, the solid surface is composed of two different wetting surfaces with contact angles θ_1 and θ_2. Reasoning similar to that used for Wenzel's law can be employed. The work necessary to displace the interface over a distance dx is

$$dW = F_x\,dx = (\gamma_{SL} - \gamma_S)_1 f_1\,dx + (\gamma_{SL} - \gamma_S)_2 f_2\,dx + \gamma_{LG}\cos\theta^*\,dx, \tag{1.44}$$

where f_1 and f_2 are respectively the surface fractions of materials 1 and 2.

Using Young's law twice and taking account of $dW/dx = 0$ yields

$$\cos\theta^* = f_1\cos\theta_1 + f_2\cos\theta_2. \tag{1.45}$$

For a composite surface made from many different materials (or one that has surface functionalization), the Cassie–Baxter law can be written

$$\cos\theta^* = \sum_i f_i\cos\theta_i, \tag{1.46}$$

with $\sum_i f_i = 1$. Suppose that one of the components of the wall is super-hydrophobic (super-lyophobic) and has a contact angle of π. Relation (1.45) becomes [38, 39]

$$\cos\theta^* = f\cos\theta - (1 - f), \tag{1.47}$$

and the contact angle θ^* becomes less hydrophilic or more hydrophobic. In fact, air is the most hydrophobic element. So, if the substrate is composed of pillars, as shown in figure 1.22(C), a droplet on top of the pillars might not touch the substrate

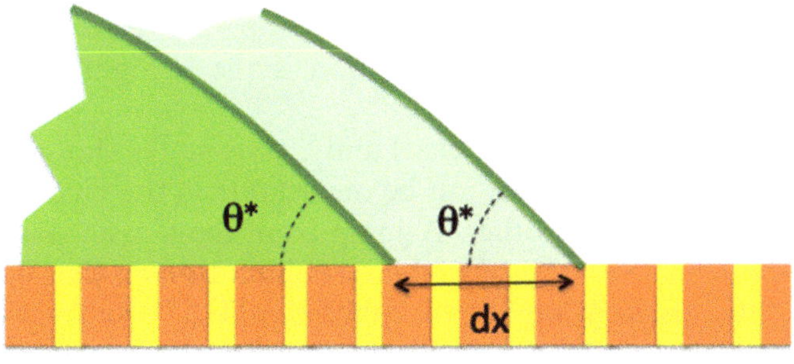

Figure 1.21. A sketch of an interface on an inhomogeneous surface.

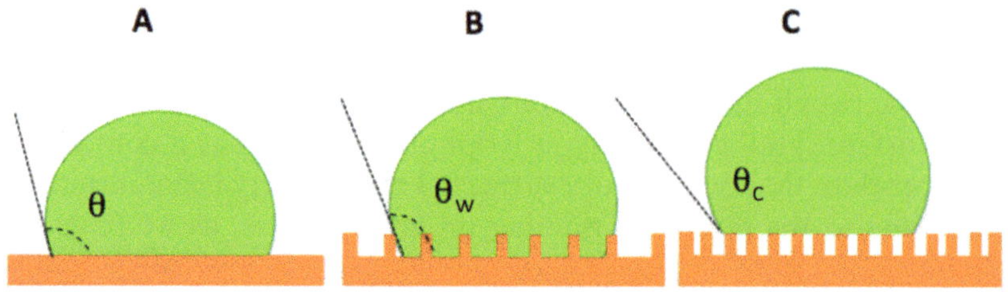

Figure 1.22. Three different morphologies of a droplet: (A) a sessile droplet on a smooth lyophobic surface; (B) a Wenzel droplet on a pillared surface (index W); (C) a Cassie droplet on a pillared surface (index C).

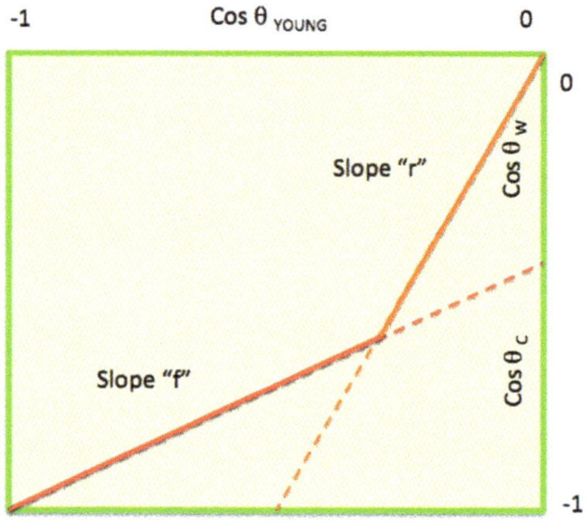

Figure 1.23. Cassie and Wenzel domains for a pillared surface: the transition between a Wenzel droplet and a Cassie droplet occurs at the intersection of the two segments [40].

bottom. This is called the 'fakir' effect, and the droplet in this case is called a Cassie droplet.

In this particular case, we have switched back to a geometrical configuration problem, where the roughness r is defined by the pillar spacing and aspect ratio. Hence, the question arises: which of the Wenzel or Cassie laws applies to a droplet on a pillared surface?

A partial answer is illustrated in figure 1.23. In the $\{\cos\theta_Y, \cos\theta_W\}$ coordinates, the Wenzel relation is the straight segment of slope r intersecting the coordinate $\{0,0\}$. On the other hand, the Cassie relation is the straight segment of slope f intersecting the coordinate $\{-1,-1\}$.

In fact, the separation between Wenzel and Cassie states shown in figure 1.23 is not always respected. The situation can be metastable in the vicinity of the domain where the two lines cross each other. A jump from the Cassie state to the Wenzel state can be observed. There is an abundance of literature on the subject because of the very important domain of super-hydrophobicity [41]. Note that surfaces that are super-hydrophobic and super-oleophobic at the same time have been developed using the Cassie concept [42].

1.1.11 Capillary rise

Capillary forces can work against gravity. This called capillary rise. The phenomenon was observed long ago. The first written observations of capillary rise are attributed to Leonardo da Vinci around the year 1480 [43]. Hauskbee and Jurin investigated the capillary ascension phenomenon during the 17th and 18th centuries. Jurin's law, which predicts the rise in a thin vertical tube immersed in a liquid at one end, was formally established in 1719 [44].

The reasoning is the following: the weight of liquid in the tube is

$$F_{\text{grav}} = \rho\ g\ \pi\ R^2 h, \tag{1.48}$$

while the capillary force is

$$F_{\text{cap}} = 2\pi\ R\ \gamma \cos\theta. \tag{1.49}$$

At equilibrium, the two forces balance each other, and Jurin's law reads

$$h = \frac{2\gamma \cos\theta}{\rho\ g\ R}. \tag{1.50}$$

The capillary rise h is inversely proportional to the tube radius R (figure 1.24). Many different geometries for the capillary rise have been investigated [2]. For example, the capillary rise in square tubes of inner dimension a is obtained using the same arguments as above

$$F_{\text{grav}} = \rho\ g\ a^2\ h = 4a\ \gamma \cos\theta = F_{\text{cap}}, \tag{1.51}$$

hence

$$h = \frac{4\gamma \cos\theta}{\rho\ g\ a}. \tag{1.52}$$

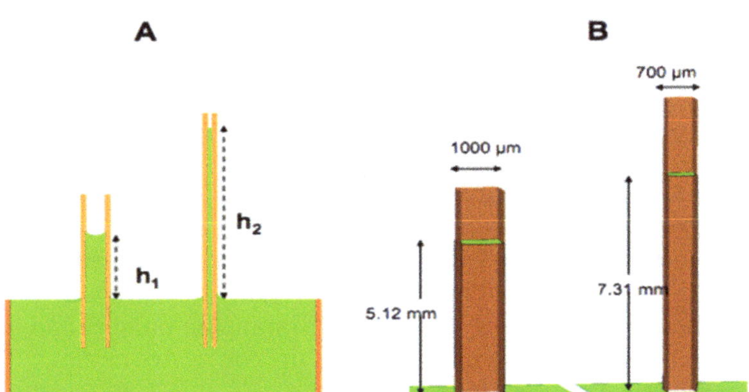

Figure 1.24. (A) Capillary rise in cylindrical tubes: the equilibrial height is inversely proportional to the diameter. (B) Capillary rise in square tubes: the equilibrial height is inversely proportional to the square side (Surface Evolver calculation).

The dynamics of the capillary rise has also been investigated, for example by O'Loughlin and coworkers [45], Tang *et al* [46], and Tsori *et al* [47] for nonparallel walls. Finally, the capillary rise can be amplified using the electrowetting effect [48].

1.1.12 Marangoni convection

Surface tension is a function of the temperature and the surfactant concentration in a liquid. Eötvös first pointed out the surface tension temperature dependency [49]. Guggenheim later derived the exact relation between surface tension and temperature [50]:

$$\gamma = \gamma_0\left(1 - \frac{T}{T_C}\right)^{\frac{11}{9}}, \tag{1.53}$$

where T_C is the critical temperature at which the interface vanishes and γ_0 is a constant. If the surface tension is known for a given temperature (e.g. room temperature T_1), a more practical expression is

$$\gamma = \gamma_1\left[1 - \beta_T(T - T_1)\right]. \tag{1.54}$$

On the other hand, in the case of added surfactants, the decrease in the surface tension is directly given by the number of molecules adsorbed at the interface, as given by the Gibbs adsorption isotherm for dilute solutions [51]:

$$\Gamma = -\frac{c}{RT}\frac{d\gamma}{dc}. \tag{1.55}$$

where Γ is the surface concentration, c the surfactant bulk concentration, and γ the surface tension (T is temperature and R the gas constant). Above a critical surfactant concentration—which is called the critical micelle concentration (CMC)—the interface is saturated and the surface tension stabilizes to a low value. Above the CMC, excess surfactants form micelles in the bulk of the fluid. In the limit of small concentration, (1.55) can be linearized:

$$\gamma = \gamma_1\left[1 - \beta_c(c - c_1)\right]. \tag{1.56}$$

A nonuniform surface temperature or a nonuniform surface surfactant concentration can trigger convective motions inside the fluid. This phenomenon is called Marangoni convection (figure 1.25), in reference to the Italian physicist Carlo Marangoni who first identified this phenomenon [52].

Two nondimensional numbers, the thermal Marangoni number and the chemical Marangoni number, indicate the importance of the convective motions. The thermal convection is governed by the number

$$Ma = \frac{d\gamma}{dT}\frac{L\Delta T}{\mu\alpha}, \tag{1.57}$$

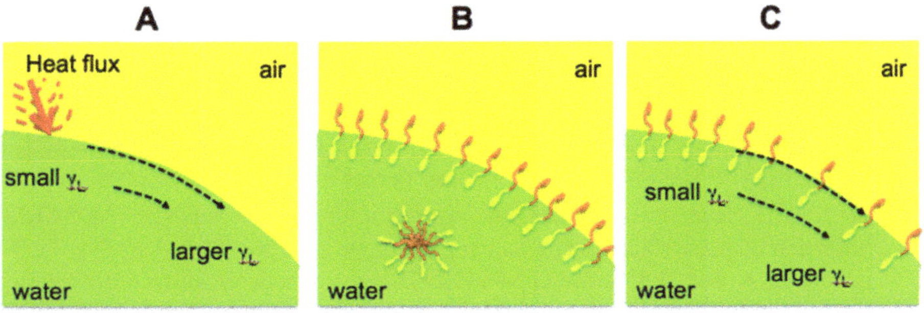

Figure 1.25. Sketches of Marangoni convection: (A) a local temperature rise at the interface reduces the surface tension, a gradient of surface tension forms at the interface, and a convective motion appears from the smaller-surface-tension area to the larger-surface-tension area. (B) A sketch of surfactants with their hydrophobic heads and hydrophilic tails migrating to the water/air interface. An excess of surfactant creates micelles inside the bulk. (C) A local uneven concentration of surface surfactants triggers convective motion from the low-surface-tension area to the larger-surface-tension area [53].

where L is a characteristic distance scale of the system and α is the thermal diffusivity. The chemical Marangoni number is

$$Ma = \frac{d\gamma}{dc}\frac{L\Delta c}{\mu\alpha}. \tag{1.58}$$

References

[1] Antonoff G N 1907 Sur la tension superficielle des solutions dans la zone critique *J. Chim. Phys.* **5** 364–71

[2] Berthier J and Brakke K A 2012 *The Physics of Microdroplets* (New York: Wiley)

[3] Baret J-C 2012 Surfactants in droplet-based microfluidics *Lab Chip* **12** 422

[4] Rodríguez-Valverde M A and Miranda M T 2011 Derivation of Jurin's law revisited *Eur. J. Phys.* **32** 49

[5] Nosonovsky M and Ramachandran R 2015 Geometric interpretation of surface tension equilibrium in superhydrophobic systems *Entropy* **17** 4684–700

[6] Finn R 2006 The contact angle in capillarity *Phys. Fluids* **18** 047102

[7] Good R J 1970 Intermolecular forces across an interface *Automotive Engineering Congress and Exposition 1970-01-12 (Detroit, MI)*

[8] Girifalco L A and Good R J 1957 A theory for the estimation of surface and interfacial energies. I. Derivation and application to interfacial tension *J. Phys. Chem.* **61** 904–9

[9] Marmur A 2006 Soft contact: measurement and interpretation of contact angles *Soft Matter* **2** 12–7

[10] Owens D K and Wendt R C 1969 Estimation of the surface free energy of polymers *J. Appl. Polym. Sci.* **13** 1741–7

[11] Israelachvili J N 2011 *Intermolecular and Surface Forces with Applications to Colloidal and Biological Systems* (Waltham, MA: Academic) 3rd edn

[12] Guisbiers G, Arscott S and Snyders R 2012 An accurate determination of the surface energy of solid selenium *Appl. Phys. Lett.* **101** 231606

[13] Yu Y S and Zhao Y P 2009 Deformation of PDMS membrane and micro-cantilever by a water droplet: comparison between Mooney–Rivlin and linear elastic constitutive models *J. Colloid Interface Sci.* **332** 467–76
[14] del Rio O I and Neumann A W 1997 Axisymmetric drop shape analysis: computational methods for the measurement of interfacial properties from the shape and dimensions of pendant and sessile drops *J. Coll. Int. Sci.* **196** 134–47
[15] Hoofar M and Neumann A W 2004 Axisymmetric drop shape analysis (ADSA) for the determination of the surface tension and contact angle *J. Adhes.* **80** 727–43
[16] Chaudhury M K and Whitesides G M 1992 Correlation between surface free energy and surface constitution *Science, New Ser.* **255** 1230–2
[17] Kano Y and Akiyama S 1992 Critical surface tension of poly (vinylidene fluoride-co-hexafluoroacetone) by the contact angle method *Polymer* **33** 1690
[18] Zisman W A 1941 The spreading of oils on water, part III. Spreading pressures and the Gibbs adsorption relation *J. Chem. Phys.* **9** 789–93
[19] Chen-Chih Tsai Y G and Kornev K G 2014 Wetting of nanofiber yarns *Colloids Surf.* A **459** 22–30
[20] Fox H W and Zisman W A 1950 The spreading of liquids on low-energy surfaces, I: polytetrafluoroethylene *J. Colloid Sci.* **5** 514–31
[21] Fox H W and Zisman W A 1952 The spreading of liquids on low-energy surfaces, II: modified tetrafluoroethylene polymers *J. Colloid Sci.* **7** 109–21
[22] Mahadik D B *et al* 2011 Effect of surface composition and roughness on the apparent surface free energy of silica aerogel materials *APL* **99** 104104
[23] Trigwell S, De S, Sharma R, Mazumder M K and Mehta J L 2006 Structural evaluation of radially expandable cardiovascular stents encased in a polyurethane film *J. Biomed. Mater. Res.* B **76** 241–50
[24] Bonn D, Eggers J, Indekeu J, Meunier J and Rolley E 2009 Wetting and spreading *Rev. Mod. Phys.* **81** 739–805
[25] Siboni S, Della Volpe C, Maniglio D and Brugnara M 2004 The solid surface free energy calculation, II. The limits of the Zisman and of the 'equation-of-state' approaches *J. Colloid Interface Sci.* **271** 454–72
[26] Ruzeng Zhu S C and Xiaosong Wang 2010 Theoretical foundation of Zisman's empirical equation for wetting of liquids on solid surfaces *Eur. J. Phys.* **31** 251–6
[27] Berthier J *et al* 2017 On the use of Zisman's function for the determination of the velocity of spontaneous capillary microflows *Proc. of the 2017 Nanotech Conf. (Washington DC, 13–15 May)*
[28] de Genes P G 1985 Wetting: statics and dynamics *Rev. Mod. Phys.* **57** 827
[29] Berthier J, Loe-Mie F, Tran V-M, Schoumacker S, Mittler F, Marchand G and Sarrut N 2009 On the pinning of interfaces on micropillar edges *J. Colloid Interface Sci.* **338** 296–303
[30] Berthier J, Brakke K A and Berthier E 2016 *Open Microfluidics* (New York: Wiley)
[31] Ondarcuhu T 1995 Total or partial pinning of a droplet on a surface with a chemical discontinuity *J. Phys. II* **5** 227–41
[32] Langbein D W 2002 *Capillary Surfaces: Shape-Stability-Dynamics, in Particular Under Weightlessness* (New York: Springer)
[33] Berthier J, Brakke K A, Grossi F, Sanchez L and Di Cioccio L 2010 Self-alignment of silicon chips on wafers: a capillary approach *J. Appl. Phys.* **108** 054905

[34] Berthier J, Mermoz S, Brakke K A, Sanchez L, Fretigny C and Di Cioccio L 2013 Capillary self-alignment of polygonal chips: a generalization for the shift-restoring force *Microfluid. Nanofluid.* **14** 845–58

[35] Fukushima T, Konno T T, Iwata E, Kobayashi R, Kojima T, Murugesan M, Bea J-C, Lee K-W, Tanaka T and Koyanagi M 2011 Self-assembly of chip-size components with cavity structures: high precision alignment and direct bonding without thermal compression for hetero-integration *Micromachine* **2** 49–68

[36] Wenzel R N 1936 Resistance of solid surface to wetting by water *Ind. Eng. Chem.* **28** 988–94

[37] Cassie A B D and Baxter S 1944 Wettability of porous surfaces *Trans. Faraday Soc.* **40** 546–51

[38] Barthlott W and Neinfuis C 1997 Purity of the sacred lotus, or escape from contamination in biological surfaces *Planta* **202** 1–8

[39] Fan J-G, Tang X-J and Zhao Y-P 2004 Water contact angles of vertically aligned SI nanorod arrays *Nanotechnology* **15** 501–4

[40] Onda T, Shibuichi S, Satoh N and Tsujii K 1996 Super water repellent surfaces *Langmuir* **12** 2125–7

[41] Patankar N A 2004 Transition between superhydrophobic states on rough surfaces *Langmuir* **20** 7097–102

[42] Nosonovsky M 2007 Multiscale roughness and stability of superhydrophobic biomimetic interfaces *Langmuir* **23** 3157–61

[43] Kemp M 2006 *Leonardo da Vinci : the marvellous works of nature and man* (Oxford: Oxford Univ. Press)

[44] Jurin J 1719 An account of some experiments shown before the Royal Society; with an enquiry into the cause of the ascent and suspension of water in capillary tubes *Phil. Trans.* **30** 739–47

[45] O'Loughlin M, Wilk K, Priest C, Ralston J and Popescu M N 2013 Capillary rise dynamics of aqueous glycerol solutions in glass capillaries: a critical examination of the Washburn equation *J. Colloid Interface Sci.* **411** 257–64

[46] Tang L H and Tang Y 1994 Capillary rise in tubes with sharp grooves *J. Phys. II France* **4** 881

[47] Tsori Y 2006 Discontinuous liquid rise in capillaries with varying cross-sections *Langmuir* **22** 8860–3

[48] Chen J H and Hsieh W H 2006 Electrowetting-induced capillary flow in a parallel-plate channel *J. Colloid Interface Sci.* **296** 276–83

[49] Eötvös R 1886 Ueber den Zusammenhang der Oberflächenspannung der Flüssigkeiten mit ihrem Molecularvolumen *Ann. Phys.* **263** 448–59

[50] Guggenheim E A 1945 The principle of corresponding states *J. Chem. Phys.* **13** 253–61

[51] Toshev B V 2006 Gibb's thermodynamics of wetting *Int. J. Mech. Eng. Educ.* **34** 256

[52] Marangoni C 1869 *Sull'espansione delle goccie d'un liquido galleggianti sulla superficie di altro liquido (On the Expansion of a Droplet of a Liquid Floating on the Surface of Another Liquid)* (Pavia, Italy: Fratelli Fusi (Fusi Brothers))

[53] Alexeev A, Gambaryan-Roisman T and Stephan P 2005 Marangoni convection and heat transfer in thin liquid films on heated walls with topography: experiments and numerical study *Phys. Fluids* **17** 062106

Chapter 2

The Lucas–Washburn–Bosanquet approach

2.1 Introduction

Capillarity—a word derived from the Latin 'capillaris' meaning hair—was observed long ago by scholars and scientists such as da Vinci, Boyle, and Hauskbee [1]. However, at this time, it remained very approximate and did not have a strong background.

A successful quantitative treatment of capillary action did not take place until 1805, when it was completed by two scientists, Young [2] and Laplace [3]. They derived the so-called Young–Laplace equation for capillary action. By 1830, the German mathematician Gauss had determined the boundary conditions governing capillary action (i.e. the conditions at the liquid–solid interface). Later, in 1894, Neumann published a study of the interaction between two immiscible liquids and generalized Young's law to immiscible liquid contact [4].

Investigation of capillary flow began around 1920 with the works of Lucas [5], Washburn [6], Rideal [7], and Bosanquet [8]. They investigated the capillary flow inside horizontal and vertical cylinders and derived laws to understand their dynamics.

In this chapter, we present the approach followed by these scientists. This approach is still the basis for our understanding of capillary flow.

2.2 The Bosanquet equation

Historically, the first work was performed for the capillary flow inside cylindrical tubes. The Bosanquet equation is the general form of a capillary flow, generalizing the Lucas–Washburn (LW) law [5, 6]. Although the general Bosanquet equation [8] was published after the investigations of Lucas and Washburn, we present it first due to the general character of this equation. Bosanquet's equation states that a change of momentum results from the difference between capillary forces F_{cap} and wall friction F_{drag}, i.e.

$$\frac{d(mV)}{dt} = F_{cap} - F_{drag}, \tag{2.1}$$

doi:10.1088/978-0-7503-5507-0ch2

where m is the mass of liquid in the capillary tube, V the average liquid velocity, and t the time. Relation (2.1) assumes that the motion is completely driven by surface tension, with no applied pressure at either end of the capillary tube (figure 2.1). It also assumes that the tube is horizontal. Later in this chapter, we shall see the extension of the Bosanquet equation to vertical tubes.

The capillary force F_{cap} is given by

$$F_{\text{cap}} = 2\pi R\, \gamma \cos\theta, \tag{2.2}$$

where R is the tube radius, γ the liquid surface tension (unit N m^{-1}), and θ the contact angle of the liquid with the wall (we shall see later the complexity of determining this contact angle).

The wall friction F_{drag} is calculated using the Poiseuille–Hagen formulation [9]. Note that this is an approximation, because the velocity field behind the advancing meniscus is quite complicated due to microscopic-scale recirculation, but the velocity field at about one hydraulic diameter behind the meniscus is already laminar. So, the Poiseuille–Hagen wall friction expression can be used. This expression is directly derived from the Stokes formulation [10]. Integration of the Stokes equation yields the velocity in a cross-section of the tube

$$v\ (r) = 2V\left(1 - \frac{r^2}{R^2}\right), \tag{2.3}$$

And the wall friction is—as an absolute value

$$\tau = \left\| \mu \frac{\partial v}{\partial r}_{(r=R)} \right\| = 4\mu\, \frac{V}{R}. \tag{2.4}$$

Hence the total wall friction (viscous force) is

$$F_{\text{drag}} = 4\mu\, \frac{V}{R}\,(2\pi R z) = 8\pi\, \mu\, zV, \tag{2.5}$$

If S denotes the cross-sectional area $S = \pi R^2$ and m the mass of liquid in the tube—which depends on the time—$m = \rho\, z\, S = \rho\, z\, \pi\, R^2$, where ρ is the liquid density, equation (2.1) can be rewritten as

$$\rho\, \pi R^2 \frac{d(zV)}{dt} = 2\pi R\, \gamma \cos\theta - 8\pi\, \mu\, zV, \tag{2.6}$$

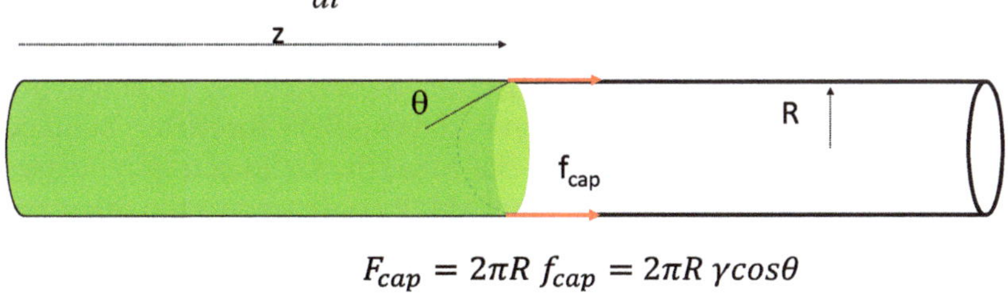

Figure 2.1. A sketch of a tube with an advancing meniscus. F_{cap} is the capillary force in the z direction (unit N) and f_{cap} the capillary force per unit length (unit N m^{-1}).

or, using $V = \frac{dz}{dt}$

$$\rho\, \pi R^2 \frac{d}{dt}\left(z \frac{dz}{dt}\right) = 2\pi R\, \gamma \cos\theta - 8\pi\, \mu\, z \frac{dz}{dt}, \tag{2.7}$$

Let us note $a = 8\mu/\rho R^2$ (unit 1 s^{-1}) and $b = 2\gamma \cos\theta/\rho R$ (unit $m^2\, s^{-2}$). The division of all the terms of (2.7) by $\rho\pi\, R^2$ yields

$$\frac{d}{dt}\left(z \frac{dz}{dt}\right) = \frac{2\gamma \cos\theta}{\rho R} - \frac{8\mu}{\rho R^2} z \frac{dz}{dt} = b - a\left(z \frac{dz}{dt}\right), \tag{2.8}$$

leading to the following differential equation in z^2:

$$\frac{d^2 z^2}{dt^2} + a \frac{dz^2}{dt} - 2b = 0. \tag{2.9}$$

The solution below was first given by Bosanquet:

$$z^2 - z_0^2 = \frac{2b}{a}\left[t - \frac{1}{a}\left(1 - e^{-at}\right)\right], \tag{2.10}$$

where $z = z_0$ at $t = 0$.

In the case of a vertical tube, let us change the notation by changing the travel distance z into h, the capillary height. This time, gravity must be taken into account. The pressure associated with gravity is

$$P_{grav} = \rho g h, \tag{2.11}$$

and (2.7) becomes

$$\rho\, \pi R^2 \frac{d}{dt}\left(h \frac{dh}{dt}\right) = 2\pi R\, \gamma \cos\theta - 8\pi\, \mu\, h \frac{dh}{dt} - \rho g h \pi\, R^2, \tag{2.12}$$

where g is gravitational acceleration. In this case, a numerical integration is required.

2.3 Simplification: inertial and viscous regimes

A simple and convenient method used to interpret the Bosanquet equation is to split the equation in two. Observing that, at the beginning of the flow, the wetted wall surface is very small, one assumes that the wall friction can be neglected in this first phase. This phase—in which the wall friction is negligible—is called the inertial regime or the inviscid regime. In this regime, the capillary forces are balanced by the inertial forces. As the liquid starts filling the tube, the wall friction increases, even if the flow velocity decreases. Inertia becomes small compared to the other two forces, and the dynamics of the flow is governed by the balance between capillary force and wall friction. This is the viscous regime, governed by the LW equation.

In this section, we successively present the two regimes.

2.3.1 The inertial evanescent regime

The inertial regime occurs when $at \ll 1$, or $t \ll 1/a = \frac{\rho R^2}{8\mu}$ [8, 11]. In this case, using the Taylor expansion

$$e^{-at} \cong 1 - at + \frac{a^2 t^2}{2}, \tag{2.13}$$

and substituting (2.13) into (2.10) and setting $z_0 = 0$, one obtains

$$z^2 = bt^2, \tag{2.14}$$

or

$$z = \sqrt{b}\ t. \tag{2.15}$$

In conclusion, in the early inertial regime, the travel distance z is proportional to the time t. The derivation of z in respect to t yields a constant velocity for the flow

$$V = \sqrt{b} = \sqrt{\frac{2\gamma \cos\theta}{\rho R}}. \tag{2.16}$$

This expression shows that the constant inertial velocity is related to the ratio between capillary forces and inertia.

2.3.2 The viscous regime: the Lucas–Washburn–Rideal law

The viscous forces (wall friction) quickly increase with the wetting of the wall. In parallel, the inertial term decreases with the decreasing velocity of the fluid. This statement is verified by the following developments. In fact, it will be shown that the term $z\frac{dz}{dt} = zV$ progressively becomes constant, so that its time derivative $\frac{d}{dt}\left(z\frac{dz}{dt}\right)$ becomes null and the inertial term disappears. The regime is now the viscous regime described by Lucas and Washburn [5, 6]:

$$F_{\text{cap}} = F_{\text{drag}}. \tag{2.17}$$

Hence, the dynamics of the flow is given by the relation

$$2\pi R\ \gamma \cos\theta = 8\pi\ \mu\ z\frac{\mathrm{d}z}{\mathrm{d}t}, \tag{2.18}$$

or

$$z\frac{\mathrm{d}z}{\mathrm{d}t} = \frac{R\ \gamma \cos\theta}{4\mu}. \tag{2.19}$$

The integration of (2.19) leads to

$$z = \sqrt{\frac{R\ \gamma \cos\theta}{2\mu}t}\ . \tag{2.20}$$

In the viscous regime, the travel distance is proportional to the square root of the time. The flow velocity is then

$$V = \frac{dz}{dt} = \sqrt{\frac{R\,\gamma\cos\theta}{8\mu}}\sqrt{\frac{1}{t}}. \tag{2.21}$$

2.3.3 Transition between the two regimes

The transition between the two regimes occurs when the linear (inertial) regime crosses the parabolic solution (viscous regime) [11]. This occurs at a time τ_{in} given by

$$\sqrt{\frac{R\,\gamma\cos\theta}{2\mu}\tau_{in}} = \sqrt{\frac{2\gamma\cos\theta}{\rho R}}\,\tau_{in}, \tag{2.22}$$

or

$$\tau_{in} = \frac{\rho\,R^2}{4\mu}. \tag{2.23}$$

The transition time τ_{in} is proportional to the square of the tube radius R: $\tau_{in} \sim R^2$. In the case of water flowing in a tube that has a 1 mm radius, $\tau_{in} = 0.\ 25\text{s}$.The transition time is usually very short. Because the viscous regime is linked to a fully developed boundary layer along the walls of the tube, the transition time τ_{in} can be physically understood as the time necessary for the viscous boundary layer to diffuse over a length on the order of the radius of the tube [11].

Note that if the transition time is usually very small, the transition distances are not so small due to the high inertial velocity. The distance at which transition occurs is

$$z_{in} = V_{in}\tau_{in} = \frac{\rho\,R^2}{4\mu}\sqrt{\frac{2\gamma\cos\theta}{\rho R}}. \tag{2.24}$$

In summary, the two regimes defined in the simplified approach are sketched in figure 2.2. We shall see later that this approach is a simplification of the reality, but it is convenient since the first inertial regime has a very short duration. Most of the time, it is even masked by the initial conditions for the capillary flow (the filling of an inlet reservoir, for example).

2.3.4 Examples

Let us consider a few liquids—aqueous and organic—and two different solids, glass and poly(methyl methacrylate) (PMMA). Aqueous liquids have a small contact angle with a glass surface, and organic liquids with PMMA. The duration of the inertial regime is very short, but the inertial length z_{in} depends greatly on the tube radius. The transition times and transition distances are given in tables 2.1 and 2.2 for different liquids and two different tubes whose radii are 1 mm and 0.25 mm.

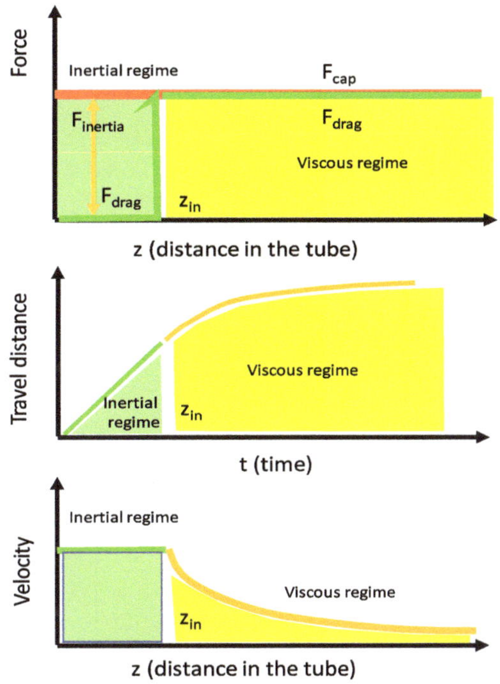

Figure 2.2. A sketch of the two regimes in the simplified approach: the inertial regime (light green) and the viscous regime (yellow). Top, a sketch of the forces taken into account; middle, travel distance vs. time; bottom, velocity vs. distance.

A plot of the transition distance as a function of the transition time for the pairs of liquids and solids of table 2.2 is shown in figure 2.3.

2.4 The full Bosanquet solution

Note that relation (2.23) yields $a\ \tau_{in} = 2$, which is far from the relation $a\, t \ll 1$ for the inertial regime. The split of the dynamics into two regimes attributes too much importance to the inertial regime. In reality, there is a transition regime between the inertial regime and the viscous regime. A more detailed analysis is necessary.

The Bosanquet equation (2.10) encompasses the two regimes. Setting $z_0 = 0$ at the tube entrance, the travel distance z is given by

$$z^2 = \frac{2b}{a}\left[t - \frac{1}{a}(1 - e^{-at})\right], \tag{2.25}$$

where $a = 8\mu/\rho R^2$, $b = 2\gamma \cos\theta/\rho R$ and $b/a = R\gamma \cos\theta\ /4\mu$.

A comparison between the piecewise simplified model (for inertial and viscous regimes) and the Bosanquet analytic solution is shown in figures 2.4 and 2.5 for a solution of isopropyl alcohol (50% w/w) [12] in a cylinder that has a radius of 1 mm. On a large scale (figures on the left), the inertial regime does not appear clearly. Zooming in to the initial time shows the difference in the case of the Bosanquet

Table 2.1. Inertial times and distances for some pairs of liquids and solids (radius $R = 1$ mm).

Liquid/solid	μ	ρ	θ	R	Gamma	a (1 s^{-1})	b (m^2 s^{-2})	t_{in} (s)	V_{in} (mm s^{-1})	z_{in} (mm)
Water/glass	1.00E−03	1000.0	0.0	1.00E−03	7.20E−02	8.0	0.14	0.25	379.5	94.9
Ethanol/glass	1.00E−03	1000.0	0.0	1.00E−03	3.60E−02	8.0	0.07	0.25	268.3	67.1
Nonanol/PMMA	1.15E−02	827.0	13.0	1.00E−03	2.95E−02	111.2	0.07	0.02	263.6	4.74
Pentanol/PMMA	3.75E−03	814.0	12.5	1.00E−03	2.54E−02	36.9	0.06	0.05	246.8	13.4
Chloroform/PMMA	6.00E−04	1490.0	17.3	1.00E−03	2.76E−02	3.2	0.04	0.62	188.1	116.7
IPA50/PMMA	3.20E−03	1100.0	18.0	1.00E−03	3.20E−02	23.3	0.06	0.09	235.2	20.2
FC40/PMMA	2.65E−03	1850.0	12.7	1.00E−03	1.60E−02	11.5	0.02	0.17	129.9	22.7
Toluene/PMMA	5.90E−04	867.0	7.8	1.00E−03	2.85E−02	5.4	0.07	0.37	255.2	93.8
Dodecane/PMMA	1.20E−03	750.0	26.7	1.00E−03	2.54E−02	12.8	0.06	0.16	246.0	38.4

Table 2.2. Inertial times and distances for some pairs of liquids and solids (radius $R = 0.25$ mm)

Liquid/solid	μ	ρ	θ	R	Gamma	a (1 s^{-1})	b (m^2 s^{-2})	t_{in} (s)	V_{in} (mm s^{-1})	z_{in} (mm)
Water/glass	1.00E−03	1000.0	0.0	0.25e−3	7.20E−02	128.00	5.76E−01	0.016	758.95	11.86
Ethanol/glass	1.00E−03	1000.0	0.0	0.25e−3	3.60E−02	128.00	2.88E−01	0.016	536.66	8.39
Nonanol/PMMA	1.15E−02	827.0	13.0	0.25e−3	2.95E−02	1779.93	2.78E−01	0.001	527.31	0.59
Pentanol/PMMA	3.75E−03	814.0	12.5	0.25e−3	2.54E−02	589.68	2.44E−01	0.003	493.67	1.67
Chloroform/PMMA	6.00E−04	1490.0	17.3	0.25e−3	2.76E−02	51.54	1.41E−01	0.039	376.14	14.60
IPA50/PMMA	3.20E−03	1100.0	18.0	10.25e−3	3.20E−02	372.36	2.21E−01	0.005	470.46	2.53
FC40/PMMA	2.65E−03	1850.0	12.7	0.25e−3	1.60E−02	183.35	6.75E−02	0.011	259.80	2.83
Toluene/PMMA	5.90E−04	867.0	7.8	0.25e−3	2.85E−02	87.10	2.61E−01	0.023	510.43	11.72
Dodecane/PMMA	1.20E−03	750.0	26.7	0.25e−3	2.54E−02	204.80	2.42E−01	0.010	491.98	4.80

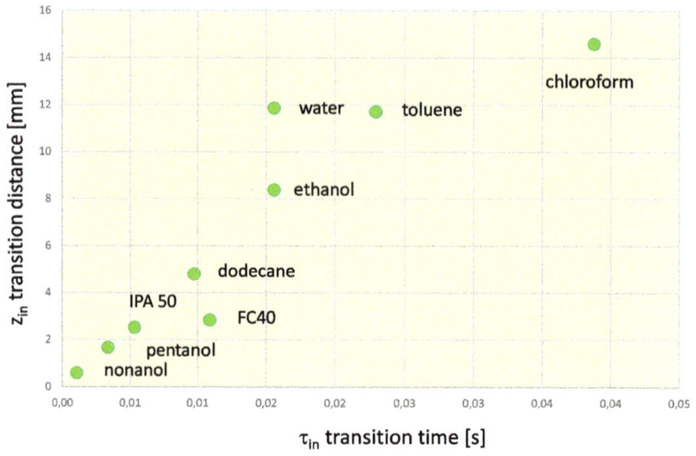

Figure 2.3. Transition distances vs. transition times for the liquid/solid pairs of table 2.2.

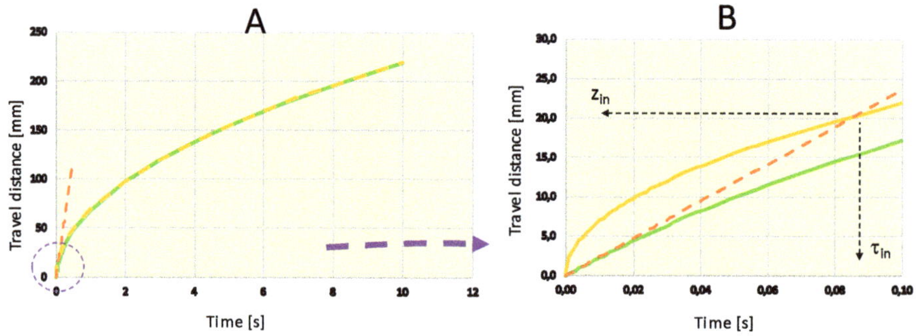

Figure 2.4. The travel distance of IPA50 inside a PMMA cylinder that has a radius of 1 mm: a comparison between the simplified inertial and viscous regimes (red dotted line and orange continuous line) and the Bosanquet analytical solution (green). (A) Timescale from 0 to 10 s; (B) timescale from 0 to 0.10 s.

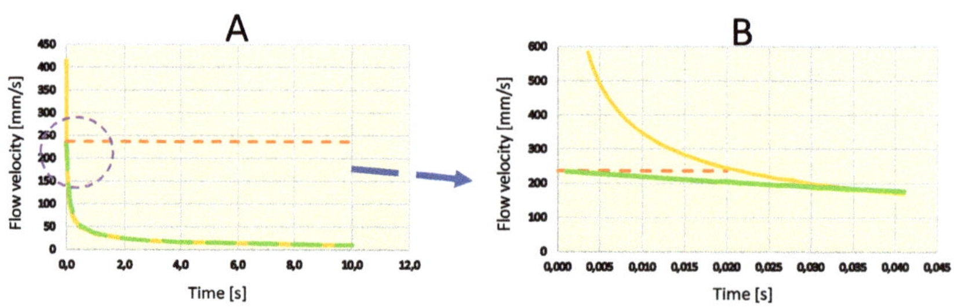

Figure 2.5. The flow velocity of IPA50 inside a cylinder that has a radius of 1 mm: a comparison between the simplified inertial and viscous regimes (red dotted line and orange continuous line) and the Bosanquet analytical solution (green). (A) Timescale from 0 to 10 s; (B) timescale from 0 to 0.04 s.

analytic solution (figures on the right). Note that the two curves corresponding to the inertial and viscous regime cross at τ_{in} and z_{in} (in figure 2.4 for travel distances), but earlier in figure 2.5 for the velocities.

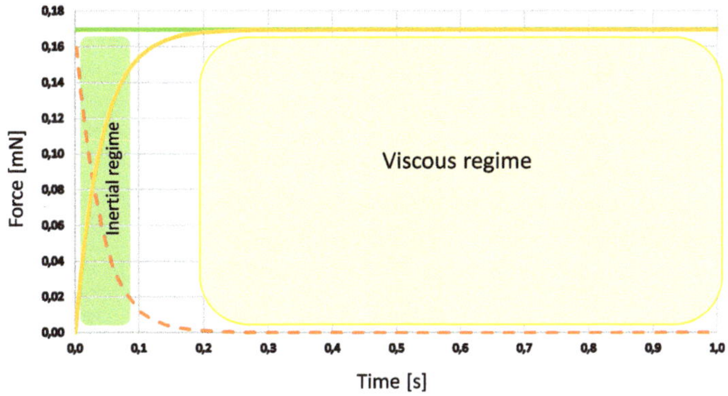

Figure 2.6. Forces governing the capillary dynamic of IPA50 flowing in the tube with a radius of 1 mm: the green horizontal line corresponds to the capillary force (F_{cap}), the orange line corresponds to the wall friction (F_{drag}), and the red dotted line to the inertial forces.

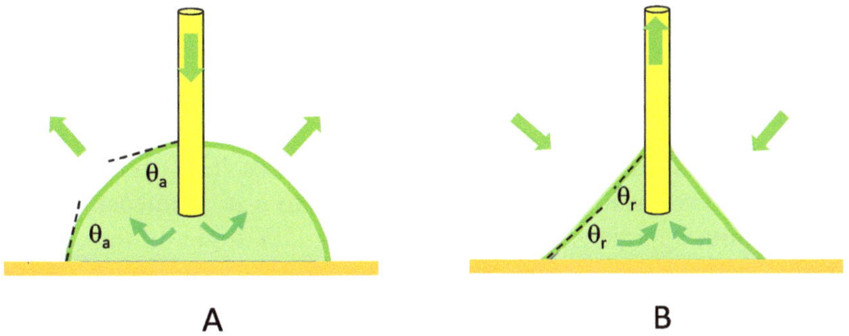

Figure 2.7. A sketch of the advancing and receding contact angles: (A) advancing contact angle; (B) receding contact angle.

Considering the forces (inertial, viscous, and capillary) the drawing of figure 2.2 must be corrected by figure 2.6. A transition appears between the two regimes (inertial and viscous).

2.5 Correcting for the dynamic contact angle

The capillary force depends on the value of the contact angle. In the preceding sections, a constant contact angle has been used. In this section, the real contact angle is investigated, and we show how it modifies the Lucas–Washburn–Bosanquet approach.

2.5.1 The dynamic contact angle

Let us now consider the contact angle, which is used in the calculation of the capillary force. It has been observed that the contact angle of a liquid advancing or receding on a solid surface does not keep its static value (figure 2.7) [13]. To distinguish it from its static value, this changing contact angle is called the dynamic

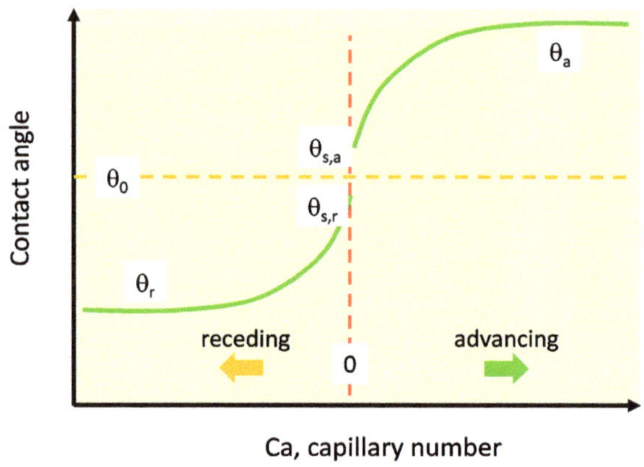

Figure 2.8. The variation of the advancing and receding contact angles as a function of the capillary number.

contact angle (DCA). Figure 2.8 shows the evolution of the contact angle as a function of the capillary number. We recall that the capillary number is defined by

$$\mathrm{Ca} = \frac{\mu\, V}{\gamma}. \tag{2.26}$$

It is the ratio between viscous and surface tension forces. The capillary number also is convenient when we wish to reduce the velocity to a nondimensional form.

At rest, the contact angle is the static Young contact angle. In the case of a rough surface, the static contact angle can take two different values, $\theta_{s,a}$ and $\theta_{s,r}$, which are somewhat different from the 'theoretical' Young value, depending on how the liquid has been immobilized [14]. In the following, we assume a perfectly smooth contact and

$$\theta_{s,\,a} = \theta_{s,\,r} = \theta_0. \tag{2.27}$$

A. The different approaches used for the DCA

The basis for our understanding of the DCA was established by de Gennes [15, 16], who demonstrated that the loss of interfacial tension forces due to the DCA $F = \gamma\,(\cos\theta_0 - \cos\theta_\mathrm{d})$, where γ is the liquid/air interfacial tension and θ_d and θ_0 are the dynamic and equilibrium contact angles respectively, must be compensated either by the viscous dissipation at a mesoscopic scale or by the viscous dissipation in the vicinity of the contact line, or in the precursor film.

The first case gave birth to the hydrodynamic (HD) model, which considers the viscous energy dissipation in the mesoscopic layer associated with the bending of the velocity profile shown in figure 2.9. A. In the HD approach, the increase of the contact angle outside the microscopic layer (L_mi) is linked to the cubic root of the capillary number [17–20]

$$\theta_\mathrm{d}^3 = \theta_0^3 + 9\chi Ca = \theta_0^3 + 9\ln\left(L_\mathrm{ma}/L_\mathrm{mi}\right) Ca, \tag{2.28}$$

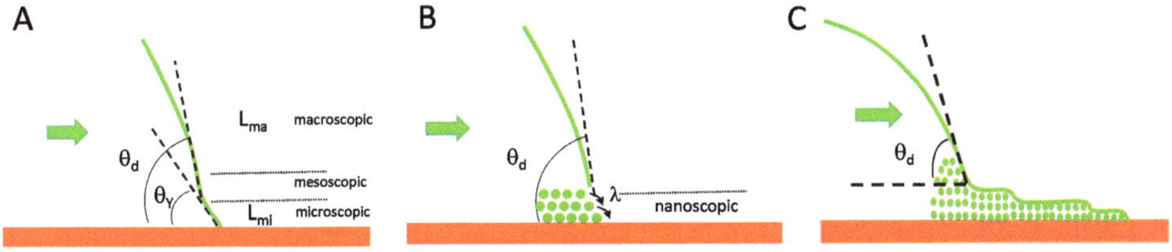

Figure 2.9. Sketches of the different models for the DCA: (A) the hydrodynamic model assumes a microscopic layer that has a static contact angle with the solid wall and a bending of the interface starting at a mesoscopic scale; (B) the molecular kinetic theory (MKT) model that considers adsorption–desorption processes at a nanoscopic scale at the triple phase zone (TPZ); the MKT-self-layering model that considers a wall viscosity associated with the formation of layers in the precursor film.

where Ca is the capillary number ($Ca = \mu V/\gamma$), and χ is a parameter given by $\chi = \ln(L_{\mathrm{ma}}/L_{\mathrm{mi}})$, L_{ma} is a macroscopic length scale which is of the order of the size of the meniscus, and L_{mi} is a microscopic length scale, which is the cutoff length below which the continuum theory breaks down.

The molecular kinetic theory (MKT) is the second approach. It considers adsorption–desorption processes at the wall at the vicinity of the triple contact line (TPCL) at a molecular scale (figure 2.9(B)) [21–23]. The velocity-dependent DCA is then

$$\cos\theta_{\mathrm{d}} = \cos\theta_0 - \frac{\zeta}{\gamma}V, \tag{2.29}$$

where ζ is a frictional coefficient that applies between the liquid and solid surfaces at the triple contact line (unit Pa s). The term $\frac{\zeta}{\gamma}V$ is a nondimensional term similar to the capillary number.

Finally, if a precursor film exists—which is often the case when the static contact angle is very small—the MKT model has been reformulated to introduce a wall viscosity corresponding to the formation of molecular layers on the wall [24–26]:

$$\cos\theta_{\mathrm{d}} = \cos\theta_0 - \frac{\mu}{\gamma}V, \tag{2.30}$$

where μ is a wall viscosity (unit Pa s). We will not go into details here. Let us just mention that the precursor film (a film of nanoscopic thickness preceding the advancing liquid) has great importance for the wetting of solid surfaces [27]. Relation (2.30) is similar to (2.29), but ζ and μ have different meanings.

Also, in connection with the HD model, some empirical equations have been proposed based on the adhesion energy. Roughly, these correlations are valid for capillary numbers in the range from 5×10^{-4} to 10^{-2}. Below the 5×10^{-4} threshold, a stick–slip motion may appear, leading to fluctuations in the DCA, especially when solid surface roughness is present [28]. At even smaller capillary numbers (less than 10^{-6} or 10^{-7}), the DCA is the static advancing or receding angle ($\theta_{s,a}$ or $\theta_{s,r}$). These correlations usually link the cosine of the advancing contact angle to the capillary number (see table 2.3). As noted by Seebergh and Berg [29], most correlations are of the form

$$H = \frac{\cos\theta_0 - \cos\theta_d}{(1 + \cos\theta_0)} = A\ Ca^B \tag{2.31}$$

At first look, the Hoffman–Tanner correlation [17, 20] seems different, but note that this correlation considers $\theta_0 = 0$, and $Ca < 0.01$, thus the DCA is small, and $\cos\theta_d$ can be approximated by its Taylor expansion $\cos\theta_d \approx 1 - \theta^2/2$. Then, (2.31) yields $H \approx \theta^2/4 = A\ Ca^B$, or $\theta \approx 2A^{1/2}\ Ca^{B/2}$. Considering Jiang's correlation [30], where $A = 4.96$ and $B = 0.702$, the DCA is $\theta \approx 4.45Ca^{0.35}$, very close to the Hoffman–Tanner correlation.

In fact, the only correlation with a different form is that of Blake and coworkers [21, 22]. In chapter 4, we will investigate the MKT approach more carefully.

B. A comparison of the different models

In figure 2.10, the different correlations are compared. The value of the DCA is plotted as a function of the capillary number using the correlations of table 2.3.

The capillary force inside a cylinder that has a radius of 1 mm is shown in figure 2.11 for the different correlations. All the correlations—except the linear correlation of Blake *et al* for $\beta = 0.05$—produce similar capillary forces.

2.5.2 A dynamic contact angle correction to the Lucas–Washburn law

According to section 2.4.1, the LW law should take into account a DCA. This remark has been made by many researchers [32, 33]. The LW law (2.18) should be written with a dynamic angle (θ_d) that depends on the capillary number and indirectly on the location z of the advancing meniscus $\theta_d(Ca(z))$.

$$2\pi R\ \gamma\cos\theta_d(z) = 8\pi\ \mu\ z\frac{dz}{dt}, \tag{2.32}$$

or

$$\frac{1}{2}\frac{dz^2}{dt} = \frac{R\ \gamma\cos\theta_d(z)}{4\mu}. \tag{2.33}$$

Table 2.3. The different correlations for the DCA.

Correlation	Expression	Remarks	References
Hoffman–Tanner	$\theta = \alpha Ca^{1/3}$	α ~4–5	[17, 20]
Jiang	$\cos\theta = \cos\theta_0 - (1 + \cos\theta_0)\tanh(4.96Ca^{0.702})$	$Ca < 0.01$	[30]
	$\cos\theta = \cos\theta_0 - (1 + \cos\theta_0)4.96Ca^{0.702}$		
Bracke	$\cos\theta = \cos\theta_0 - 2(1 + \cos\theta_0)\ Ca^{\frac{1}{2}}$		[31]
Seebergh	$\cos\theta = \cos\theta_0 - 2.29(1 + \cos\theta_0)\ Ca^{0.54}$		[29]
Blake and Wu	$\cos\theta = \cos\theta_0 - \frac{\beta}{\mu}\ Ca = \cos\theta_0 - \frac{\beta}{\gamma}\ V$	β ~0.02–0.2	[21–26]

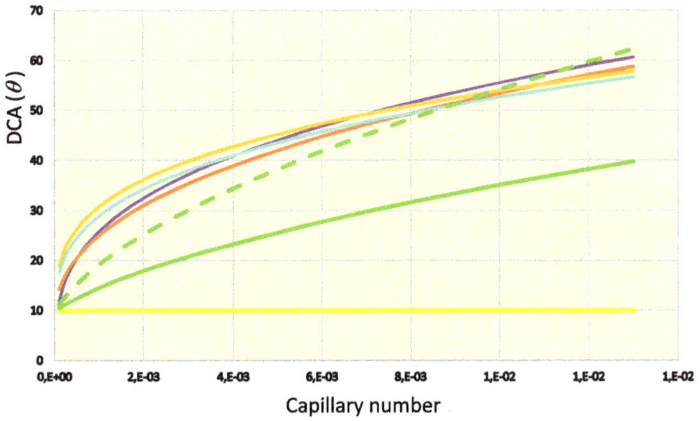

Figure 2.10. The DCA vs. capillary number for a static angle $\theta_0 = 10°$. Yellow line: θ_0; orange: Bracke's correlation; light blue line: Seebergh's correlation; violet line: Tanner–Hoffman correlation; red line: Jiang's correlation; continuous green line: Blake/Wu's correlation ($\beta = 0.05$); dotted green line: Blake/Wu's correlation ($\beta = 0.14$).

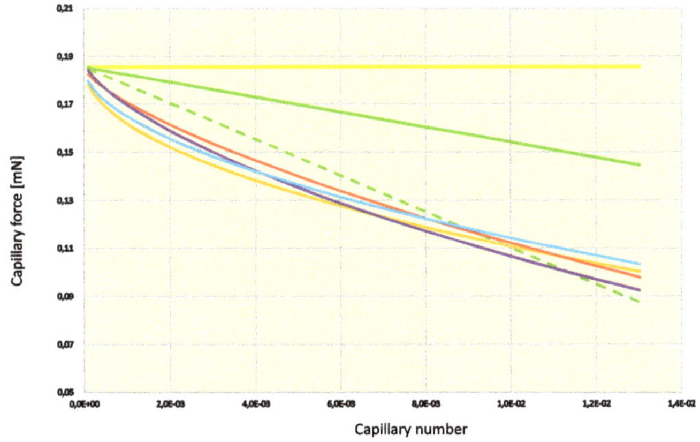

Figure 2.11. Capillary force vs. capillary number for a static angle $\theta_0 = 10°$. Yellow line: θ_0; orange: Bracke's correlation; light blue line: Seebergh's correlation; violet line: Tanner–Hoffman correlation; red line: Jiang's correlation; continuous green line: Blake's correlation ($\beta = 0.05$); dotted green line: Blake's correlation ($\beta = 0.14$).

Then, integrating (2.33) gives

$$z^2 = \frac{\gamma R}{2\mu} \int_0^t \cos\theta_d \, dt \tag{2.34}$$

Upon introducing the dynamic correction $\cos\theta_d = \cos\theta_0 - \mathscr{C}(z)$, where $\mathscr{C}(z) = H(1 + \cos\theta_0) = (1 + \cos\theta_0)\, A\, Ca^B$ or $\mathscr{C}(z) = (\beta/\mu)\, Ca$ depending on the form of the correlation considered in the calculation, we obtain

$$z^2 = \frac{\gamma R}{2\mu} \int_0^t (\cos\theta_0 - \mathscr{C}) dt = \frac{\gamma R}{2\mu} \cos\theta_0 - \frac{\gamma R}{2\mu} \int_0^t \mathscr{C} dt\,. \tag{2.35}$$

Noting that $\mathscr{C}$ can be expressed as $\mathscr{C} = K\ V^n = K\ (\mathrm{d}z/\mathrm{d}t)^n$, where K is a constant depending on the liquid and geometry, equation (2.35) becomes

$$z^2 = \frac{\gamma\ R}{2\mu}\cos\theta_0 t - \frac{\gamma\ R}{2\mu}K\int_0^t (\mathrm{d}z/\mathrm{d}t)^n \mathrm{d}t\ . \tag{2.36}$$

The case of Blake's correlation is particularly interesting because $n = 1$ (when the surrounding fluid is air) and $K = \beta/\gamma$. We shall see that a closed-form solution exists in this case. In the other cases, such as Bracke's correlation, Joos *et al* and Martic *et al* used numerical integration. If $n = 1$, relation (2.36) can be integrated

$$z^2 = \frac{\gamma\ R}{2\mu}\cos\theta_0 t - \frac{\beta\ R}{2\mu}z. \tag{2.37}$$

If z_0 denotes the solution of the LW relation using the static angle θ_0: $z_0 = \sqrt{\frac{\gamma\ R}{2\mu}\cos\theta_0 t}$, then

$$z^2 = z_0^2 - \frac{\beta\ R}{2\mu}z. \tag{2.38}$$

The solution for the DCA correction is then the solution of a quadratic equation. Finally,

$$z = -\frac{\beta\ R}{4\mu} + \sqrt{\left(\frac{\beta\ R}{4\mu}\right)^2 + z_0^2}\ . \tag{2.39}$$

The corresponding velocity is obtained using the time derivation of (2.39):

$$V = \frac{V_0}{\sqrt{\left(\frac{\beta\ R}{4\mu\ z_0}\right)^2 + 1}}. \tag{2.40}$$

An example of the importance of the DCA correction is shown in figure 2.12, for $\beta = 0.2$, $\gamma = 30$ mN m^{-1}, $\mu = 3$ mPa s, and $R = 1$ mm (which are approximate data for

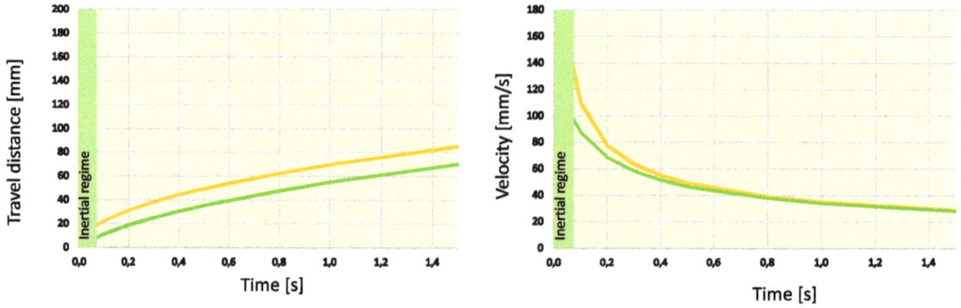

Figure 2.12. Left: travel distance vs time; right: velocity vs time. The orange curves correspond to the LW law with the static angle and the green curves to the LW law using the DCA. The velocity plot does not apply for the inertial regime occurring before 0.05 s.

a 50% solution of IPA in a PMMA tube). After 1 s, the velocity corrected for a DCA V is the same as the velocity V_0 obtained using the static contact angle. This is explained by the disappearance of the term $\left(\frac{\beta R}{4\mu z_0}\right)^2$ in the denominator of (2.40).

2.5.3 Graphical representation in $1/V$

In the viscous regime, the velocity can be expressed as a function of the travel distance instead of the time. Eliminating the time between (2.20) and (2.21) yields

$$V = \frac{R\gamma\cos\theta}{4\mu}\frac{1}{Z}. \tag{2.41}$$

First, the contact angle may be considered to be constant. In this case, note that expression (2.41) corresponds to an inertialess regime, because zV is constant. Relation (2.41) can also be written as

$$\frac{1}{V} = \frac{4\mu}{R\gamma\cos\theta}z. \tag{2.42}$$

The plot of the inverse of the velocity $1/V$ as a function of the travel distance z is then a straight line (figure 2.13(A)). The inertial regime appears at the tube inlet with an inflexion of the Bosanquet solution. The effect of the DCA is shown in figure 2.13(B).

Now, let us take into account the DCA as determined by Blake's correlation. Using relation (2.39) for the travel distance z and relation (2.40) for the velocity V, one finds

$$Vz = \frac{V_0}{\sqrt{\left(\frac{\beta R}{4\mu z_0}\right)^2 + 1}}\left[-\frac{\beta R}{4\mu} + \sqrt{\left(\frac{\beta R}{4\mu}\right)^2 + z_0^2}\right] = V_0 z_0\left[1 - \frac{\frac{\beta R}{4\mu z_0}}{\sqrt{\left(\frac{\beta R}{4\mu z_0}\right)^2 + 1}}\right]. \tag{2.43}$$

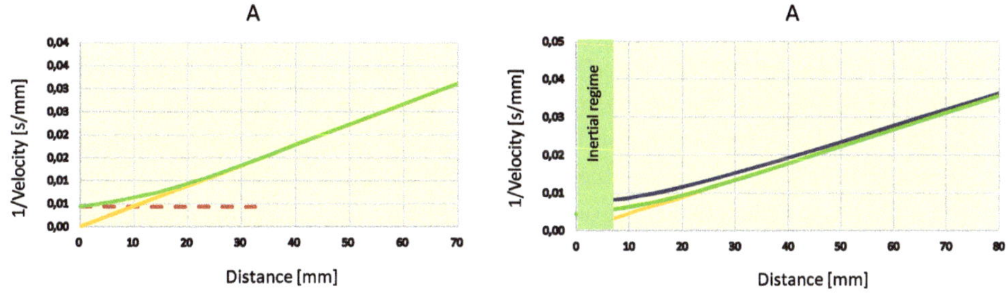

Figure 2.13. (A) A plot of $1/V$ vs z for a capillary-driven flow of IPA50 inside a tube that has a radius of 1 mm. The inertial regime is seen in the tube inlet (red dotted line), then the straight sloping line corresponds to the fully viscous regime (orange line), and the green line is the Bosanquet solution. (B) A comparison between the Bosanquet solution using a constant contact angle (green line) and the viscous regime using a DCA (Blake's correlation, $\beta = 0.2$ Pa s). The orange line represents the viscous solution with a constant contact angle. Here, $\theta_0 = 25°$, $\mu = 3$ mPa s, and $\gamma = 30$ mN m^{-1}.

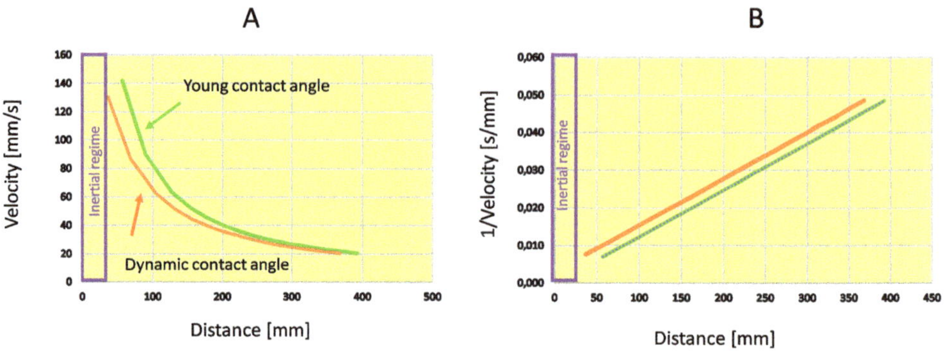

Figure 2.14. (A) A comparison between the velocities obtained using a Young contact angle and a DCA; (B) the same comparison for the inverse velocities.

Then

$$\frac{1}{V} = \frac{z}{V_0 z_0 \left[1 - \frac{\frac{\beta R}{4\mu z_0}}{\sqrt{\left(\frac{\beta R}{4\mu z_0}\right)^2 + 1}}\right]}. \tag{2.44}$$

A comparison between the inverse velocities given by a constant (Young) contact angle and a DCA is shown in figure 2.14. We find that the $1/V$ plot is still nearly linear when the DCA is used.

The $1/V$ representation will appear to be very convenient in the case of open-capillary flows, in particular for the notion of average friction length that will be presented in chapter 4.

2.6 Conclusions

In this chapter, the state of the art of capillary-driven flows has been presented. This state of the art concerns capillary-driven flows inside cylindrical tubes. Fundamental laws, such as the LW law, are the basis for our understanding of capillary flows.

However, in recent years, the development of new, more sophisticated microsystems has promoted the development of the theory of capillarity in more complicated channels or along solid structures. Solid surfaces and composite walls with geometrical complex shapes have appeared, and devices based on open-surface capillary flows are undergoing fast development. The rest of this book is dedicated to their study.

References

[1] Hauksbee F 1712 An account of an experiment touching the ascent of water between two glass planes, in an hyperbolick figure *Philos. Trans. Royal Soc. Lond.* **27** 539–40

[2] Young T 1805 An essay on the cohesion of fluids *Philos. Trans. Royal Soc. Lond.* **95** 65–87

[3] Simon marquis de Laplace P 1805 *Traité de Mécanique Céleste* **vol 4** (Paris: Courcier))

[4] Neumann F and Wangerin A (ed) 1894 *Vorlesungen über die Theorie der Capillarität [Lectures on the Theory of Capillarity]* (Leipzig: B. G. Teubner))
[5] Lucas R 1918 Ueber das zeitgesetz des kapillaren aufstiegs von flüssigkeiten *Kolloid-Z* **23** 15–22
[6] Washburn E W 1921 The dynamics of capillary flow *Phys. Rev.* **17** 273
[7] Rideal E K 1922 On the flow of liquids under capillary pressure *Philos. Mag. Ser. 6* **64** 1152–9
[8] Bosanquet C 1923 On the flow of liquids into capillary tubes *Philos. Mag. Ser. 6* **45** 525–31
[9] Bejan A 1984 *Convection Heat Transfer* (New York: Wiley)
[10] Stokes G G 1845 On the theories of the internal friction of fluids in motion, and of the equilibrium and motion of elastic solids *Trans. Camb. Phil. Soc.* **8** 287–341
[11] Quéré D 1997 Inertial capillarity *Europhys. Lett.* **39** 533–8
[12] Park J-G, Lee S-H, Ryu J-S, Hong Y-K, Kim T-G and Busnaina A A 2006 Interfacial and electrokinetic characterization of IPA solutions related to semiconductor wafer drying and cleaning *J. Electrochem. Soc.* **153** G811–814
[13] Blake T D 2012 Forced wetting of a reactive surface *Adv. Colloid Interface Sci.* **179–82** 22–8
[14] Marmur A 1994 Thermodynamic aspects of contact angle hysteresis *Adv. Colloid Interface Sci.* **50** 121–41
[15] de Gennes P G 1985 Wetting statics and dynamics *Rev. Mod. Phys.* **5** 827–63
[16] De Gennes P-G, Brochart-Wyart F and Quéré D 2004 *Capillarity and Wetting Phenomena Drops, Bubbles, Pearls, Waves* (Berlin: Springer)
[17] Hoffman R L 1975 A study of the advancing interface. I. Interface shape in liquid–gas systems *J. Colloid Interface Sci.* **50** 228–541
[18] Voinov O 1976 Hydrodynamics of wetting *Fluid Dyn.* **11** 714–21
[19] Dussan E B 1976 On the difference between a bounding surface and a material surface *J. Fluid Mech.* **75** 609–23
[20] Tanner L H 1979 The spreading of silicone oil drops on horizontal surfaces *Phys. D: Appl. Phys.* **12** 1473
[21] Blake T D and Haynes J 1969 Kinetics of liquid–liquid displacement *J. Colloid Interface Sci.* **30** 421–3
Blake T D and De Coninck J I 2002 The influence of solid–liquid interactions on dynamic wetting *Adv. Colloid Interface Sci.* **96** 21–36
[22] Blake T D 2006 The physics of moving wetting lines *J. Colloid Interface Sci.* **299** 1–13
[23] Duvivier D, Blake T D and De Coninck J I 2013 Toward a predictive theory of wetting dynamics *Langmuir* **29** 10132–40
[24] Wu P, Nikolov A and Wasan D 2017 Capillary dynamics driven by molecular self-layering *Adv. Colloid Interface Sci.* **243** 114–20
[25] Wu P, Nikolov A D and Wasan D T 2017 Capillary rise: validity of the dynamic contact angle models *Langmuir* **33** 7862–72
[26] Hamraoui A, Thuresson K, Nylander T and Yaminsky V 2000 Can a dynamic contact angle be understood in terms of a friction coefficient? *J. Colloid Interface Sci.* **226** 199–204
[27] Sah A 2014 Precursor film: a key driver to determine wetting behavior in the vicinity of surface heterogeneity *Soft Mater.* **10** 3890
[28] Schwartz M and Tejada S B 1972 Studies of dynamic contact angles on solids *J. Colloid Interface Sci.* **38** 359–75
[29] Seebergh J E and Berg J C 1992 Dynamic wetting in the low capillary number regime *Chem. Eng. Sci.* **47** 4455–64

[30] Jiang T S, Oh S G and Slattery J C 1979 Correlation for dynamic contact angle *J. Colloid Interface Sci.* **69** 74–7

[31] Bracke M, De Voeght F and Joos P 1989 The kinetics of wetting the dynamic contact angle *Prog. Colloid Polym. Sci.* **79** 142–9

[32] Joos P, van Remoortere P and Bracke M 1990 The kinetics of wetting in a capillary *J. Colloid Interface Sci.* **136** 189–97

[33] Martic G, Gentner F, Seveno D, Coulon D, De Coninck J and Blake T D 2002 A molecular dynamics simulation of capillary imbibition *Langmuir* **18** 7971–6

IOP Publishing

Open-Channel Microfluidics (Second Edition)
Fundamentals and applications
Jean Berthier, Ashleigh B Theberge and Erwin Berthier

Chapter 3

Condition for capillary flow in open channels

In this chapter, we present the conditions for the establishment of spontaneous capillary flow (SCF) in open channels. SCF corresponds to the case in which the value of the pressure at the inlet is zero. First, the flow condition for monolithic channels is presented (i.e. the same wetting properties of the walls lead to a unique value of the contact angle), then this condition is extended to composite walls. Finally, the case of enhanced capillary flow is discussed: in such a case, a small additional pressure—usually at the inlet—contributes to the establishment of the flow.

3.1 Spontaneous capillary flow in a monolithic channel

Let us consider the open channel sketched in figure 3.1.

In an open channel with monolithic walls, the condition for SCF is [1, 2]

$$\cos\theta > \frac{p_{\mathrm{F}}}{p_{\mathrm{w}}} \tag{3.1}$$

where p_{W} and p_{F} are respectively the wetted and free perimeters in a cross-section. This condition simply derives from the axial orientation of the capillary force. It can be shown [1, 3, 4] that the capillary force along the triple line (the contact line between liquid, solid, and air) is $F_{\mathrm{cap,\ wall}} = p_w \gamma \cos\theta$. The free surface reduces the capillary force—it takes energy to extend the free surface—and the total capillary force is

$$F_{\mathrm{cap}} = p_w \gamma \cos\theta + p_{\mathrm{F}} \gamma \cos\pi = \gamma \cos\theta\, p_{\mathrm{w}} - \gamma\, p_{\mathrm{F}}. \tag{3.2}$$

Note that the negative contribution of the free surface is $\gamma \cos\pi\, p_{\mathrm{F}} = -\gamma\, p_{\mathrm{F}}$. The condition for SCF is then $F_{\mathrm{cap}} = \gamma \cos\theta\, p_{\mathrm{w}} - \gamma\, p_{\mathrm{F}} > 0$. For a totally closed channel, $p_{\mathrm{F}} = 0$, and condition (3.2) reduces to the usual condition $\cos\theta > 0$, or $\theta < \pi/2$.

Condition (3.1) has been verified using a numerical approach by the Surface Evolver software for many different channel shapes [5, 6]. Take the example of suspended channels, i.e. channels delimited by two parallel vertical walls (figure 3.2).

doi:10.1088/978-0-7503-5507-0ch3

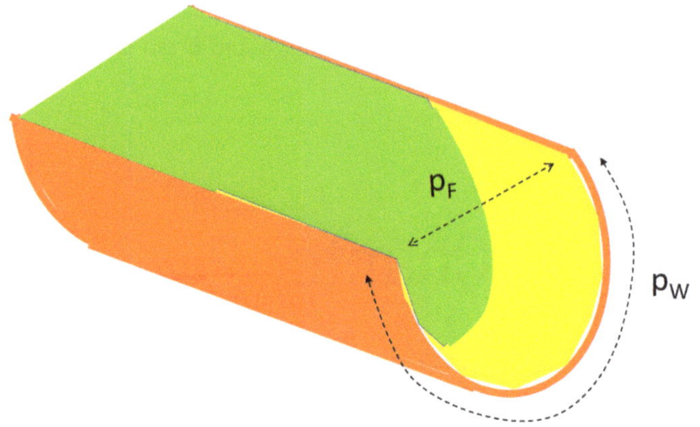

Figure 3.1. A sketch of a monolithic open channel.

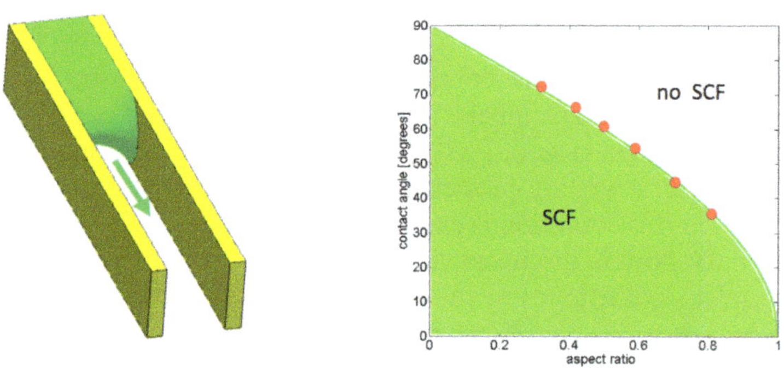

Figure 3.2. Left: a suspended channel (Evolver picture); right: comparison between theory and Evolver for the SCF; the contact angle that permits SCF as a function of the aspect ratio (channel width/channel height).

3.2 Spontaneous capillary flow in composite open channels: the generalized Cassie condition

In modern microfluidic systems, channel walls are often composed of different materials. A generalization of relation (3.1) has been derived that considers Gibbs' free energy [5]. Using the notations of figure 3.3, the SCF condition is

$$\cos \theta^* = \frac{\sum_i p_i \cos \theta_i}{\sum_i p_i} > 0, \tag{3.3}$$

where the index i denotes all the walls and air. Here, θ_i are the different contact angles. Note that the contact angle with air is π, then (3.3) can be rewritten as

$$\cos \theta^* = \frac{\sum_j p_j \cos \theta_j - p_F \cos \theta_F}{\sum_j p_j + p_F} > 0, \tag{3.4}$$

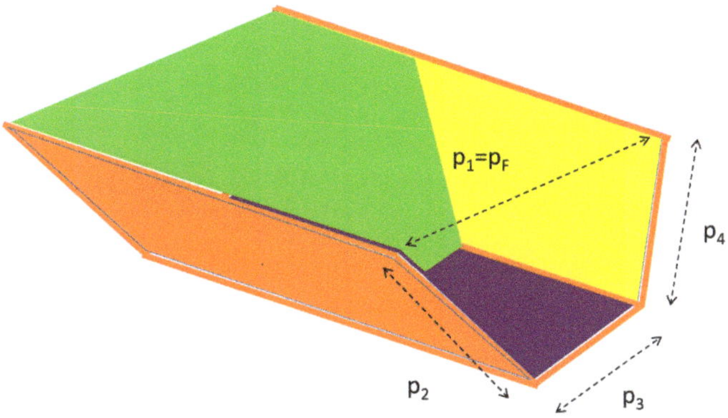

Figure 3.3. A sketch of an open, composite channel. The perimeters p_2, p_3, p_4 are along the walls and p_1 is the free perimeter.

where the index j denotes only the walls. This condition resembles the Cassie contact angle for a droplet on a composite surface. Hence, the condition is called the generalized Cassie condition.

Condition (3.4) states that the total capillary force exerted by the walls on the liquid is positive, because

$$F_{\mathrm{cap}} = \sum_i \gamma\, p_i \cos\theta_i = \sum_j \gamma\, p_j \cos\theta_j - \gamma\, p_{\mathrm{F}} \cos\theta_{\mathrm{F}} > 0, \tag{3.5}$$

Condition (3.4) has been verified using a numerical approach by the Surface Evolver software [3]. Figure 3.4 shows the SCF condition in composite suspended channels and in rounded open channels.

3.3 Common geometries

Table 3.1 indicates the SCF conditions deduced from relation (3.4) for the most common open-channel geometries. In all cases, the theoretical conditions agree with the Evolver results.

A few remarks can be made about table 3.1. First, the presence of rounded inner corners lowers the constraint on the Young angle θ for obtaining an SCF. Indeed, in the case of a triangular open channel, the condition $\cos\theta > w/\frac{2h}{\cos\alpha} = \sin\alpha$ for SCF in a sharp triangular channel (#1) is less strict than the condition $\cos\theta > w/\left(\frac{2h}{\cos\alpha} - \frac{2R}{\tan\alpha}\right)$ for a rounded-tip triangular channel (#2). The same remark can be made for the rectangular U-channel (#3), for which the condition $\cos\theta > w/(2h + w)$ is less strict than the condition $\cos\theta > w/[2h + w + R(\pi - 4)]$ for the rounded-corner U-channel (#4). The presence of rounded corners decreases the wetted perimeter, while the free perimeter remains unchanged. We shall see later that another difference between sharp and rounded corners pertains to the notion of capillary filaments (figure 3.5). Capillary filaments are detailed in chapter 6.

Second, the presence of additional micro-grooves on the channel bottom also relaxes the SCF condition. In particular, this is the case for a rectangular

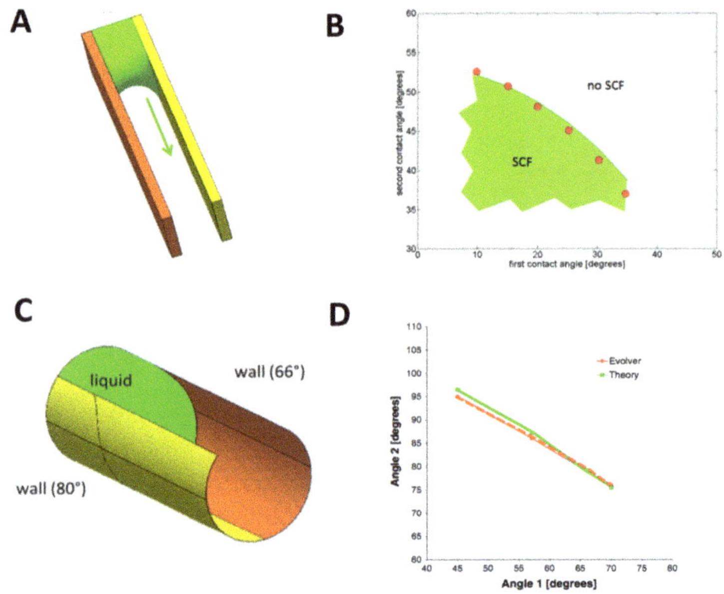

Figure 3.4. (A) Evolver simulation of the SCF in composite suspended channels; (B) a comparison between theory and the Evolver results for SCF onset; (C) an Evolver simulation of the SCF in composite round open channels; (D) a comparison between theory and the Evolver results for SCF onset.

Table 3.1. A summary of the different conditions for liquid spreading under capillary action in different open microfluidic geometries. Images were obtained using the Surface Evolver software.

Geometry	SCF condition	View	
Triangular	$\cos\theta > \frac{w}{\frac{2h}{\cos\alpha}} = \sin\alpha$	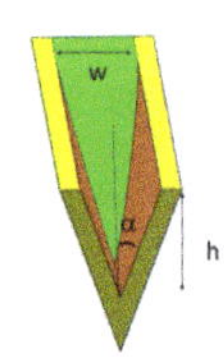	#1
Triangular with a rounded tip	$\cos\theta > \frac{w}{\frac{2h}{\cos\alpha} - \frac{2R}{\tan\alpha}}$	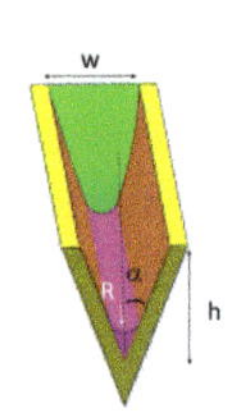	#2
Rectangular U-channel	$\cos\theta > \frac{w}{2h + w}$	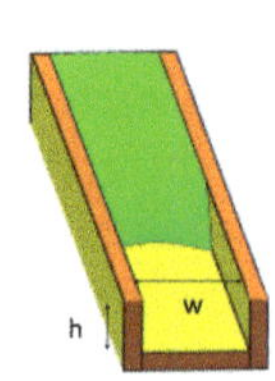	#3

Rounded-corner U-channel	$\cos\theta > \frac{w}{2h + w + R(\pi - 4)}$	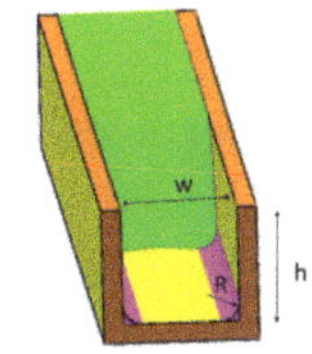	#4
Rectangular U-channel with a grooved bottom	$\cos\theta > \frac{w}{2h + w + 2na}$	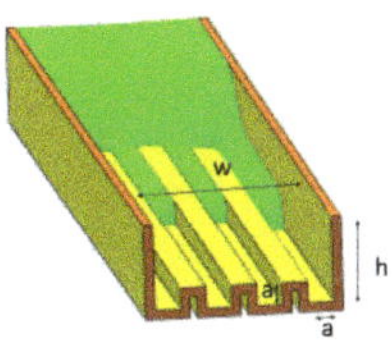	#5
Rectangular U-channel with a triangularly grooved bottom	$\cos\theta > \frac{w}{2h + \frac{w}{\sin\alpha}\left(1 - \frac{\cos\alpha}{n}\right)}$	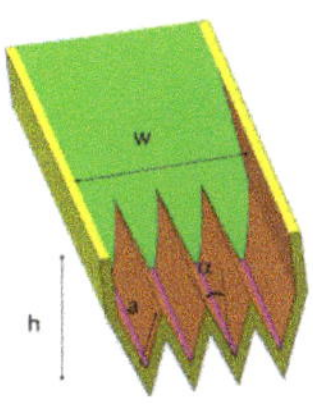	#6
Trapezoidal channel	$\cos\theta > \frac{w_1}{w_2 + 2\sqrt{h^2 + \left(\frac{w_1 - w_2}{2}\right)^2}}$	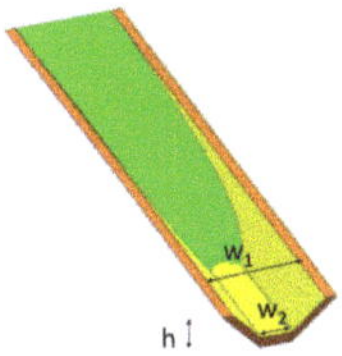	#7
Half-pipe	$\cos\theta > \frac{2}{\pi}$	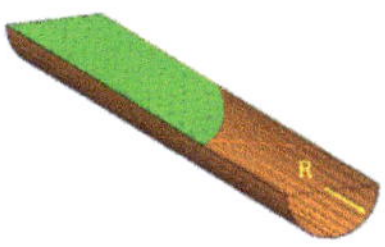	#8
Horizontal rails	$\cos\theta > \frac{h}{w}$	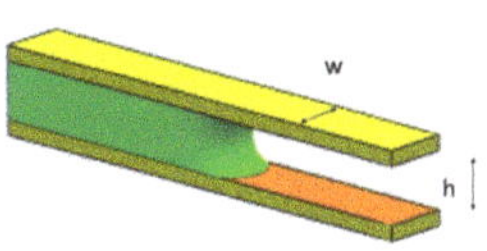	#9

(*Continued*)

Table 3.1. (*Continued*)

Geometry	SCF condition	View	
Horizontal rails (vanes)	$\cos\theta > \frac{h}{\sin\theta(w + h\cos\theta)}$	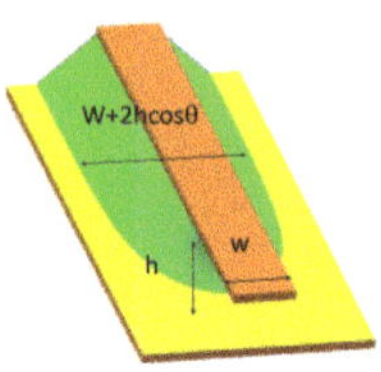	#10
Vertical fin (vanes)	$\cos\theta > \sqrt{\frac{w}{w+h}} = \frac{\pi}{4}$	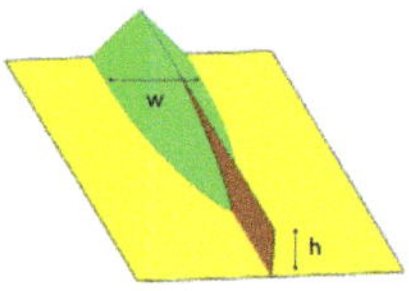	#11
Suspended channel	$\cos\theta > \frac{w}{h}$	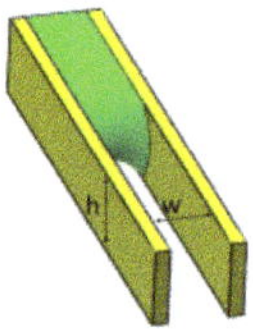	#12
Embossed channel	$\cos\theta > \frac{w + 2R}{w + 2R(1 + \alpha)}$	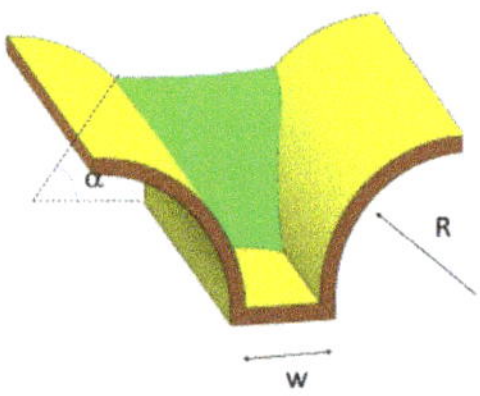	#13
Bundle of fibers	$\cos\theta > \frac{b}{n\,a}$	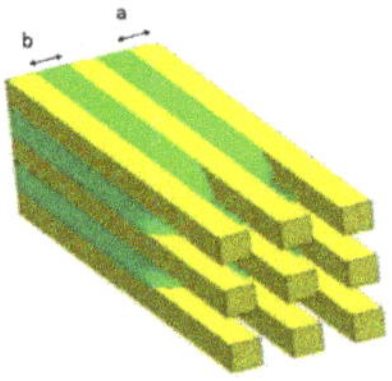	#14

U-channel, for which the condition $\cos\theta > w/(2h + w)$ in the case in which there are no grooves on the channel bottom (#3) is stricter than the condition $\cos\theta > w/(2h + w + 6a)$ in the presence of rectangular grooves (#5) and the condition $\cos\theta > w/\left[2h + \frac{w}{\sin\alpha}\left(1 - \frac{\cos\alpha}{n}\right)\right]$ in the presence of micro V-grooves

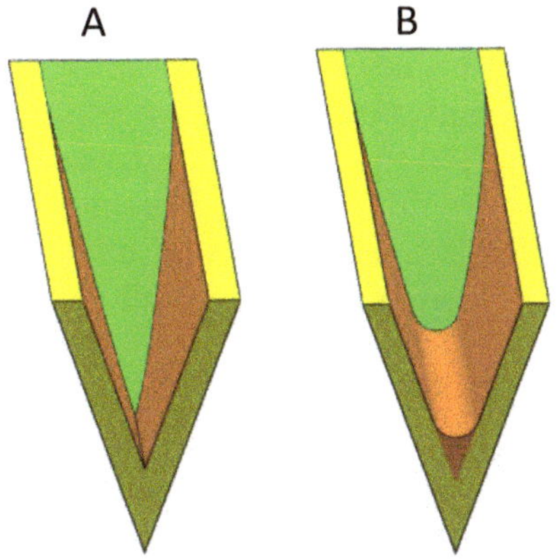

Figure 3.5. Two different aspects of the meniscus: (A) a sharp triangular groove; (B) a rounded triangular groove. In case (A), a capillary filament forms in the wedge. Images obtained using Surface Evolver software.

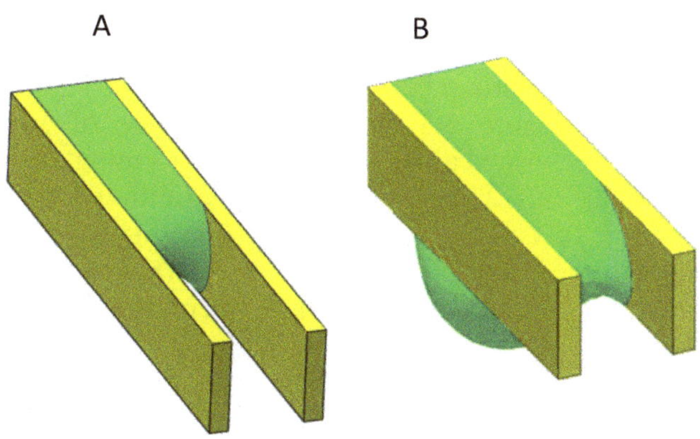

Figure 3.6. Two different aspects of a liquid in a suspended channel: (A) for a Bond number smaller than one; (B) for a Bond number larger than one. Images obtained using Surface Evolver software.

(#6). Grooves on the bottom considerably increase the wetted perimeter, while the free perimeter is unchanged.

Third, the SCF conditions for horizontal rails (#10) and suspended channels (#12) are basically similar—it is the same geometry after a rotation—under the condition that gravity can be neglected. We recall that gravity can be neglected when the channel width w or the channel height h of the suspended channel geometry is less than the capillary number, i.e. max $(w, h) < \sqrt{\gamma/\rho g}$ or identically that the Bond number is smaller than one: $Bo = \rho g \max(w, h)^2/\gamma < 1$ [7, 8]. Figure 3.6 shows the difference between a suspended channel in which gravity is negligible (A) and one in which it is not (B).

Finally, the SCF condition in a bundle of parallel fibers (#14) is relaxed when the number of fibers increases ($n \to \infty$). Indeed, when n increases, the condition on the contact angle $\cos\theta > b/(n\,a)$ becomes less strict. At the limit, $b/(n\,a) \to 0$ and one retrieves the condition $\cos\theta > 0$.

The SCF condition is also valid for composite channels, i.e. channels with walls of different wettability. Some examples are given in table 3.2.

Note that it was assumed that the open surface was flat (zero pressure), which is valid when the liquid is pinned along the upper edges of the channel. In some cases (see chapter 7), the liquid may not be pinned on the channel edges, as shown in figure 3.7. In such cases, Mitteramskogler *et al* have shown—using lengthy geometrical developments—that condition (3.3) is still valid [9].

3.4 Enhanced open-capillary flows

In contrast to SCFs, in which the inlet pressure is zero, 'enhanced' open-capillary flow may be obtained using an additional inlet pressure or an additional front-end capillary force. Additional pressure in the reservoir can arise from hydrostatic pressure, such as that produced by an elevated reservoir [10] or a pipette [11], or Laplace pressure, such as that produced by a bulging reservoir. Additional capillary force at the front end of the flow can be obtained via electrowetting [12].

3.4.1 Fluid walls

It was shown by Ed Walsh's group [13–15] that an immobile liquid path can behave as a channel bottom. Figure 3.8 shows the principle of such a system. A fluorocarbon (FC40) is deposited along a straight line, forming a liquid path starting below the tip of a needle (inlet) and ending in a sink (outlet). Outside this path, the surface is hydrophobic (plastic material). Slowly injecting aqueous liquid through the needle tip triggers a slow flow along the FC40 path (figure 3.8(A)). In this case, a small pressure exists at the needle tip, which decreases towards the sink. The circulating aqueous fluid has a convex shape that is more pronounced closer to the inlet than to the outlet (figures 3.8(B) and (C)), because the pressure is higher there. A small layer of viscous FC40 remains at rest while aqueous flow takes place on top of it. The drag force exerted by the circulating fluid is insufficient to move the FC40.

This example shows that a (small) additional pressure can be used to trigger or boost an open-capillary flow. In the following sections, the conditions for enhanced capillary flows are presented.

It is also interesting to compare the preceding 'fluid walls' approach and the phenomenon of coarsening using precursor films (figure 3.9). In this last case (figure 3.9(A)), the liquid of the smallest droplet moves via the precursor film towards the largest droplet. In the case of 'fluid walls,' a path is established between the droplets which replaces the precursor film (figure 3.10(B)). The same effect is achieved: the largest droplet pumps the liquid of the smallest droplet.

3.4.2 Constant additional inlet pressure

In practice, it is difficult to create a constant inlet pressure, but it is of academic interest to investigate the effect of a constant additional pressure. In the case of a positive reservoir pressure P_{in}, the condition for obtaining a capillary flow is

$$P_{in} > P_{cap}, \tag{3.6}$$

Table 3.2. The SCF condition in composite channels. Images were obtained using the Surface Evolver software.

Geometry	SCF condition	View	
Closed channel (composite)	$2h\cos\theta_3 + w\,(\cos\theta_1 + \cos\theta_2) > 0$	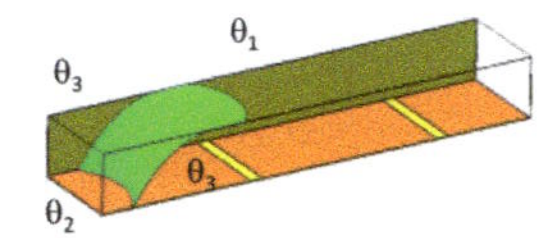	#1
Rectangular U-channel (composite)	$h\,(\cos\theta_1 + \cos\theta_1) + w\cos\theta_0 > w$	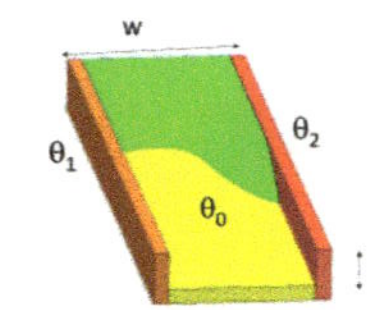	#2
Trapezoidal channel (composite)	$w_2\cos\theta_0 + \sqrt{h^2 + \left(\frac{w_1 - w_2}{2}\right)^2}\cos\theta_1 + \sqrt{h^2 + \left(\frac{w_1 - w_2}{2}\right)^2}\cos\theta_2 > w_1$	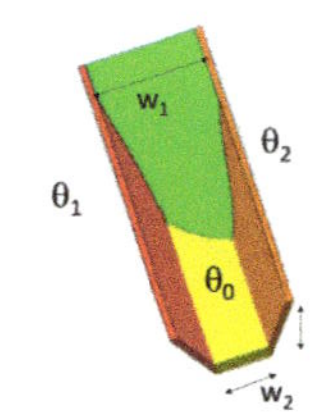	#3
Suspended channel (composite)	$h\,(\cos\theta_1 + \cos\theta_1) > w$	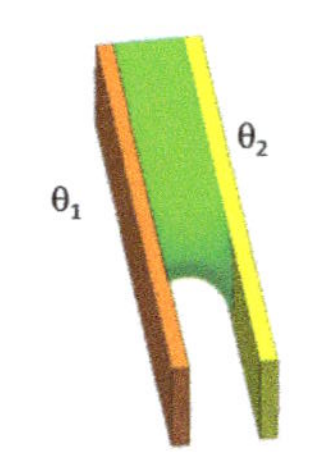	#4

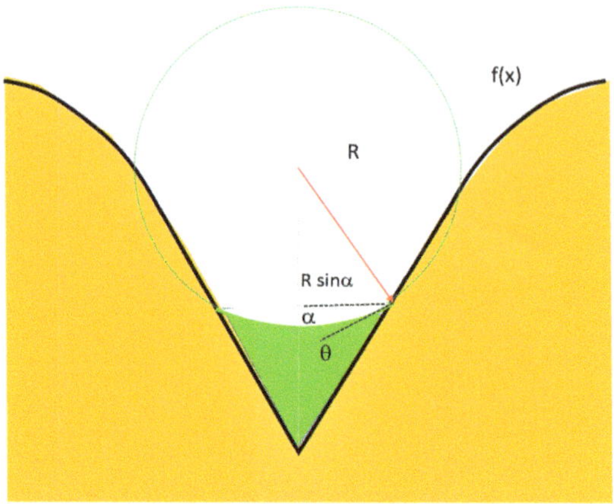

Figure 3.7. A sketch of an open-capillary flow in a V-grove when the free boundary is not pinned to an edge [9].

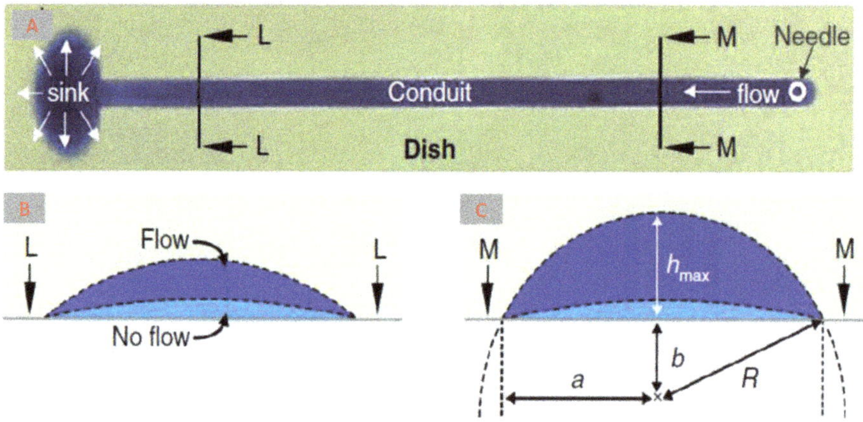

Figure 3.8. (A) The top view of the FC40 path; (B) the liquid profile close to the outlet; (C) the liquid profile close to the inlet. Reprinted from [15]. CC BY 4.0.

where P_{cap} is the capillary pressure of the front meniscus, which can be expressed as

$$P_{\text{cap}} = -\frac{\gamma\, p_{\text{w}} \cos\theta - \gamma\, p_{\text{F}}}{S}, \tag{3.7}$$

where S is the flow area in a cross-section. The minus sign in (3.7) indicates a negative pressure when the capillary force is oriented downstream. Note that it can also be positive under condition (3.6). Substituting (3.7) into (3.6) yields

$$\frac{S\, P_{\text{in}}}{\gamma} > p_{\text{F}} - p_{\text{w}} \cos\theta. \tag{3.8}$$

Alternatively,

$$\cos\theta > \frac{p_{\text{F}}}{p_{\text{w}}} - \frac{S\, P_{\text{in}}}{p_{\text{w}}\, \gamma}. \tag{3.9}$$

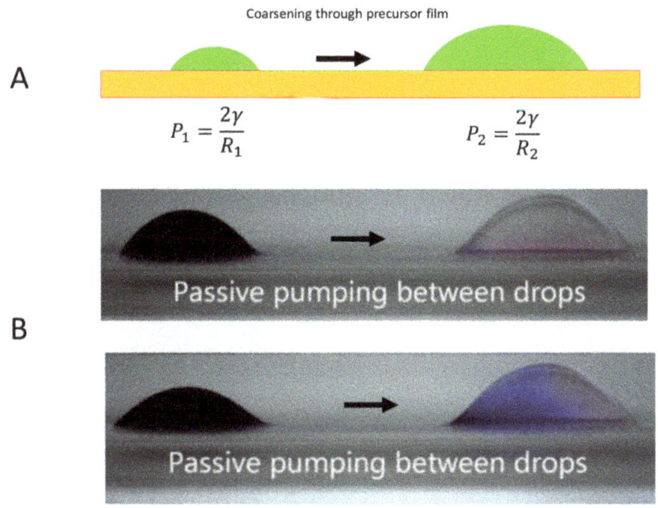

Figure 3.9. (A) A sketch of coarsening produced by a precursor film; (B) coarsening produced by a 'fluid wall' path, reproduced from [13]. CC BY 4.0.

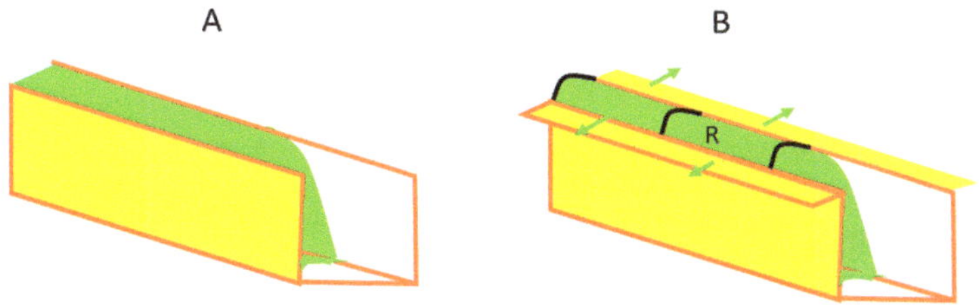

Figure 3.10. (A) An SCF in a rectangular U-channel; (B) a slightly pressurized capillary flow in the open rectangular U-channel: the free surface presents a curvature r that slowly decreases with distance. If the pinning condition is not met, liquid overflows laterally.

This condition is evidently less restrictive than the SCF condition $\cos\theta > p_F/p_w$, because the term $S\,P_{in}/p_w\gamma$ is positive. There is, however, an additional restrictive condition: the pressure in the open channel must be smaller than the Laplace pressure of the free surface at all locations in the channel, otherwise overflow occurs due to loss of pinning and liquid spills out of the channel (figure 3.10). In the case of a uniform U-groove channel, the location where the Laplace pressure is highest is at the inlet. If the channel does not have a uniform cross-section, then overspill may occur elsewhere, for example in restricted cross-sectional areas.

Hence, the condition for enhanced capillary flow in an open channel is

$$P_{Laplace} > P_{in} > P_{cap}, \tag{3.10}$$

We recall that pinning on an edge [1, 16, 17] depends on the contact angle with the walls adjacent to the edge and the shape of the liquid surface (figure 3.11). For an

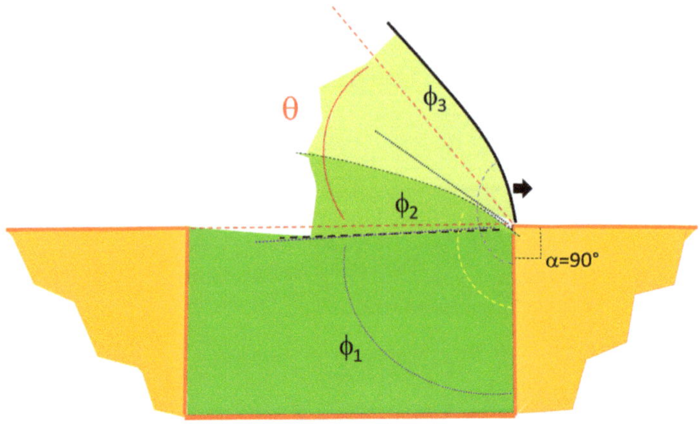

Figure 3.11. Pinning of an open flow on the edge of a rectangular open channel: the green surface corresponds to the SCF (nearly horizontal surface), with an angle of ϕ_1; the light green surface at an angle of ϕ_2 corresponds to an enhanced flow which does not overflow; when the angle of the free surface (black line) exceeds $\alpha+\theta$, as in the case of ϕ_3, for example, overflow occurs.

edge of geometric angle α and a liquid that makes a Young angle of θ with the walls on both sides of the edge, the pinning condition is

$$\phi < \theta + \alpha, \tag{3.11}$$

where ϕ is the angle between the free surface and the channel wall.

Hence, for a uniform cross-section channel with a boundary of width w in contact with the surrounding air,

$$P_{\text{Laplace}} = \gamma\kappa = \gamma\left(\frac{1}{R_1} + \frac{1}{R_2}\right), \tag{3.12}$$

where κ is the maximum pinning curvature of the free surface, with corresponding principal curvature radii R_1 and R_2. In the case of a channel that has a uniform cross-section, the curvature radius along the channel axis is negligible, such that $R_2 \sim 0$, and the curvature is $\kappa = 1/R_1$, where R_1 is the curvature radius in the transverse direction. The condition (3.10) then becomes

$$\frac{1}{R_1} > \frac{P_{\text{in}}}{\gamma} > \frac{p_{\text{F}} - p_{\text{w}} \cos\theta}{S}. \tag{3.13}$$

Let us now consider an aqueous flow in two possible configurations shown in figure 3.12. First, the top plate, on both sides of the channel, is hydrophobic and the channel interior is hydrophilic [8]; second, the whole domain—channel and plate—is hydrophilic.

In the first case, the maximum curvature before pinning is lost corresponds to a semicylindrical bulging surface [3]. Indeed, if we consider a spherical interface, for example, instead of a cylindrical interface, the maximum Laplace pressure is obtained at the minimum value of the curvature. For a spherical interface, this

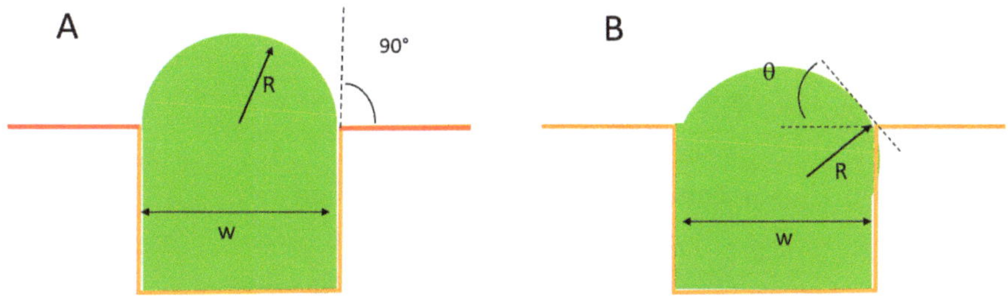

Figure 3.12. (A) An open channel consisting of a hydrophobic (lyophobic) top plate and a hydrophilic (lyophilic) interior. The maximum protrusion from the liquid domain makes an angle of 90° with the top plate. (B) A wholly hydrophilic (lyophilic) device. The pinning condition is a contact angle with the plate less than θ.

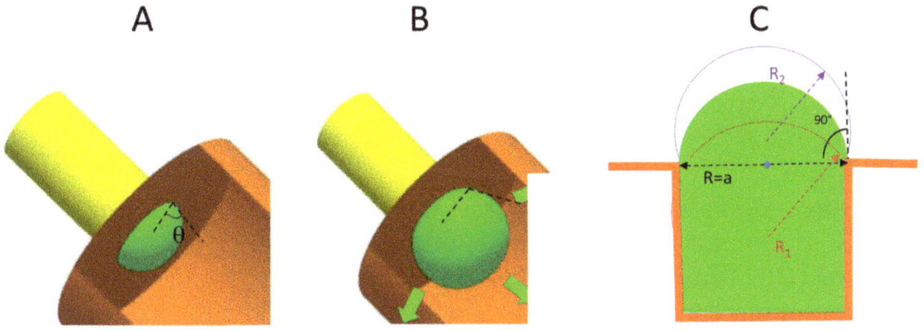

Figure 3.13. (A) A stable liquid interface with $\theta < 90°$; (B) when the droplet radius is larger than that of the tube, the interface breaks down. A and B are the results of an Evolver calculation. (C) The semicircular interface corresponds to the maximum Laplace pressure.

takes place when $R_1 = R_2 = R$, which corresponds to the case of a half-sphere and a contact angle of 90°. When the interface contact angle with the plane exceeds 90°, the interface breaks down, as shown in figure 3.13. The same reasoning applies for the case of a cylindrical interface.

The stability condition is deduced from (3.11) by taking into account the maximum Laplace pressure: $\phi < \theta + \min(\alpha, 90°)$. Then,

$$\kappa = \frac{1}{R_1} = \frac{2}{w}, \tag{3.14}$$

while in the second case, the pinning is lost when the curvature exceeds

$$\kappa = \frac{1}{R_1} = \frac{2\sin\theta}{w}. \tag{3.15}$$

Successively substituting (3.14) and (3.15) into (3.13) yields, for the hydrophilic/hydrophobic case,

$$\frac{2}{w} > \frac{P_{in}}{\gamma} > \frac{w - (2h + w)\cos\theta}{S}, \tag{3.16}$$

where the liquid cross-sectional area S is

$$S = h\,w + \frac{\pi R^2}{2} = h\,w + \frac{\pi w^2}{8}. \tag{3.17}$$

The condition on P_{in} for enhanced capillary flow is then

$$\frac{2}{w} > \frac{P_{\text{in}}}{\gamma} > \frac{w - (2h + w)\cos\theta}{h\,w + \frac{\pi w^2}{8}}. \tag{3.18}$$

This condition has been plotted in figure 3.14(A). We have distinguished three 'regimes.' The first one is SCF, which occurs for $P_{\text{in}} = 0$ and even if P_{in} is negative, as long as the capillary pressure of the meniscus is able to drive the flow. The second one, called 'enhanced SCF,' corresponds to the case of a negative capillary pressure at the meniscus and a positive inlet pressure. The third one, which we can call 'forced capillary flow,' corresponds to the case in which the Laplace pressure of the advancing meniscus becomes positive, but the inlet pressure is larger and sufficient to drive the flow.

Hence, three domains appear in the figure: the usual SCF for $P_{\text{in}} \leqslant 0$, an enhanced SCF domain for $P_{\text{in}} < 2\gamma/w$ and $\cos\theta > w/(2h + w)$, and a 'forced' capillary flow (CF) for $\cos\theta < w/(2h + w)$.

Similar reasoning can be applied to a wholly hydrophilic device. In such a case, the maximum liquid cross-sectional area S depends on the contact angle θ:

$$\begin{aligned} S &= \min\left[h\,w + \frac{R^2}{2}(2\theta - \sin 2\theta),\; h\,w + \frac{\pi w^2}{8}\right] \\ &= \min\left[h\,w + \frac{w^2}{8\sin^2\theta}(2\theta - \sin 2\theta),\; h\,w + \frac{\pi w^2}{8}\right]. \end{aligned} \tag{3.19}$$

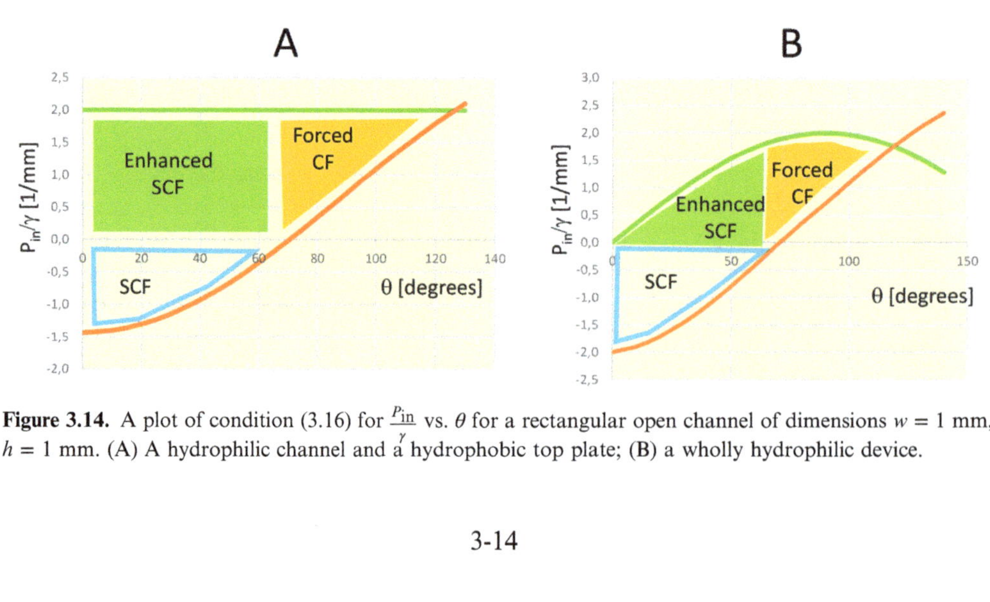

Figure 3.14. A plot of condition (3.16) for $\frac{P_{\text{in}}}{\gamma}$ vs. θ for a rectangular open channel of dimensions $w = 1$ mm, $h = 1$ mm. (A) A hydrophilic channel and a hydrophobic top plate; (B) a wholly hydrophilic device.

The condition for enhanced capillary flow is then

$$\frac{2\sin\theta}{w} > \frac{P_{in}}{\gamma} > \frac{w-(2h+w)\cos\theta}{\min\left\{h\,w+\frac{w^2}{8\sin^2\theta}(2\theta-\sin 2\theta),\ h\,w+\frac{\pi w^2}{8}\right\}}. \tag{3.20}$$

This condition has been plotted in figure 3.14(B). Again, three domains appear in the figure: the usual SCF for $P_{in} \leqslant 0$, an enhanced SCF domain for $P_{in} < 2\gamma/w$ and $\cos\theta > w/(2h+w)$, and a 'forced' CF for $\cos\theta < w/(2h+w)$.

3.4.3 Overfilled reservoir: initial additional Laplace pressure

Consider a rectangular open channel exiting from a large reservoir (figure 3.15(A)). Let us take the example of an initially bulging reservoir obtained by pipetting a large volume of liquid. It is supposed here that there is pinning of the liquid at the reservoir edges, so that no overflow occurs. It is also assumed that no additional liquid is added to the reservoir during the capillary flow. A volume of liquid V_0 is first pipetted into the reservoir of radius a and depth h (figure 3.15(B)).

If we assume that the reservoir radius is of the order of the capillary length, the free surface has the shape of a spherical cap. Then the initial bulge H_0 of the free surface in the reservoir is

$$V_0 = \pi\,a^2h + \frac{1}{6}\pi H_0(3a^2 + H_0{}^2). \tag{3.21}$$

When the flow advances in the channel and reaches a distance z, the bulge height H decreases and may even become negative, according to

$$V_0 - wh\,z = \pi\,a^2h + \frac{1}{6}\pi H(3a^2 + H^2), \tag{3.22}$$

where h and w are the depth and width of the rectangular open channel, respectively.

Relation (3.22) is a third-order polynomial parameterized by the term $-whz$: $V_0 - wh\,z = f(H^3)$. If $a \gg H$, the third-order polynomial reduces to a first-order polynomial whose solution is

$$H \cong 2\frac{V_0 - wh\,z}{\pi\,a^2} - 2h. \tag{3.23}$$

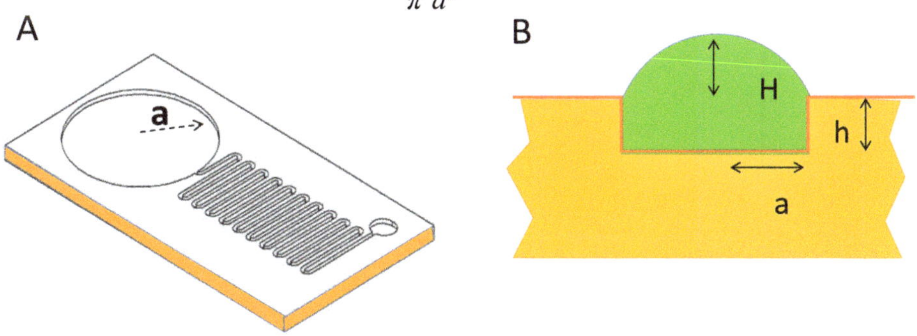

Figure 3.15. (A) A sketch of the capillary channel and the reservoir; (B) initial filling of the reservoir.

Relation (3.23) indicates that the bulge of the reservoir's free surface decreases with the increasing travel distance of the capillary flow in the open channel. At the beginning of the flow, the travel distance $z = 0$ and the bulge height H_0 is given by

$$H_0 \cong 2\frac{V_0}{\pi\, a^2} - 2h. \tag{3.24}$$

Let us determine the travel distance at which the bulge vanishes. Substituting $H = 0$ into (3.23) yields the distance z_1:

$$z_1 \cong \frac{V_0 - \pi\, a^2 h}{wh}. \tag{3.25}$$

The values H_0 and z_1 are indicated in figure 3.16.

Because no liquid is added in the reservoir, after some time, the flow stops. Let z_{stop} and H_{stop} respectively denote the stopping travel distance and the height in the reservoir when the flow stops. The flow stops when the balance between the Laplace pressure of the reservoir and the capillary pressure at the front meniscus equilibrates, i.e.

$$P_{res} = P_{cap}. \tag{3.26}$$

The Laplace pressure of the reservoir is

$$P_{res} = 2\frac{\gamma}{R}, \tag{3.27}$$

where R is the curvature radius of the free surface of the reservoir. On the other hand, the capillary pressure is

$$P_{cap} = -\frac{\left(\gamma \cos\theta\, p_w - \gamma\, p_F\right)}{wh}. \tag{3.28}$$

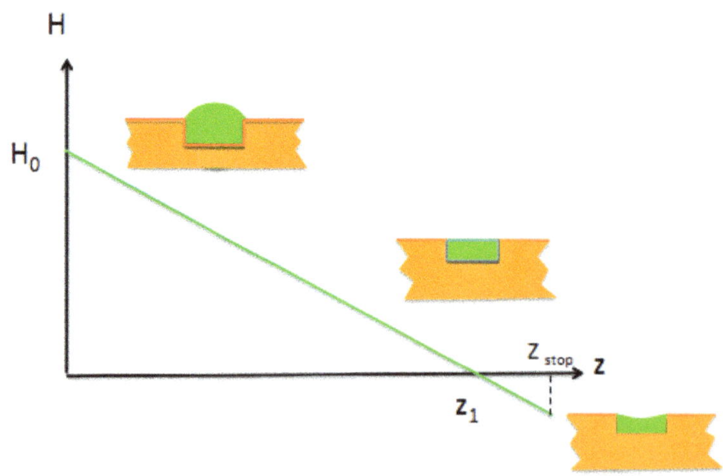

Figure 3.16. Evolution of the bulge height H with the travel distance z.

Equation (3.26) now becomes

$$2\frac{\gamma}{R} = -\frac{\left(\gamma \cos\theta p_{\mathrm{w}} - \gamma p_{\mathrm{F}}\right)}{wh}. \tag{3.29}$$

Under the assumption that the liquid surface in the reservoir has a spherical shape, the curvature radius R of the surface is the following function of H:

$$R = \frac{a^2 + H^2}{2H}. \tag{3.30}$$

Note that H can be negative or positive in (3.30), depending on the concave or convex shape of the interface, and similarly, R can be negative or positive. Substituting into (3.29), one obtains

$$4\frac{H_{\mathrm{stop}}}{\left(a^2 + H_{\mathrm{stop}}{}^2\right)} = \frac{1}{wh}\left[w - (2h + w)\cos\theta\right]. \tag{3.31}$$

Again, if $a \gg H_{\mathrm{stop}}$, H_{stop}^2 can be neglected in (3.31) and relation (3.31) reduces to

$$H_{\mathrm{stop}} \cong \frac{a^2}{4wh}\left[w - \cos\theta(2h + w)\right]. \tag{3.32}$$

Relation (3.32) gives the bulge height (positive or negative) which corresponds to the pressure equilibrium, i.e. to the maximum travel distance in the channel. If the SCF condition

$$\cos\theta > \frac{p_{\mathrm{F}}}{p_{\mathrm{w}}} = \frac{w}{2h + w} \tag{3.33}$$

is met, H_{stop} is negative, else H_{stop} is positive. Substitution of the bulge height H_{stop} derived from (3.32) into relation (3.23) gives the maximum travel distance or equivalently the maximum spread of the droplet z_{stop}, given by the relation

$$z_{\mathrm{stop}} \cong \frac{V_0 - \pi a^2\left(h + \frac{H_{\mathrm{stop}}}{2}\right)}{wh} \cong \frac{V_0 - \pi a^2\left(h + \frac{a^2}{8wh}\left[w - (2h + w)\cos\theta\right]\right)}{wh}. \tag{3.34}$$

Note that if the pipetted volume is just that of the flat reservoir, $V_0 = \pi a^2 h$, then

$$z_{\mathrm{stop}} \cong -\frac{\pi a^2 H_{\mathrm{stop}}}{2wh} \cong \frac{\pi a^4\left[(2h + w)\cos\theta - w\right]}{8w^2h^2}. \tag{3.35}$$

In this case of an initial flat surface in the reservoir, there is only penetration into the channel if the SCF condition is met. Note that if $a \gg w$ and $a \gg h$, then $a^2/h^2 \gg 1$ and $a^2/w^2 \gg 1$. Thus, z_{stop} may be large when the wetting condition is favorable, i.e. when $\theta \approx 0$ or $(2h + w)\cos\theta - w \approx 2h$. As an example, assume that $a = 3h = 3w$, and $\theta \approx 0$. Then, (3.35) yields $z_{\mathrm{stop}} \cong 20\pi\ h \cong 63h$.

Figure 3.17 shows the relative maximum distance (z_{stop}/h) as a function of the relative reservoir dimension (a/h) for two contact angles 10° and 30°. Figure 3.18,

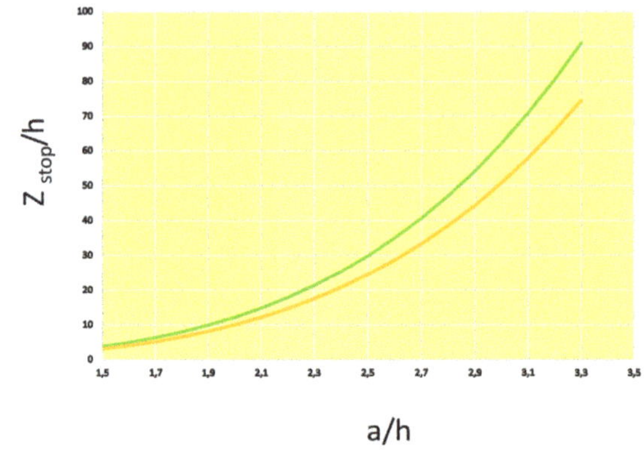

Figure 3.17. The relative maximum distance (z_{stop}/h) as a function of the relative reservoir dimension (a/h) for two contact angles, 10° (green line) and 30° (orange line).

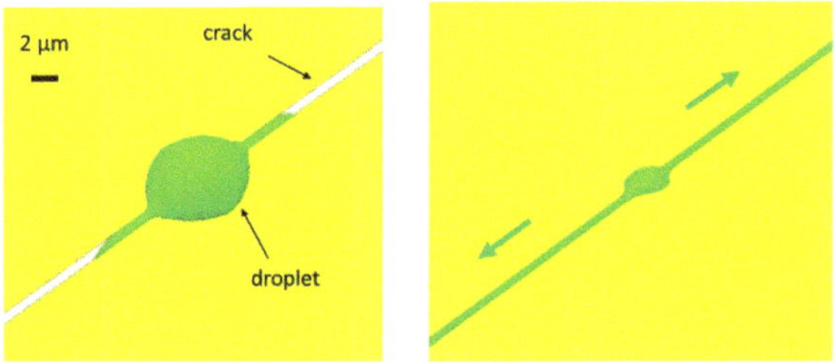

Figure 3.18. A droplet spreading in a crack. The capillary motion combines the capillary forces in the crack and the Laplace pressure of the droplet (produced via Surface Evolver calculation).

obtained using the software Surface Evolver, shows the vanishing of a droplet in a crack (a narrow channel in the plate). In this particular case, $a \sim 8\ h$ and the maximum distance is of the order of $z_{\mathrm{stop}} \approx 1000h$. The droplet completely disappears in the crack.

3.5 Conclusions

In this chapter, the condition for observing SCF in open channels of arbitrary shapes, composite or not, has been presented. This condition was shown to depend uniquely on the contact angle of the liquid and on the channel geometry. A generalized Cassie angle was derived which took into account the different materials contacting the fluid, including air. This angle determines the possibility of SCF. If it is smaller than 90°, SCF occurs.

It was shown that surface tension does not appear in the condition, meaning that all fluids with the same contact angle θ behave similarly relative to the onset of capillary flow. However, as we shall see later, their dynamics is not the same.

By increasing the inlet pressure or the capillary force, an 'enhanced capillary flow' can be obtained, allowing for capillary flow even if the SCF condition is not met.

This additional pressure must be small enough not to ruin the pinning along the open edges of the channel.

References

[1] Berthier J and Brakke K A 2012 *The Physics of Microdroplets* (New York: Wiley)
[2] Casavant B P *et al* 2013 Suspended microfluidics *Proc. Natl Acad. Sci.* **110** 10111–6
[3] Berthier J, Brakke K A and Berthier E 2016 *Open Microfluidics* (New York: Wiley)
[4] Yang D, Krasowska M, Priest C, Popescu M N and Ralston J 2011 Dynamics of capillary-driven flow in open microchannels *J. Phys. Chem.* **115** 18761–9
[5] Berthier J, Brakke K A and Berthier E 2014 A general condition for spontaneous capillary flow in uniform cross-section microchannels *Microfluid. Nanofluid.* **16** 779–85
[6] Brakke K A 1992 The surface evolver *Exp. Math.* **1** 141–65
[7] Massey B S 1986 *Measures in Science and Engineering: Their Expression, Relation and Interpretation* (Hemel Hempstead: : Ellis Horwood Limited)
[8] Berthier J and Silberzan P 2010 *Microfluidics for Biotechnology* 2nd edn (Boston, MA: Artech House)
[9] Mitteramskogler T, Hingerl K and Jakoby B 2022 A condition for spontaneous capillary flow in open microgroove *Acta Mech.* **233** 3923–39
[10] Waghmare P R and Mitra S K 2010 Finite reservoir effect on capillary flow of microbead suspension in rectangular microchannels *J. Colloid Interface Sci.* **351** 561–9
[11] Li C, Boban M and Tuteja A 2017 Open-channel, water-in-oil emulsification in paper-based microfluidic devices *Lab Chip* **17** 1436–41
[12] Satoh W, Hosono H and Suzuki H 2005 On-chip microfluidic transport and mixing using electrowetting and incorporation of sensing functions *Anal. Chem.* **77** 6857–63
[13] Walsh E J, Feuerborn A, Wheeler J H R, Tan A N, Durham W M, Foster K R and Cook P R 2017 Microfluidics with fluid walls *Nat. Commun.* **8** 816
[14] Soitu C, Feuerborn A, Deroy C, Castrejón-Pita A A, Cook P R and Walsh E J 2019 Raising fluid walls around living cells *Sci. Adv.* **5** 8002
[15] Deroy C, Stovall-Kurtz N, Nebuloni F, Soitu C, Cook P R and Walsh E J 2021 Predicting flows through microfluidic circuits with fluid wall *Microsyst. Nanoeng.* **7** 3
[16] Berthier J, Loe-Mie F, Tran V-M, Schoumacker S, Mittler F, Marchand G and Sarrut N 2009 On the pinning of interfaces on micropillar edges *J. Colloid Interface Sci.* **338** 296–303
[17] Tung C-K, Krupa O, Apaydin E, Liou J-J, Diaz-Santana A, Kim B J and Wu M 2013 A contact line pinning based microfluidic platform for modelling physiological flows *Lab Chip* **13** 3876

IOP Publishing

Open-Channel Microfluidics (Second Edition)

Fundamentals and applications

Jean Berthier, Ashleigh B Theberge and Erwin Berthier

Chapter 4

Flow dynamics in open channels of uniform cross-section

In chapter 2, we presented capillary flow in closed cylindrical tubes. In this chapter, the dynamics of spontaneous capillary flows is presented, first for closed—composite and non-composite—channels, then for open channels. The capillary flow inside closed channels of arbitrary cross-section is first investigated. The approach appears to be useful as a prerequisite for the comprehension of capillary flow in open channels.

It is shown that the Lucas–Washburn–Rideal (LWR) law—valid for cylindrical ducts—is only approximate for noncircular cross-sections which can be circumscribed by a circle, and presents an important bias for flat channels. The use of the hydraulic diameter does not solve this difficulty. In the case of closed channels of arbitrary cross-section, it is shown that the 'average friction length' must replace the hydraulic diameter of circular or nearly circular channels. The notion of 'average friction length' is also the background for the dynamics of capillary flow in open channels.

Because biologic fluids often have a non-Newtonian rheology, the last section is devoted to the dynamics of capillary flow in non-Newtonian fluids such as alginates and whole blood.

4.1 Spontaneous capillary flow in composite, closed channels of arbitrary uniform cross-section

Chapter 2 presented capillary flow inside cylindrical tubes. In the case of closed channels of arbitrary cross-section, the dynamics of the flow requires the notion of 'average friction length,' which is presented in this section.

4.1.1 The Bosanquet equation and the average friction length

The capillary force in a cylinder is given by

$$F_{\text{cap, cyl}} = 2\pi R\, \gamma \cos\theta, \tag{4.1}$$

doi:10.1088/978-0-7503-5507-0ch4

where R is the cylinder's interior radius and θ the Young contact angle. Consider now a closed channel of arbitrary cross-section (figure 4.1).

We have seen in chapter 1 that in such channels, the capillary force is given by

$$F_{\text{cap}} = p\,\gamma\cos\theta, \tag{4.2}$$

where p the total perimeter of the cross-section. Because the channel has a uniform cross-section, the perimeter p and the cross-sectional area S_c are constant along the whole device.

Let us now use the Bosanquet equation (see chapter 2) [1]. The mass of fluid m is proportional to the penetration distance z

$$m = \rho\, S_C z, \tag{4.3}$$

and the Bosanquet equation can be rewritten as

$$\rho\, S_C \frac{\mathrm{d}(zV)}{\mathrm{d}t} = p\,\gamma\cos\theta - p\,z\,\bar{\tau}, \tag{4.4}$$

where $\bar{\tau}$ is the wall friction averaged between the tip of the flow and inlet. In order to solve (4.4), the wall friction has to be expressed as a function of V [2, 3]. The local wall friction can be expressed as

$$\tau = \mu\frac{\partial v}{\partial \hat{n}} = \mu\frac{V}{\lambda}, \tag{4.5}$$

where λ is the friction length and $\hat{n}$ the unit vector perpendicular to the wall (figure 4.2(A)). In a cross-section, the friction length λ can be averaged by

$$\tilde{\tau} = \frac{1}{p}\int_0^p \tau\,\mathrm{d}s = \frac{\mu V}{p}\int_0^p \frac{1}{\lambda}\mathrm{d}s = \mu\frac{V}{\bar{\lambda}}, \tag{4.6}$$

where s is the curvilinear coordinate along the perimeter and

$$\frac{1}{\bar{\lambda}} = \frac{1}{p}\int_0^p \frac{1}{\lambda}\mathrm{d}s. \tag{4.7}$$

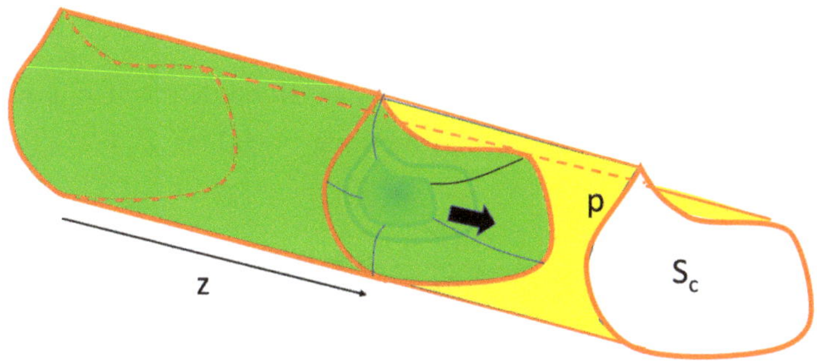

Figure 4.1. A closed channel of uniform cross-section. The travel distance is z, the cross-sectional area Sc and the perimeter p. Note the approximation for the travel distance z, due to the shape of the convex = concave meniscus.

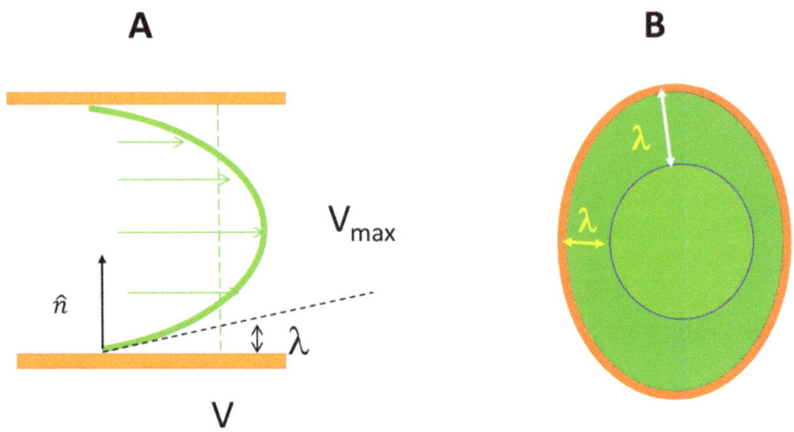

Figure 4.2. (A) A sketch of the velocity profile and the local friction length; (B) a sketch of the different friction lengths for an oval closed channel.

Because the cross-section is constant, i.e. independent of the distance in the channel, the average wall friction over a distance z is therefore

$$\bar{\tau} = \frac{1}{z}\int_0^z \tilde{\tau}\,\mathrm{d}z = \frac{\mu\, V}{\bar{\lambda}}\frac{1}{z}\int_0^z \mathrm{d}z = \mu\frac{V}{\bar{\lambda}}. \tag{4.8}$$

The distance $\bar{\lambda}$ is called the average friction length. This definition of the average friction length produces the solution given in (4.4) following the introduction of the velocity V into the friction term.

Equation (4.4) then becomes the first-order differential equation

$$\rho\, S_C\frac{\mathrm{d}(z\,\dot{z})}{\mathrm{d}t} + \frac{\mu\, p\, z\, \dot{z}}{\bar{\lambda}} - p\,\gamma\cos\theta = 0, \tag{4.9}$$

with the notation $\dot{z} = \mathrm{d}z/\mathrm{d}t = V$. In the case of a cylindrical channel ($p = 2\pi R$, $S_C = \pi R^2$, $\bar{\lambda} = R/4$, $\theta^* = \theta$) and assuming that inertia is negligible, equation (4.9) reduces to the well-known LWR equation (chapter 2). Equation (4.9) can be cast in the form

$$\frac{\rho\, S_C}{2}\frac{\mathrm{d}^2 z^2}{\mathrm{d}t^2} + \frac{\mu\, p}{2\bar{\lambda}}\frac{\mathrm{d}z^2}{\mathrm{d}t} - p\,\gamma\cos\theta = 0, \tag{4.10}$$

and we are left with a differential equation in z^2. If we define $A = \mu\, p/\left(\bar{\lambda}\rho\, S_C\right)$ and $B = 2\,p\,\gamma\cos\theta/\left(\rho\, S_C\right)$ and use the intermediate variables $Z = z^2$, the differential equation (4.10) becomes

$$\frac{\mathrm{d}^2 Z}{\mathrm{d}t^2} + A\frac{\mathrm{d}Z}{\mathrm{d}t} - B = 0, \tag{4.11}$$

Introducing $U = \mathrm{d}Z/\mathrm{d}t$, equation (4.11) can be integrated twice and we find

$$z^2 = \frac{1}{A}\left[B\,t + C\left(1 - e^{-At}\right)\right], \tag{4.12}$$

where C is an integration constant which is difficult to determine due to the indetermination of the velocity at $t = 0$. The first term on the right-hand side of (4.12), i.e. $z^2 = B\,t/A$ is the LWR law corresponding to the inertialess regime or the viscous regime. The second term corresponds to the inertial regime [4]. In the following section, the inertial regime which occurs at the beginning of the flow is presented.

4.1.2 The inertial regime

Due to the negative exponential term in (4.12), the duration τ_{inertial} of the inertial regime at the beginning of the motion is very short. The inertial term is negligible when $t \gtrsim 1/2A$. So τ_{inertial} can be approximated by

$$\tau_{\text{inertial}} \sim \frac{1}{2A} = \frac{\bar{\lambda}\,\rho\,S_C}{2\mu\,p} = \frac{\bar{\lambda}\,\rho\,D_H}{8\mu}, \tag{4.13}$$

where D_H is the hydraulic diameter ($D_H = 4S_c/p$). For a cylindrical duct whose diameter is 1 mm and a water flow, the duration of the inertial regime is $\tau_{\text{inertial}} \sim 1/30$ s. This is the reason why scientists at the time of Washburn did not observe the inertial regime. Relation (4.13) shows that the inertial regime may be neglected for very viscous liquids or in the case of nanoscale channels [5, 6]. Let us determine a maximum for τ_{inertial} in the general case. First, the average friction length is, per definition, a fraction of the hydraulic diameter $\bar{\lambda} \sim D_H/\varepsilon$, where ε is of the order of or less than eight (for the case of a circular cross-section). Second, capillarity-based systems usually have dimensions smaller than the capillary length $\sqrt{\gamma/\rho g}$. Then, using (4.13), the duration of the inertial regime is less than

$$\tau_{\text{inertial}} < \frac{1}{64g}\frac{\gamma}{\mu}. \tag{4.14}$$

Clearly, only mercury has an inertial regime of the order of one second. The inertial regime is a fraction of a second for the liquids currently used in biotechnology and biology. However, the travel distance during the inertial regime is not totally negligible. Following Quéré [4], at the very beginning of the fluid motion, i.e. during the inertial regime, friction can be neglected (because the wetted length is negligible) and equation (4.9) for the fluid motion simplifies to

$$\rho\,S_C\frac{d(z\,\dot{z})}{dt} = p\,\gamma\cos\theta. \tag{4.15}$$

Integration of the differential equation (4.15) yields

$$z = \sqrt{\frac{2p\,\gamma\cos\theta}{\rho\,S_C}}\;t = \sqrt{B}\,t. \tag{4.16}$$

Note the linearity of relation (4.16), signifying that the inertial velocity is constant. The travel distance at the end of the inertial regime is

$$z_{\text{inertial}} \simeq \sqrt{B}\,t_{\text{inertial}}. \tag{4.17}$$

A maximum value for z_{inertial} can be obtained by considering a hydraulic diameter equal to the capillary length and a friction length of $\bar{\lambda} \sim D_H/8$. Introducing these maximum values into (4.13) and (4.17) yields

$$z_{\text{inertial}} < \frac{\gamma}{32\mu}\sqrt{\frac{2D_{\text{H}}}{g}}. \tag{4.18}$$

For water, the inertial distance is of the order of a few millimeters, depending on the dimension of the channel.

4.1.3 The viscous regime

After the initial evanescent inertial regime, the expression for the travel distance z is the solution of the simplified equation (4.9):

$$\frac{\mu\, p\, z\, \dot{z}}{\bar{\lambda}} - p\,\gamma\cos\theta = 0. \tag{4.19}$$

The integration of (4.19) yields

$$z = \sqrt{\frac{\gamma}{\mu}2\bar{\lambda}\cos\theta t}. \tag{4.20}$$

The travel distance varies as the square root of the time, in agreement with the LWR model of capillary flows inside cylinders [7–9]. The liquid velocity can be readily derived from (4.20):

$$V = \frac{\mathrm{d}z}{\mathrm{d}t} = \sqrt{\frac{\gamma}{\mu}2\bar{\lambda}\cos\theta}\,\frac{1}{2\sqrt{t}}. \tag{4.21}$$

In this form, the fluid velocity is the product of the square root of an 'intrinsic' velocity γ/μ—related to the physical properties of the materials—by the square root of a 'geometrical' velocity $\sqrt{\bar{\lambda}/2t}$.

On the other hand, eliminating the time t between the two relations (4.20) and (4.21) produces a relation between the velocity and the travel distance:

$$V = \frac{\gamma}{\mu}\frac{\bar{\lambda}\cos\theta}{z}. \tag{4.22}$$

We can read relation (4.22) differently by multiplying both sides by the perimeter p and dividing by the cross-sectional area S_c:

$$\mu\frac{V}{\bar{\lambda}}\frac{pz}{S_c} = \frac{p\,\gamma\cos\theta}{S_c} = P_{\text{cap}}. \tag{4.23}$$

The right-hand side of (4.23) is the capillary force F_{cap} divided by the cross-section S_c, i.e. the capillary pressure at the advancing interface, P_{cap}. The inlet pressure being zero, the pressure drop between the channel inlet and the advancing meniscus is

$$\Delta P = \mu\frac{V}{\bar{\lambda}}\frac{pz}{S_c} = \frac{p\,\gamma\cos\theta}{S_c}. \tag{4.24}$$

The flow being laminar, the Navier–Stokes equation governs the physics of the flow; the pressure drop is proportional to the flow velocity [10–12] and has a coefficient R_H called the hydraulic resistance. Here, $\widetilde{R}_H$ denotes the hydraulic resistance per unit length ($\widetilde{R}_H z = R_H$), and the pressure drop in relation (4.24) can be expressed as a function of $\widetilde{R}_H$:

$$\Delta P = \mu \frac{V}{\bar{\lambda}} \frac{pz}{S_c} = \widetilde{R}_H z\ V. \tag{4.25}$$

Note that the flow behavior (Navier–Stokes law) is the same for laminar forced flows inside ducts and capillary flows. The difference pertains to the boundary conditions: forced microflows usually have usually a pressure or a flow rate imposed at the back end of the flow, while capillary flows are driven by capillary forces at the front end of the flow.

An interesting application of formula (4.25) is the direct deduction of the average friction length $\bar{\lambda}$—and consequently the capillary velocity—when the hydraulic resistance is known. The friction length $\bar{\lambda}$ is deduced from (4.25) as follows:

$$\bar{\lambda} = \mu \frac{p}{\widetilde{R}_H S_c}. \tag{4.26}$$

The friction length $\bar{\lambda}$ can then be related to the hydraulic diameter ($D_H = 4S_c/p$). Substitution of the hydraulic diameter into (4.26) yields

$$\bar{\lambda} = \frac{4\mu}{\widetilde{R}_H D_H}. \tag{4.27}$$

In the literature, tables have been published for hydraulic resistances R_H as functions of the shapes and dimensions of the channels, for many different confined channels [13–16]. If Re denotes the Reynolds number and f the Fanning coefficient, the pressure drop is given by

$$\Delta P = \widetilde{R}_H z\ V = \frac{\mu f \,\mathrm{Re}}{2D_H^2} z\ V. \tag{4.28}$$

In this form, we can equate the pressure drop between relation (4.25)—using the average friction length $\bar{\lambda}$—and relation (4.28) —using the hydraulic diameter:

$$\Delta P = \widetilde{R}_H z\ V = \frac{\mu f \,\mathrm{Re}}{2D_H^2} z\ V = \mu \frac{V}{\bar{\lambda}} \frac{pz}{S_c}. \tag{4.29}$$

The hydraulic resistance $\widetilde{R}_H$ is then eliminated, and the average friction length is related to the hydraulic diameter by

$$\bar{\lambda} = \frac{8\ D_H}{f\ \mathrm{Re}} = \frac{2\ D_H}{Po}, \tag{4.30}$$

where Po is the Poiseuille number ($Po = f\,Re/4$). For a closed channel, substituting relation (4.30) into the expression for the travel distance (4.20) yields

$$z = 4\sqrt{\frac{\gamma \cos\theta}{\mu} \frac{D_H}{f\ \mathrm{Re}} t} = \sqrt{\frac{4D_H}{Po} \frac{\gamma \cos\theta}{\mu} t}, \tag{4.31}$$

and the velocity of the flow is

$$V = 2\sqrt{\frac{\gamma\cos\theta}{\mu}\frac{D_{\mathrm{H}}}{f\,\mathrm{Re}}}\sqrt{\frac{1}{t}} = \sqrt{\frac{\gamma\cos\theta}{\mu}\frac{D_{\mathrm{H}}}{Po}}\sqrt{\frac{1}{t}}. \tag{4.32}$$

The following tables 4.1 and 4.2 are extracted from the literature. For simple geometries, they give the value of the friction length from the expression of the hydraulic resistance—using relation (4.27)—or from the expression of the Poiseuille number ($Po = f\,Re/4$)—using relation (4.30). Tables 4.1 and 4.2 are derived from references [13–20].

In the case of very complicated geometries, neither the hydraulic resistance nor the Poiseuille number has an analytical expression, and a numerical approach is the only solution that can be used to determine the average friction length.

Table 4.1. Values of the average friction length for different geometries.

Cross-section	Hydraulic resistance (per unit length)	Area and perimeter *Sc, p*	Average friction length $\bar{\lambda}$
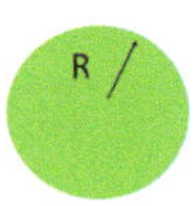	$\widetilde{R}_{\mathrm{H}} = \frac{8\mu}{R^2}$	$S_c = \pi R^2$ $L = 2\pi R$	$\bar{\lambda} = \frac{R}{4}$
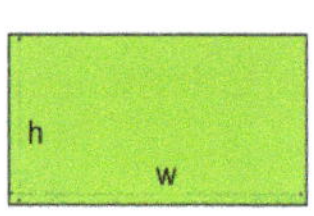	$\widetilde{R}_{\mathrm{H}} = \frac{4\mu}{h^2}\frac{1}{q(\varepsilon)}$, $\varepsilon = h/w$ $q(\varepsilon) = \frac{1}{3} - \frac{64}{\pi^5}\varepsilon\tanh\left(\frac{\pi}{2\varepsilon}\right)$	$S_c = h\,w$ $L = 2\,(w + h)$	$\bar{\lambda} = \frac{\varepsilon\, q(\varepsilon)\,(w+h)}{2}$
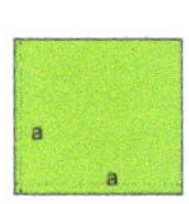	$\widetilde{R}_{\mathrm{H}} = \frac{28.3\,\mu}{a^2}$	$S_c = a^2$ $L = 4\,a$	$\bar{\lambda} = \frac{a}{7.1}$
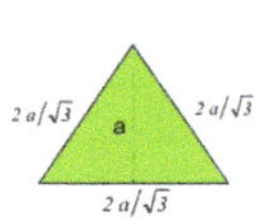	$\widetilde{R}_{\mathrm{H}} = \frac{60\,\mu}{a^2}$	$S_c = a^2/\sqrt{3}$ $L = 6\,a/\sqrt{3}$	$\bar{\lambda} = \frac{a}{15\sqrt{3}}$
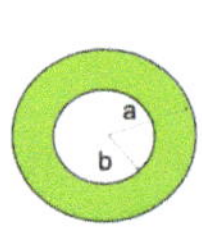	$\widetilde{R}_{\mathrm{H}} = \frac{8\mu}{a^2}\frac{1}{\beta}$ $\varepsilon = b/a$ $\beta = \varepsilon^2 - 1 + \frac{2\ln(1/\varepsilon) + \varepsilon^2 - 1}{\ln(1/\varepsilon)}$	$S_c = \pi\,(a^2 - b^2)$ $L = 2\pi\,(a + b)$	$\bar{\lambda} = \frac{\beta}{4}\frac{a^2}{(a-b)}$

Table 4.2. Values of the average friction length for different geometries.

Cross-section	*fRe*	Hydraulic diameter	Average friction length
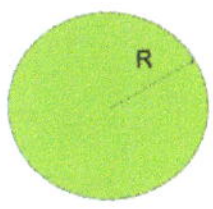	64	$D_H = 2\,R$	$\bar{\lambda} = \frac{R}{4}$
 $h << w$	96	$D_H \approx 2\,h$	$\bar{\lambda} \approx \frac{h}{6}$
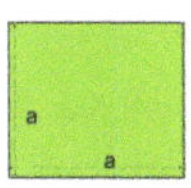	56.92	$D_H = a$	$\bar{\lambda} = \frac{a}{7.1}$
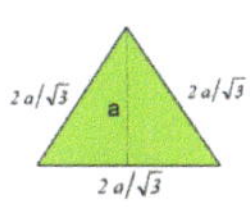	120	$D_H = \frac{a}{\sqrt{3}}$	$\bar{\lambda} = \frac{a}{15\sqrt{3}}$
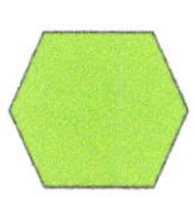	60	$D_H = a\,\sqrt{3}$	$\bar{\lambda} = \frac{a}{4.33}$
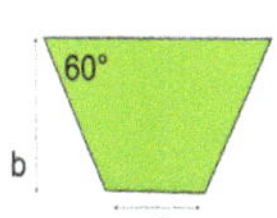	b/a *fRe* 4 55.66 2 55.22 1 56.60 0.5 62.77	$D_H = \frac{2b\,(a\,\sqrt{3}+b)}{\sqrt{3}\,(a+b\,\sqrt{3})}$	$\bar{\lambda} = D_H/6.96$ $\bar{\lambda} = D_H/6.90$ $\bar{\lambda} = D_H/7.07$ $\bar{\lambda} = D_H/7.85$

4.1.4 On the use of the hydraulic diameter

Note that the LWR expression (4.20)—in which the radius R is replaced by the hydraulic radius—is often wrongly used in the literature to calculate the travel distance. The difference between the travel distance obtained by this method (hydraulic diameter) and that obtained using expression (4.30) is

$$\Delta z = \sqrt{\frac{\gamma \cos\theta}{\mu}}\sqrt{t}\left(\sqrt{\frac{16 D_H}{f\,\mathrm{Re}}} - \sqrt{\frac{D_H}{4}}\right) = \sqrt{\frac{\gamma \cos\theta}{\mu}}\sqrt{t}\sqrt{\frac{16 D_H}{f\,\mathrm{Re}}}\left(1 - \sqrt{\frac{f\,\mathrm{Re}}{64}}\right). \quad (4.33)$$

[For 'regular' shapes, such as squares or hexagons, or even some trapezoids, the value of *fRe* is close to 64 (table 4.2). However, for 'irregular' shapes, such as slender/flat rectangles or triangles, a systematic error is made. This systematic relative error is

$$E = \left| \frac{\Delta z}{z} \right| = \left| 1 - \sqrt{\frac{f \, \mathrm{Re}}{64}} \right|. \tag{4.34}$$

For the values $fRe = \{56.92, 60, 64, 96, 120\}$, respectively corresponding to square, hexagonal, circular, rectangular, and triangular shapes, the relative errors are $E = \{6\%, 3\%, 0\%, 22\%, 37\%\}$. The error is large for very irregular shapes.

4.2 Spontaneous capillary flow in open channels of arbitrary uniform cross-section

Consider now the case of open channels. The approach is similar to that of the preceding section, with two exceptions. First, replace the Young contact angle θ by the generalized Cassie angle θ^* defined in chapter 3. For a monolithic open channel,

$$\cos\theta^* = \frac{p_\mathrm{w}\cos\theta - p_\mathrm{F}}{p_\mathrm{w} + p_\mathrm{F}}, \tag{4.35}$$

and for a composite open channel, $\cos\theta^* = \frac{\sum_i p_{\mathrm{w},\,i}\cos\theta_i - p_\mathrm{F}}{\sum_i p_{\mathrm{w},\,i} + p_\mathrm{F}}$, where the index i refers to the wetting properties of the different walls. The capillary force is then

$$F_\mathrm{cap} = \gamma\, p\cos\theta^* = \gamma\left(p_\mathrm{w}\cos\theta - p_\mathrm{F}\right) \tag{4.36}$$

The second condition is relative to the friction force. In the case of an open channel, the wall friction is given by

$$F_\mathrm{drag} = \bar{\tau}\, p_\mathrm{W}\, z = z\int_{\Gamma_W}\mu\frac{V}{\lambda}\,\mathrm{d}l = \mu\frac{V}{\bar{\lambda}}\,p\,z. \tag{4.37}$$

where l is the curvilinear coordinate along the wetted contour Γ_W and the friction length is defined as

$$\frac{1}{\bar{\lambda}} = \frac{1}{p}\int_{\Gamma_\mathrm{W}}\frac{1}{\lambda}\,\mathrm{d}l, \tag{4.38}$$

Note that the integration is now only on the wetted contour Γ_W. Using (4.35), (4.37), and (4.38), this definition of the average friction length agrees with the viscous regime travel distance $z = \sqrt{\frac{\gamma}{\mu}2\bar{\lambda}\cos\theta^* t}$.

4.2.1 The Bosanquet equation

In the case of an open-capillary flow, the Bosanquet equation is similar to equation (4.9), except that the contact angle θ is replaced by θ^*:

$$\rho\, S_\mathrm{C}\frac{\mathrm{d}(z\,\dot{z})}{\mathrm{d}t} + \frac{\mu\, p\, z\,\dot{z}}{\bar{\lambda}} - p\,\gamma\cos\theta^* = 0. \tag{4.39}$$

In relation (4.39), the cross-sectional area S_c is delimited by the wetted walls and the flat free surface (zero pressure). The notion of the average friction length detailed in equations (4.6) and (4.7) still holds. The difference stems from the free surface that does not contribute to friction. Hence, for the same cross-section, the friction length is larger in the case of an open boundary (the wall friction is less when there is an open boundary). The difficulty is the determination of this friction length. An approximation of the friction length for simple geometries is presented in chapter 5.

4.2.2 The inertial regime

The same considerations can be made for the inertial regime as for closed channels. The travel distance in the inertial regime still has the same expression

$$z = \sqrt{\frac{2p\,\gamma\cos\theta^*}{\rho\,S_C}}\;t = \sqrt{B}\;t, \tag{4.40}$$

with the difference that the coefficient B is now $B = 2\,p\,\gamma\cos\theta^*/\left(\rho\;S_C\right)$. The inertial velocity V_{in} is now constant:

$$V_{in} = \sqrt{\frac{p\,\gamma\cos\theta^*}{\rho\,S_C}} = \sqrt{B}. \tag{4.41}$$

Compared with the same closed-channel geometry, as friction does not participate in the inertial regime, the inertial velocity does not depend on the friction length $\bar{\lambda}$ but only on the Cassie angle θ^* representing the capillary force. On the other hand, it is to be expected that the transition to the viscous regime depends on the average friction length. The end of the inertial regime is given by

$$\tau_{\text{inertial}} \sim \frac{\bar{\lambda}\,\rho\,S_C}{2\mu\,p} = \frac{\bar{\lambda}\,\rho\,D_H}{8\mu}. \tag{4.42}$$

The travel distance at the end of the inertial regime is then

$$z_{\text{inertial}} \simeq \sqrt{B}\;t_{\text{inertial}}, \tag{4.43}$$

and it depends on both the friction length and the Cassie angle. Note that the transition distance z_{inertial} between inertial regime and the viscous regime is smaller in the case of an open channel of the same cross-section because the inertial velocity V_{in} is proportional to the square root of $\cos\theta^*$ or $\cos\theta$, and $\cos\theta^* < \cos\theta$.

In figure 4.3, we have indicated the values of the inertial distance for three open rectangular channels and six different liquids.

4.2.3 The viscous regime

After the time τ_{inertial}, friction is important and inertia vanishes quickly. Then

$$z = \sqrt{\frac{B}{A}t} = \sqrt{\frac{\gamma}{\mu}2\bar{\lambda}\cos\theta^* t}\,. \tag{4.44}$$

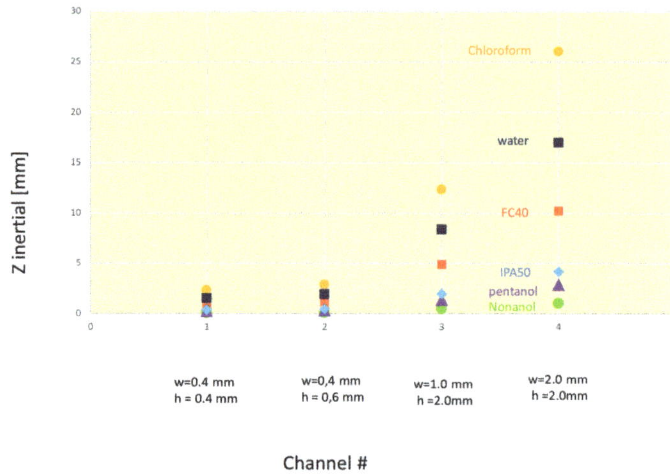

Figure 4.3. Different inertial distances for four rectangular open channels and six liquids.

Relation (4.42) is the LWR law as extended to open channels. The liquid velocity is then

$$V = \frac{\mathrm{d}z}{\mathrm{d}t} = \sqrt{\frac{\gamma}{\mu}\frac{\bar{\lambda}}{2}\cos\theta^*}\,\frac{1}{\sqrt{t}}. \tag{4.45}$$

Eliminating the time between the two preceding relations yields the relation between V and z

$$V = \frac{\gamma}{\mu}\frac{\bar{\lambda}\cos\theta^*}{z}. \tag{4.46}$$

Note that the term zV is constant: then, as expected, inertia is null because the inertia $I = \frac{\mathrm{d}(mV)}{\mathrm{d}t} = \rho S_c \frac{\mathrm{d}(zV)}{\mathrm{d}t} = 0$.

4.2.4 An example

Plots of the travel distance as functions of the square root of time for six liquids (toluene, chloroform, dodecane, pentanol, FC40, and nonanol) in an open U-groove are shown in figure 4.4. The proportionality between the travel distance and the square root of time is clearly apparent. Note that the inertial time is so short that the inertial regime cannot be seen in the figure.

4.2.5 A comparison of the average friction lengths in closed and open channels

Let us take the example of a rectangular geometry and consider two types of channels: open rectangular and closed rectangular (figure 4.5). An expression for the average friction length of the closed channel is listed in table 4.1. Approximations for the average friction length of rectangular open channel are detailed later, in chapter 5.

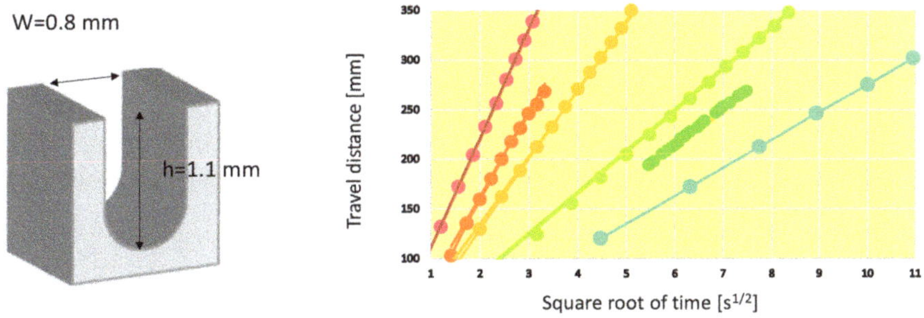

Figure 4.4. Left: a sketch of the channel cross-section ($\bar{\lambda} \sim 0.22$ mm) milled in PMMA; right: the travel distance as a function of the square root of time for six liquids; from left to right: toluene, chloroform, dodecane, pentanol, FC40, and nonanol. The dots correspond to experiments and the straight lines to relation (4.44).

Figure 4.5. (A) A sketch of an open rectangular channel and (B) a closed rectangular channel.

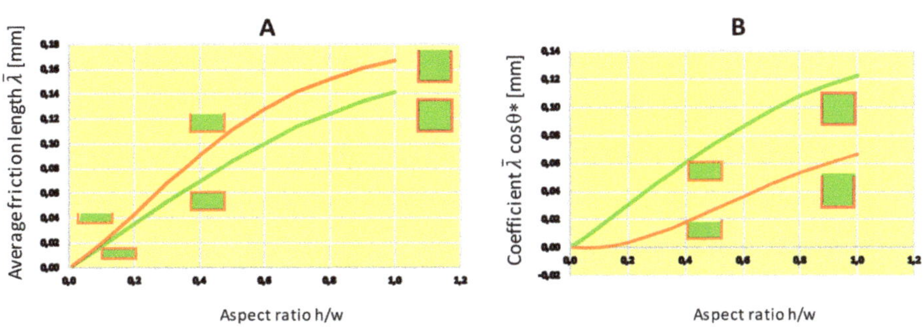

Figure 4.6. A: average friction length for a width $w = 1$ mm for the two types of channels, open and closed; B: coefficient $\bar{\lambda}\cos\theta^*$ driving the flow for the three channels.

In the case of the U-groove, $\bar{\lambda} \approx \frac{1}{6}\frac{(w+2h)}{\left(\frac{w}{h}+\frac{2h}{w}\right)}$. The different average friction lengths (for closed and open channels) are plotted against the channel aspect ratio in figure 4.6(A).

The friction length is higher for the open rectangular U-groove, signifying that the friction is less in such a geometry—which is due to a smaller contact area with the walls—but the driving term of relation (4.42) $\bar{\lambda}\cos\theta^*$ is higher for the closed-channel geometry, due to the term $\cos\theta^*$.

4.2.6 A numerical approach

Numerical approaches to open-capillary flow are complicated due to the ever-increasing free surface. Powerful computers are required to account for the very fine meshes at the vicinity of the advancing meniscus and the increasing number of meshes of the extending free surface. Different approaches are possible, such as volume of fluid (VOF) or level set methods (see [21]).

4.2.6.1 The early stages of the flow

We briefly show here an approach that illustrates the early inertial regime and its transition to the viscous regime using the software FLOW-3D (Flow Science, Santa Fe, USA) to calculate the onset of the spontaneous capillary flow of dodecane in an open U-groove 0.8 mm wide and 1 mm deep. A view of the flow is shown in figure 4.7(A) and a velocity profile is shown in figure 4.7(B).

A plot of the travel distance as a function of time is shown in figure 4.8. The transition between the inertial regime and the viscous regime takes place between 0.10 and 0.16 s.

4.2.6.2 Determination of the average friction length

The most difficult point is the determination of the average friction length. In simple geometries, an approximation of the friction length $\bar{\lambda}$ can be made. These cases are presented in chapter 6. In the general case, either an experimental or numerical approach can be used to determine the friction length. When choosing a numerical method, an easy approach—although approximate—consists in calculating the flow field in a portion of the closed channel using a non-friction condition for the free surface, then deriving the value of the local shear rate at the wall, and finally averaging the shear rate $\dot{\gamma}$ on a cross-section ($\dot{\gamma} = V/\bar{\lambda}$). An example is shown in figure 4.9.

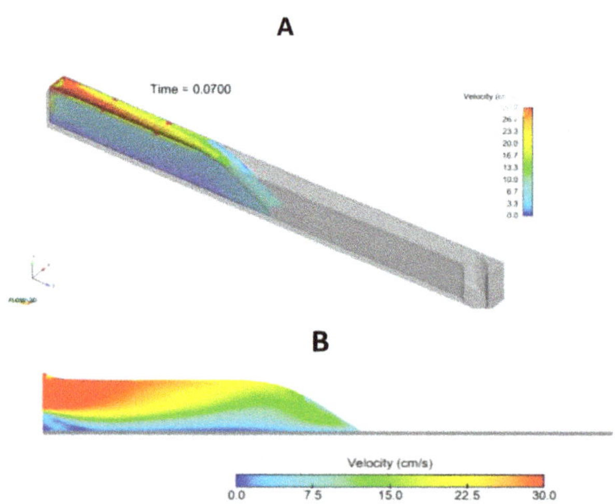

Figure 4.7. The onset of spontaneous capillary flow in dodecane (in the inertial regime): (A) a FLOW-3D view of the capillary flow; (B) the transient velocity profile showing the maximum velocity at the free boundary. Calculation courtesy of I Karampelas, Flow Science (https://www.flow3d.com/flow-science).

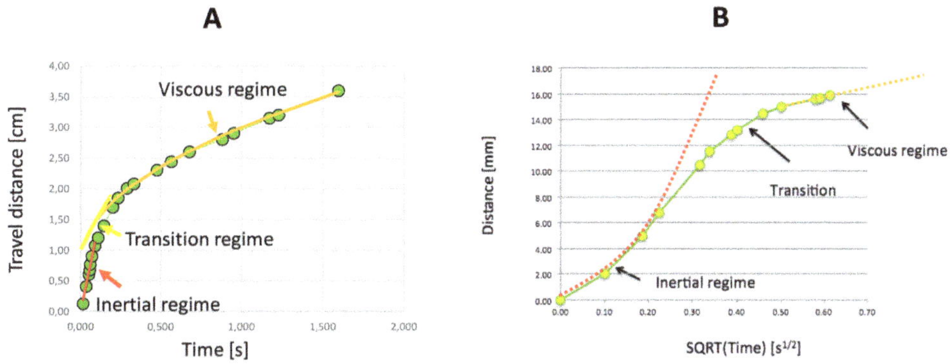

Figure 4.8. (A) The travel distance vs. time calculated using FLOW-3D for dodecane, showing the inertial regime (red line) and the viscous regime (yellow line); (B) the travel distance vs. the square root of time for the inertial regime (red dots) and the viscous regime (yellow dots). A short transition regime links the two regimes. An estimate of the inertial time using (4.42) is 0.15 s. Calculation courtesy of I Karampelas, Flow Science (https://www.flow3d.com/flow-science).

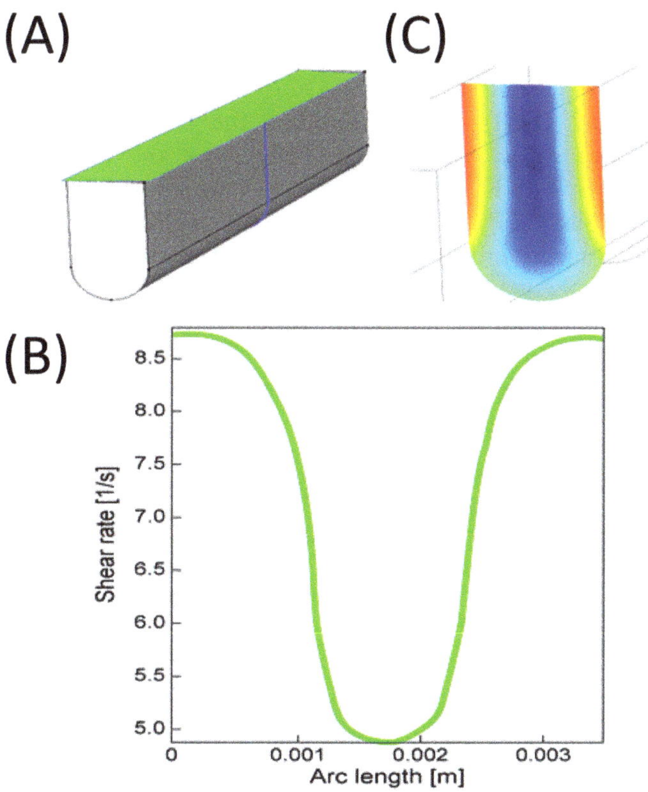

Figure 4.9. An example of numeric determination of the friction length (COMSOL: https://www.comsol.fr/comsol-multiphysics): (A) the geometry of the channel with an open boundary (green) corresponding to the symmetry condition (no friction); (B) the shear rate contour in a cross-section; (C) the wall shear rate as a function of the curvilinear distance. Integration of this shear rate along the contour yields the average shear rate and consequently the average friction length.

Finally, when the physical properties of a liquid flowing in an open channel are well known, a fit with experimental results can be a convenient way of determining the friction length.

4.3 The dynamic contact angle

The real value of the contact angle with the solid surface is a notable problem in capillary microflows. The LWR law—which dates from 1920—assumes a constant contact angle during the fluid motion. However, more recently, experimental data have shown somewhat lower velocities than that predicted by the LWR equation. This observation is attributed to the fact that the contact angle between the flowing liquid and the walls differs from its static (Young) value. Inspection of the LWR behavior in closed channels shows that a dynamic contact angle (DCA) larger than the static contact angle should be taken into account, depending on the value of the flow velocity.

The analysis of this phenomenon dates from 1979, when the paradox of the moving triple line was raised: at the same time, the meniscus at the tip of the flow advances, while the velocity of the triple line in contact with the solid is zero (there is no slip at the wall) [22–24]. The hypothesis of a very small layer of immobilized liquid molecules has been proposed to resolve this paradox.

Three main theoretical models for the DCA were successively proposed: (1) the hydrodynamic model (HM), which considers the viscous energy dissipation in the mesoscopic layer associated with the bending of the velocity profile; (2) the molecular kinetic theory (MKT), which considers the energy dissipation in the vicinity of the triple contact line (TPCL); (3) the MKT self-layering approach, in which the energy is dissipated in the precursor film [25, 26]. Also, in connection with the HD model, some empirical equations have been proposed based on the adhesion energy.

In chapter 2, we presented the different approaches to the determination of the DCA. We recall here the main conclusions: in the HD approach (figure 4.10(A)), the increase in the contact angle outside the microscopic layer (L_{mi}) is linked to the cubic root of the capillary number [27–30]:

$$\theta_d^3 = \theta_0^3 + 9\chi Ca = \theta_0^3 + 9\ln\left(L_{ma}/L_{mi}\right) Ca, \tag{4.47}$$

where θ_d is the advancing DCA, θ_0 the equilibrium (Young) contact angle, Ca the capillary number ($Ca = \mu V/\gamma$), and χ is a parameter given by $\chi = \ln\left(L_{ma}/L_{mi}\right)$. L_{ma}

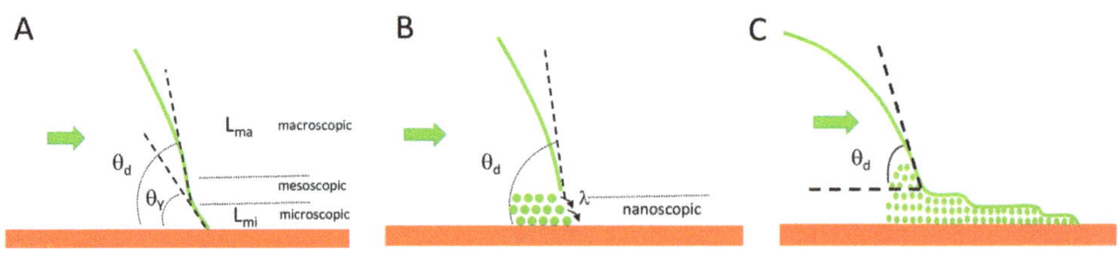

Figure 4.10. Sketches of the different models for the DCA: (A) the HM assumes a microscopic layer that has a static contact angle with the solid wall, together with a bending of the interface starting at a mesoscopic scale; (B) the MKT model considers adsorption–desorption processes at a nanoscopic scale at the TPZ (triple phase zone); (C) the MKT-self-layering model considers a wall viscosity associated with the formation of layers in the precursor film.

is a macroscopic length scale which is of the order of the size of the meniscus and L_{mi} is a microscopic length scale that is the cutoff length below which the continuum theory breaks down. The difficulty with correlation (4.47) is the determination of χ, which depends on the length L_{mi}. Even though some studies have attempted to determine χ [31], this determination is not straightforward and one generally uses a fit with the experimental results. Correlation (4.47) states that the change of the contact angle during motion is linked to the cubic root of the capillary number—or velocity.

Empirical correlations were later proposed by Bracke *et al* [32], Seebergh and Berg [33], and Jiang *et al* [34]. These correlations also make use of the capillary number. They are of the form

$$\cos\theta_{\mathrm{d}} = \cos\theta_0 - A(1 + \cos\theta_0)f(Ca), \tag{4.48}$$

where A is a nondimensional coefficient and f a monotonously increasing function. Often, the function f is taken to be $f(Ca) = Ca^B$. Then,

$$\cos\theta_{\mathrm{d}} = \cos\theta_0 - A(1 + \cos\theta_0)Ca^B. \tag{4.49}$$

The correlation constant A ranges between 2 and 4.96 and the constant B ranges between 0.42 and 0.7 [35, 36]. The term $1 + \cos\theta_0$ represents the adhesion energy. In all these correlations, the inertial forces are assumed to be negligible compared to the viscous forces, so that the regime is the viscous regime.

More recently, Blake and coworkers proposed a new approach based on MKT for viscous flows in which the DCA results from the adsorption–desorption dynamics of liquid molecules on a solid surface near the TPCL (figure 4.10(B)). This kind of rolling motion can be understood in terms of a friction coefficient [37–40]. Hence, MKT introduces a value for the viscosity of the layered liquid which depends on the liquid and the solid wall. The velocity-dependent DCA is then

$$\cos\theta_{\mathrm{d}} = \cos\theta_0 - \frac{\beta}{\gamma}V = \cos\theta_0 - \frac{\beta}{\mu}Ca, \tag{4.50}$$

where γ is the surface tension, V is the fluid velocity, and β is a coefficient of wetting-line friction in the case of the MKT approach or a frictional coefficient between the liquid and solid surfaces at the TPCL in the case of self-layering. In both cases, the unit of β is Pa s.

In the case of a precursor film, Wu *et al* [25, 26] adapted the MKT approach using the concept of self-layering of the thin precursor film (figure 4.10(C)), in which a few layers of molecules of the liquid are immobilized at the wall near the triple line.

In both approaches, the wall viscosity β (unit Pa s) is linked to the adsorption–desorption process at the TPCL by either one of these relations

$$\frac{\beta}{\mu} = \frac{\nu}{\lambda^3}\exp\left(\frac{\lambda^2\gamma(1 + \cos\theta_0)}{k_{\mathrm{B}}T}\right) \text{ for MKT, or}$$

$$\frac{\beta}{\mu} = \exp\left(\frac{2\gamma}{n\,k_{\mathrm{B}}T}\right) = \exp\left(\frac{2\gamma\,\sigma^2}{k_{\mathrm{B}}T}\right) \text{ for MKT self-layering,} \tag{4.51}$$

where μ is the viscosity of the flowing liquid, γ the surface tension, λ the nanoscopic distance over which a molecule 'jumps,' ν the molecular volume, k_B the Boltzmann constant, T the temperature, and n the number of molecules per unit area in the precursory thin film (which is related by $n = 1/\sigma^2$ to σ, the effective diameter of the molecules in the thin film).

4.3.1 A model for the relation between the travel distance and a varying dynamic contact angle

In the open-channel geometry, the extended LWR law yields the relation between the velocity V and the travel distance stated by relation (4.46). This relation can be rewritten in terms of the capillary number

$$Ca = \frac{\mu}{\gamma}V = \frac{\bar{\lambda} \cos \theta^*}{z}. \tag{4.52}$$

The DCA depends on the value of the velocity represented by the capillary number. Relation (4.52) shows that, if the travel distance is sufficient and/or if the channel cross-section is sufficiently small ($\bar{\lambda}/z \ll 1$), the use of the static contact angle is justified. Conversely, if these conditions are not met, a DCA is present that changes the dynamics of the flow.

Blake *et al*, Wu *et al*, and Hamraoui *et al* [25, 26, 37, 40] interpreted the DCA in terms of adsorption processes at the wall at the TPCL or in the precursor film. If θ_0 and θ_d correspond to the static angle and the dynamic angle, respectively, and V is the velocity, the decrease in the capillary force associated with the DCA can be written as

$$\Delta F_{\mathrm{cap}} = \gamma\,(\cos \theta_{\mathrm{d}} - \cos \theta_0) = -\beta\ V, \tag{4.53}$$

where β is a parameter that depends on the liquid and the wall. Note that the coefficient β has the dimension of viscosity i.e. Pa s. In the work of Hamraoui and coworkers, the coefficient β was found to vary between 0.02 and 0.2 for ethanol mixtures and a glass capillary tube.

Thus, the DCA can be written as

$$\cos \theta_{\mathrm{d}} = \cos \theta_0 - \frac{\beta}{\gamma}\ V = \cos \theta_0 - \frac{\beta}{\mu}\ Ca. \tag{4.54}$$

In the viscous regime, the capillary force balances the wall friction force (whether the channel is closed or open). Then

$$F_{\mathrm{cap}} = p\ \gamma \cos \theta^* = F_{\mathrm{drag}} = p\ z\ \mu\frac{V}{\bar{\lambda}}. \tag{4.55}$$

where z is the marching—or travel—distance, p the total perimeter of a cross-section (including the liquid/air boundary, if any), $\bar{\lambda}$ the average friction length [2, 3], and θ^* the generalized Cassie angle [3, 41]. Substituting $V = \mathrm{d}z/\mathrm{d}t$, relation (4.55) can be rewritten as

$$zV = \frac{1}{2}\frac{\mathrm{d}z^2}{\mathrm{d}t} = \frac{\gamma}{\mu}\bar{\lambda} \cos \theta^*. \tag{4.56}$$

or

$$z^2 = \frac{\gamma}{\mu} 2\bar{\lambda} \int_0^t \cos\theta^* \mathrm{d}t. \tag{4.57}$$

If θ^* is constant, the usual extended Lucas–Washburn law is retrieved. Let us recall that the generalized Cassie angle is given by

$$\cos\theta^* = \frac{p_W}{p}\cos\theta_d - \frac{p_F}{p}, \tag{4.58}$$

where p_W and p_F are the wetted and free perimeters in a cross-section and $p = p_W + p_F$. Then, (4.57) becomes

$$z^2 = \frac{\gamma}{\mu} 2\bar{\lambda} \int_0^t \left(\frac{p_W}{p}\cos\theta_d - \frac{p_F}{p}\right)\mathrm{d}t. \tag{4.59}$$

Let us use correlation (4.54) to express the DCA:

$$\begin{aligned} z^2 &= \frac{\gamma}{\mu} 2\bar{\lambda} \int_0^t \left[\frac{p_W}{p}\left(\cos\theta_0 - \frac{\beta}{\gamma} V\right) - \frac{p_F}{p}\right]\mathrm{d}t \\ &= \frac{\gamma}{\mu} 2\bar{\lambda} \int_0^t \left[\left(\frac{p_W}{p}\cos\theta_0 - \frac{p_F}{p}\right) - \frac{p_W}{p}\frac{\beta}{\gamma}\frac{\mathrm{d}z}{\mathrm{d}t}\right]\mathrm{d}t \end{aligned}, \tag{4.60}$$

or

$$z^2 = \frac{\gamma}{\mu} 2\bar{\lambda}\cos\theta_0^* - \frac{\gamma}{\mu} 2\bar{\lambda} \int_0^t \frac{p_W}{p}\frac{\beta}{\gamma}\frac{\mathrm{d}z}{\mathrm{d}t}\mathrm{d}t. \tag{4.61}$$

The first term in (4.61) is the travel distance for the static contact angle $z_0^2 = \frac{\gamma}{\mu} 2\bar{\lambda}\cos\theta_0^*$ and the second term is a correction due to the DCA. Relation (4.61) can be simplified to

$$z^2 = \frac{\gamma}{\mu} 2\bar{\lambda}\cos\theta_0^* - 2\bar{\lambda}\frac{p_W}{p}\frac{\beta}{\mu} z. \tag{4.62}$$

The real travel distance—accounting for the DCA correction—is then the solution of the quadratic equation

$$z^2 + 2\bar{\lambda}\frac{p_W}{p}\frac{\beta}{\mu} z - z_0^2 = 0. \tag{4.63}$$

In (4.63), we have to keep in mind that z and z_0 are functions of the time t. Finally, we obtain

$$z = -\bar{\lambda}\frac{p_W}{p}\frac{\beta}{\mu} + \sqrt{\left(\bar{\lambda}\frac{p_W}{p}\frac{\beta}{\mu}\right)^2 + z_0^2}. \tag{4.64}$$

This verifies that at $t = 0$ (we neglect the evanescent inertial motion [4]), $z_0 = 0$ and relation (4.64) yields $z = 0$. For a long channel, z_0 becomes large, the corrective term $\bar{\lambda}\frac{p_W}{p}\frac{\beta}{\mu}$ can be neglected, and $z \rightarrow z_0$. The flow velocity is easily deduced from (4.64) by time derivation:

$$V = \frac{V_0}{\sqrt{1 + \frac{\left(\bar{\lambda}\frac{p_W}{p}\frac{\beta}{\mu}\right)^2}{z_0^2}}}, \tag{4.65}$$

where V_0 is the velocity obtained using the static contact angle: $V_0 = \mathrm{d}z_0/\mathrm{d}t$. When the travel distance increases, z_0 increases and the real velocity V converges towards the velocity V_0. Note that when z_0 goes to zero, relation (4.65) indicates that V goes to infinity. In reality, relation (4.65) holds when the inertial regime ends, e.g. when $z_0 > z_{\text{inertial}}$.

4.3.2 Experiments showing the dynamic contact angle

The DCA can be experimentally observed during an open-capillary flow. Consider the device shown in figure 4.11, in which a nearly rectangular open channel has been milled in a poly(methyl methacrylate) (PMMA) plate. Note that the bottom interior corners are rounded in order to avoid the formation of capillary filaments (see chapter 6) [42, 43]. The channel bottom is estimated to have a few microns of roughness, which is one magnitude below the roughness values that were observed to produce fluctuations in velocity by Lade *et al* [44].

Experiments were performed using four different open rectangular channels milled in PMMA. The characteristics of the channels are listed in table 4.3.

Different solvents were used: aqueous isopropyl alcohol (IPA) was used at a concentration of 50% (wt)—denoted by IPA50—and colored with 0.60% yellow or 1.2% blue food coloring. Organic compounds such as nonanol, pentanol, FC40, and

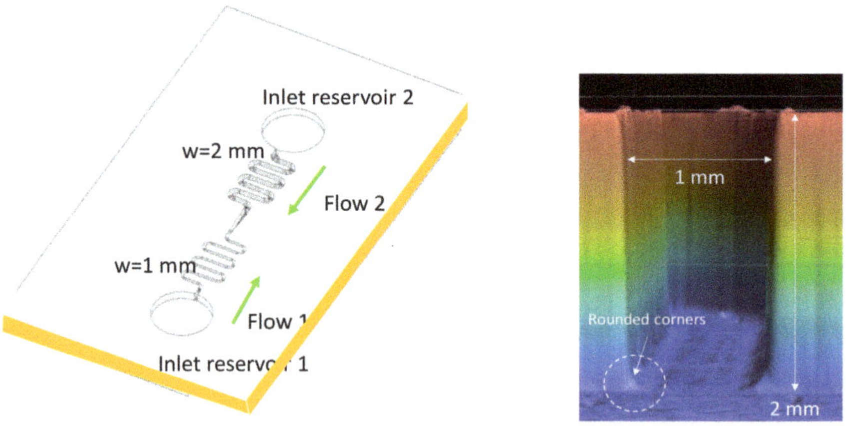

Figure 4.11. Left: a view of two open channels milled in a PMMA plate; right: a profilometric view of the open channel (courtesy of J Tokihiro and Wan-Chen Tu, University of Washington).

Table 4.3. Characteristic dimensions of the channels.

Channel	Width	Height	Wetted perimeter	Free perimeter	Friction length
Symbol/units	w (mm)	h (mm)	p_W (mm)	P_F (mm)	$\bar{\lambda}$ (mm)
Channel #1	1	2	5	1	0.20
Channel #2	2	2	6	2	0.40
Channel #3	0.4	0.6	1.6	0.4	0.09
Channel #4	0.4	0.4	1.2	0.4	0.07

Table 4.4. Properties of the liquids.

	γ (mN m^{-1})	μ (mPa s)	θ	w (mm)	h (mm)	$\cos\theta^*$	$V = \gamma/\mu$ (m s^{-1})
Water	70	1.1	46	0.4	0.6	0.30	63.6
Chloroform	28	0.6	10	1	2	0.64	46.6
IPA50	30	3.0	18	0.4	0.4	0.46	10.0
IPA50	30	3.0	18	1	2	0.63	10.0
IPA50	30	3.0	18	2	2	0.46	10.0
FC40	16	2.5	17	1	2	0.63	6.4
Pentanol	25	3.7	13	2	2	0.48	6.7
Nonanol	27	11.5	13	2	2	0.48	2.3

chloroform were colored with either Solvent Yellow 7 or with Solvent Green 3 (Sigma-Aldrich) at concentrations of 0.50 and 1.43 mg ml^{-1}, respectively. FC40 was too difficult to tint, but it was still possible to follow its travel in the channel, and the water was colored by food coloring. The physical data for these liquids are listed in table 4.4. In the case of water, the PMMA channel was O_2 plasma treated shortly before the experiment.

The values of the physical properties of the liquids were taken from physical tables and literature [45, 46], and the contact angles with PMMA were given in [44, 47]. Videos of the progression of the flow of the solvent in the device were recorded using a Nikon D5300 ultrahigh-resolution single lens reflex (SLR) camera.

4.3.3 Experimental results and comparison with the model

Let us categorize the different liquids by their intrinsic velocities $V_i = \gamma/\mu$ (the rightmost column in table 4.5). The 'fast' fluids correspond to high values of V_i and 'slow' fluids to low values of V_i. Water and chloroform are in the first category and nonanol is in the second category.

Figure 4.12 shows the dynamics of these two fluids in channels of different cross-sections. Clearly, the LWR expression that uses the static (Young) contact angle does not explain the flow velocity in every situation. A DCA is required to fit the experimental results, especially for the 'fast' fluids, i.e. water and chloroform.

Table 4.5. The ratio β/μ, the work of adhesion, and the average distance of each molecular displacement at the triple line λ for different fluids and substrates. The results are deduced from the fitted values of β.

	γ (mN m^{-1})	μ (Pa s)	β (Pa s)	β/μ	W_a (mN m^{-1})	λ (Å)
Water/PMMA (treated)	72.0	0.001	0.35	350	140	4.0
Chloroform/PMMA	28.0	0.0006	0.22	367	53	5.0
IPA50/PMMA	35.0	0.003	0.15	46.9	62	5.6
Nonanol/PMMA	28.0	0.011	0.08	7.3	57	4.2
FC40/PMMA	16.0	0.12	0.12	48	31	6.2
Pentanol/PMMA	27.0	0.004	0.12	32	54	4.5

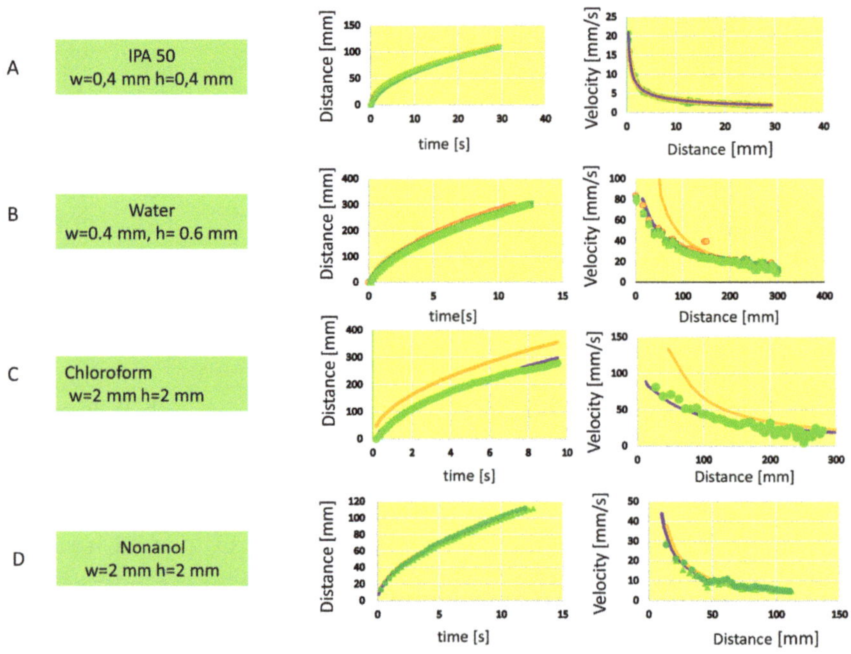

Figure 4.12. A comparison between static and dynamic angles for travel distance (left) and velocity (right) for four materials. (A) IPA50 in channel #4 (w = 0.4 mm, h = 0.4 mm); (B) water in channel #3 (w = 0.4 mm, h = 0.6mm); (C) chloroform in channel #2 (w = 2 mm, h = 2 mm); (D) nonanol in channel #2 (w = 2 mm, h = 2 mm). Green dots correspond to experimental results, the orange continuous line corresponds to the LWR extended law with a constant Young contact angle, and the purple line corresponds to the model that uses the DCA.

This DCA is obtained using viscosity coefficients of β = 0.32 Pa s for water and 0.22 Pa s for chloroform.

The 50% solution of IPA (IPA50) is located in the 'intermediate' fluids, as it is characterized by a moderate $V_i = \gamma/\mu$ value. In a small channel (w = 0.4 mm, h = 0.4 mm), its contact angle is the static angle, but this is not the case in a large channel (not shown in the figure).

Liquids with small intrinsic velocity, such as nonanol and FC40, flow slowly and do not require the use of a DCA.

4.3.4 A comparison with other correlations (Hoffman–Tanner, Bracke, Jiang)

In this section, we briefly present other correlations for the DCA and compare them to the MKT correlation. Hoffman and Tanner's [27–30] correlation for the DCA is

$$\theta \approx 4.5Ca^{\frac{1}{3}}. \tag{4.66}$$

On the other hand, Bracke *et al*'s correlation [32] links the cosine of the DCA to the capillary number

$$\cos\theta = \cos\theta_0 - 2(1 + \cos\theta_0)Ca. \tag{4.67}$$

This correlation is very similar to that of Seebergh and Berg [33], while Jiang *et al*'s correlation [34] yields

$$\cos\theta = \cos\theta_0 - (1 + \cos\theta_0)4.96\tanh\left(Ca^{0.702}\right). \tag{4.68}$$

A comparison of the different correlations is shown in figure 4.13(A) as a function of the capillary number. All the correlations produce similar DCAs except for the MKT correlation when β >0.3 Pa s.

A numerical scheme can be set up to incorporate these correlations into the LWR law for open channels. In the present case, a numerical approach was taken using the Matlab software (https://www.mathworks.com). A comparison of the different correlations is shown in figure 4.13(B). It appears that, in the case of open microflows, only the MKT correlation for the DCA fits the experimental data.

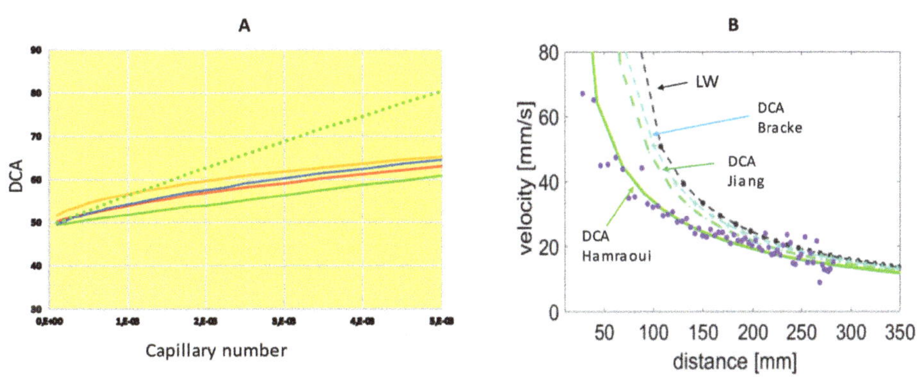

Figure 4.13. (A) A comparison between the Cassie dynamic angles obtained using the different correlations described in the text as a function of the capillary number. The two green lines corresponds to Blake's correlations for $\beta = 0.12$ (green line) and 0.32 Pa s (green dotted line). The yellow line corresponds to Bracke's correlation, the blue line to the Hoffman–Tanner correlation and the red line to Jiang's correlation. (B) A comparison of the velocities of water flows in channel #3 for different experiments and correlations. The purple dots correspond to experimental results, the continuous green line to the MKT correlation, and the dotted black line to the LWR law (with the static contact angle). Reprinted with permission from [48] © American Chemical Society.

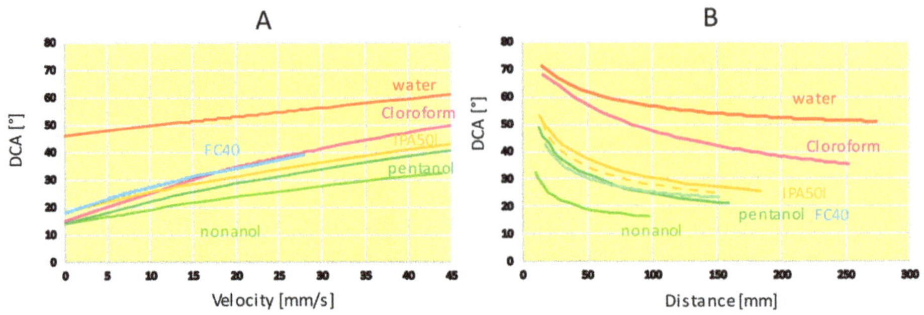

Figure 4.14. (A) The evolution of the DCA with velocity in a rectangular open channel; (B) the evolution of the DCA over distance in a rectangular open channel.

In summary, 'fast' flowing liquids characterized by a high intrinsic velocity ($V_{\text{int}} = \gamma/\mu$) require the use of a DCA for their velocities. In the case of 'slowly' flowing liquids in open channels of moderate or small cross-section, the Young (static) angle produces an adequate value for the flow velocity. The velocity difference between the DCA and the static contact angle associated to the velocity of the fluid (figure 4.14(A)) is mostly seen in the early stages of the viscous regime (figure 4.14(B)). Over long distances, the DCA and static models correspond to the same asymptotic value (the velocity is too small to change the contact angle). The MKT approach seems the most reliable way to account for the DCA in rectangular open channels.

It is of interest to notice that only the MKT correlation with the fitted viscosity β produces results that correspond to the experimental values. The value of λ—the average distance of each molecular displacement at the triple line—is deduced from the implicit relation

$$\frac{\beta}{\mu} = \frac{\nu}{\lambda^3} \exp\left(\frac{\lambda^2\gamma(1 + \cos\theta_0)}{k_B T}\right), \tag{4.69}$$

Using a numerical scheme, one can determine the value of λ. These values are indicated in table 4.5 for the six liquids (water, IPA50, chloroform, pentanol, FC40, and nonanol) and a PMMA substrate (O_2 plasma treated in the case of water). These confirm that the different values of λ are of the order of the dimensions of the molecules.

Note that, using (4.54), the MKT correlation for the DCA can be written as

$$\cos\theta = \cos\theta_0 - \frac{\beta}{\mu} Ca = \cos\theta_0 - Ca^* = \cos\theta_0 - \frac{\nu}{\lambda^3} \exp\left(\frac{\lambda^2\gamma(1 + \cos\theta_0)}{k_B T}\right) Ca, \tag{4.70}$$

where Ca^* can be viewed as a 'wall' capillary number ($Ca^* = \beta V/\gamma$).

4.4 Rough walls

Depending on the fabrication process, the walls of open channels may present some surface roughness. For example, in the case of a milled channel, depending on the milling tool and the care applied in milling, 'micro-waves' can be present on the

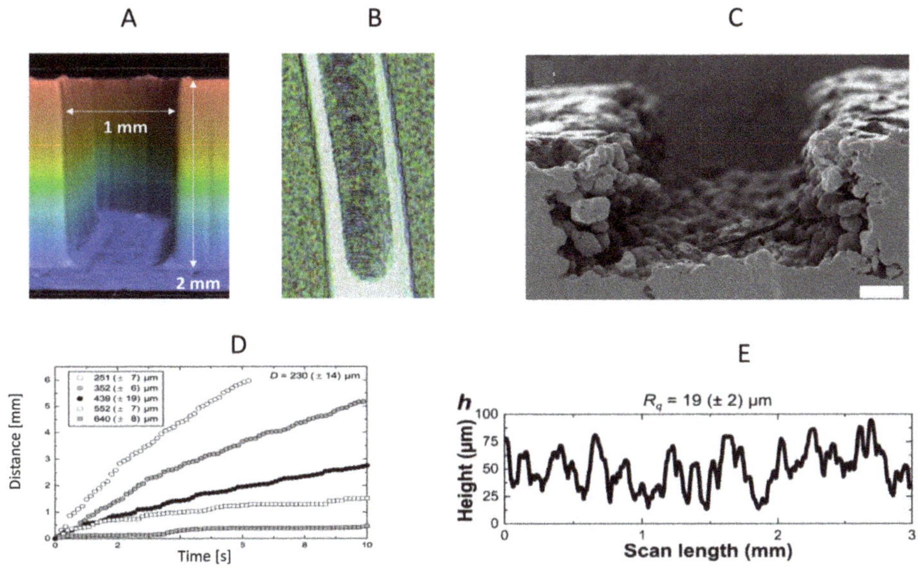

Figure 4.15. (A) striations on the channel walls caused by the milling process, photos Jodi Tokihiro, University of Washington; (B) roughness on the floor of a milled channel, photo David Gosselin, CEA-Leti; (C) Rough walls obtained by laser sintering 3D printing; (D) fluctuations of the travel distance caused by important wall roughness of a 3D printed open-channel; (E) roughness profile corresponding to channel C. Photos C,D,E, reprinted with permission from [44], © 2017 ACS.

bottom and side walls, as shown in figure 4.15(A) and (B). Roughness is also observed for 3D printed open channels, as shown in figure 4.15(C). Pronounced roughness result in a non-smooth marching distance (figure 4.15(D)) associated to velocity fluctuations (figure 4.15(E)).

4.4.1 Capillary force

The presence of roughness modifies the velocity of the capillary flow [49, 50]. The roughness is characterized by the coefficient $r = L_r/L_s$, where L_r is the length along the rough surface and L_s is the horizontal straight length (smooth surface). Wenzel's law (chapter 2) [3] states that the contact angle on the rough surface θ_r is

$$\cos\theta_r = r\cos\theta. \tag{4.71}$$

Then, in such cases, Young's law becomes

$$\gamma_{SA} = \gamma_{SL} + \gamma_{LA}\cos\theta_r = \gamma_{SL} + r\,\gamma_{LA}\cos\theta, \tag{4.72}$$

where γ_{SA}, γ_{SL}, andγ_{LA} are the surface tensions at solid/air, solid/liquid, and liquid/air interfaces, respectively. On the other hand, the TPCL is affected by roughness, and the wetted perimeter becomes rp_w instead of p_w.

The generalized Cassie angle is now

$$\cos\theta_r^* = \frac{r^2(2h + w)\cos\theta - w}{2r(h + w)}, \tag{4.73}$$

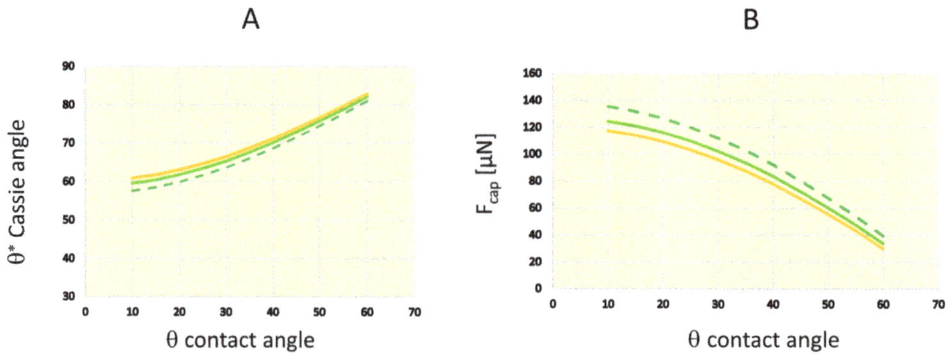

Figure 4.16. A comparison between smooth and uniformly rough walls: (A) Cassie angle, (B) capillary force. The orange line corresponds to smooth walls, the continuous green line to a roughness of $r = 1.02$, and the dashed dark green line to a roughness of $r = 1.1$.

and the capillary force in a rough-surface channel is

$$F_{\text{cap}} = \gamma p_{\text{w, r}} \cos \theta_{\text{r}} - \gamma p_{\text{F}} = \gamma (r p_{\text{w}}) (r \cos \theta) - \gamma p_{\text{F}} = \gamma r^2 p_{\text{w}} \cos \theta - \gamma p_{\text{F}}, \quad (4.74)$$

When $r = 1$ (smooth wall), relation (4.73) retrieves the usual capillary force. When $r>1$, the capillary force can be significantly increased by the presence of roughness. Hence, the wall friction is increased in the presence of wall roughness, but the increase depends on the orientation of the protrusions [44]. The effect of rough walls on the flow dynamics is complicated. Moreover, if the protrusions on the channel bottom are too tall, partial pinning might occur, as in the case of the 3D printed open channel shown in figures 4.14(D) and (E).

Figure 4.16 illustrates the effect of uniform roughness on the capillary force in a rectangular open channel that has an aspect ratio of one. Figure 4.16(A) shows the change of generalized Cassie angle and figure 4.16(B) shows the change of capillary force for a channel that has a width of 2 mm and a depth of 2 mm for roughnesses of 1.02 and 1.05 and a liquid that has a surface tension of $\gamma = 30 \text{ mN m}^{-1}$.

4.4.2 Wall friction

Open microflows are driven by capillary forces and opposed by wall friction. Let us examine wall friction for the case of rough surfaces. It is known that, at a macroscopic scale, the wall friction of a laminar flow is independent of wall roughness. The usual wall friction for a laminar flow is given by the Darcy friction factor [51, 52], defined as:

$$f_{\text{D}} = \frac{64}{\text{Re}}, \quad (4.75)$$

where Re is the Reynolds number, defined as

$$\text{Re} = VD/\nu. \quad (4.76)$$

where V is the average velocity in a cross-section, D the tube (or characteristic) diameter, and ν the kinematic viscosity. Friction in pipes is given by the Moody diagram as a function of wall roughness (figure 4.17). Friction varies with roughness only in the case of turbulent flows.

The explanation is that, in laminar flow, the fluid layer near the pipe wall has a velocity $V = 0$. Because there is no flow near the pipe wall, the roughness does not affect pressure loss.

However, this is only a legitimate approximation at the macroscale and not at the microscale. Friction has a non-negligible effect on laminar flow [53–56]. As stated by Gloss *et al*: 'With an experimental setup which allows for changes in the relative roughness of a channel, it can be shown that there is a non-negligible influence of wall roughness even for laminar flows.'

The Poiseuille number *Po*—which is the ratio between the pressure gradient $\Delta p/L$ and the average velocity *V*—defined by

$$Po = f_D \, \mathrm{Re}, \tag{4.77}$$

cannot be considered constant in a microchannel. For example, this is in contrast to $Po = f_D Re = 64$ for a circular pipe, as stated in (4.75) by the Moody diagram.

The relative roughness R_a—the ratio between the average height of surface irregularities and the channel's characteristic dimension—is the relevant parameter for laminar friction. Gloss and Herwig [54] have conducted experiments in channels with calibrated roughness, and an approximate relation for the increase in the pressure drop is

$$\Lambda = \frac{\Delta P_r}{\Delta P_s} \cong 1 + n\, R_a. \tag{4.78}$$

where *n* is a nondimensional constant in the range from three to five, and *s* and *r* are the indices for smooth walls and rough walls, respectively. Translated in terms of friction length [2, 3], we obtain

$$\Lambda = \frac{\Delta P_r}{\Delta P_s} = \frac{\bar{\tau}_r}{\bar{\tau}_s} = \frac{\bar{\lambda}_s}{\bar{\lambda}_r} \cong 1 + n\, R_a. \tag{4.79}$$

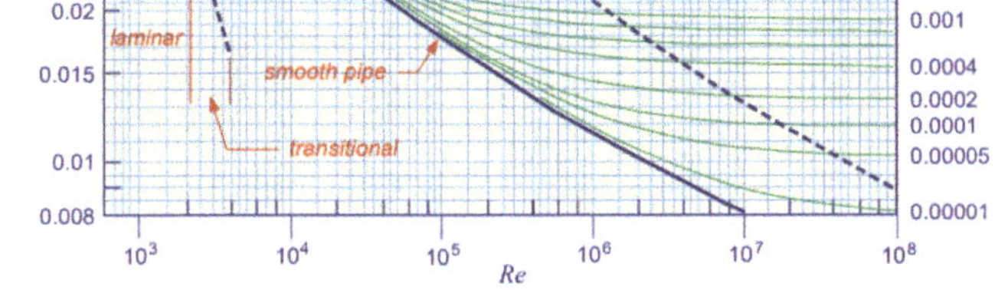

Figure 4.17. The Moody diagram for wall friction. *f* is the friction factor, *Re* the Reynolds number and *e*/*D* the relative roughness.

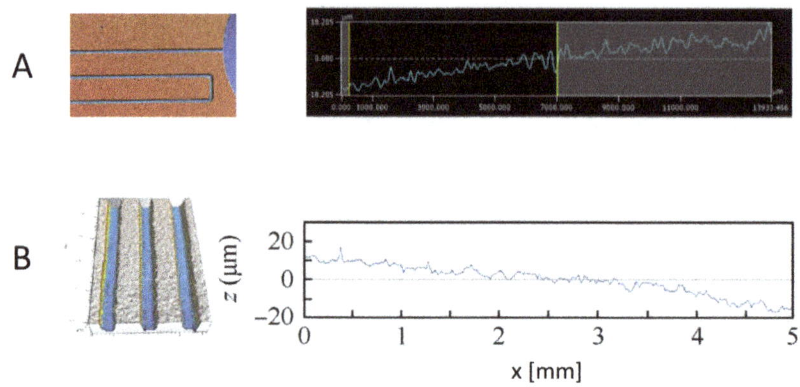

Figure 4.18. (A) An open rectangular channel milled in PMMA and its bottom roughness as measured by a profilometer; (B) the channels used by Xing *et al* [56] and their roughness. In both cases, the roughness is comparable. Figure 4.18(B) is reprinted from [55], Copyright (2016), with permission from Elsevier Ltd. on behalf of the Chinese Society of Aeronautics and Astronautics.

where $\bar{\tau}_r$ and $\bar{\tau}_s$ are the rough and smooth channel frictions and $\bar{\lambda}_r$ and $\bar{\lambda}_s$ the rough and smooth friction lengths. Figure 4.18 shows typical open channels and their roughness profiles. The first figure (4.18(A)) corresponds to the current analysis, while the second (figure 4.18(B)) corresponds to the case investigated by Xing and coworkers [56].

Hence, we obtain the friction length ratio

$$\bar{\lambda}_r \cong \frac{\bar{\lambda}_s}{1 + n\, R_a}. \tag{4.80}$$

So, $\bar{\lambda}_r < \bar{\lambda}_s$ and the friction $\bar{\tau}_r = \mu V / \bar{\lambda}_r > \bar{\tau}_s = \mu V / \bar{\lambda}_s$, so that the roughness decreases the friction length and increases the wall friction.

4.4.3 Conclusions

In the viscous regime, in a rough open channel, the force balance determining the motion is

$$F_{cap} = \gamma\left(r^2 p_w \cos\theta - p_F\right) = F_{drag} = pS\, \mu\frac{V}{\bar{\lambda}_r} = pS\, \mu\frac{V}{\bar{\lambda}}(1 + n\, R_a), \tag{4.81}$$

where $r = L_r / L_s$ is the wall's superficial tortuosity and $R_a = \Delta y / y$ is the ratio of the roughness's average height to the channel's characteristic cross-dimension.

At the time of writing, no experimental report on the global effect of roughness has yet been published. To our knowledge, only Eijo and coworkers [57] have started to investigate capillary flow in rough channels, but they have only considered the very beginning of the capillary flow, in which inertial, DCA, and roughness aspects are mixed.

4.5 A summary of the dynamics of capillary flow in an open channel

Flow dynamics in open channels is still a subject of research. Different approaches have been proposed, first for triangular grooves [58–62], then for rectangular open channels [2, 63–65] and suspended channels [66]. The model presented here has the advantage that it is valid for many different geometries as soon as the average friction length is known. All models show that, in the viscous regime, the square-root law for the travel distance as a function of time is respected for low velocities and/or very small channel cross-sections when the contact angle is equal or very close to the static contact angle.

Let us recall the main features of channel-based capillary dynamics in figure 4.19. Downstream from the reservoir, there is an inertial regime region where the travel distance is a linear function of time $z \sim t$. The inertial velocity is nearly constant and may be as high as 20–40 mm s^{-1}. The viscous regime then appears, first with a DCA, then with the static Young contact angle. In this last region, the travel distance is proportional to the square root of time $z \sim \sqrt{t}$ and the velocity varies according to $V \sim 1/z$.

4.6 The capillary dynamics of non-Newtonian fluids

By definition, a Newtonian fluid has a viscosity that is independent of the shear rate associated with the velocity field. The viscosity of Newtonian fluids may depend on external factors, such as temperature. Most of the time, during experiments, Newtonian viscosity is constant. However, some biological and biotechnological fluids are non-Newtonian, meaning that their viscosity depends on the shear rate of the flow. Liquid colloids and polymers generally have non-Newtonian behaviors. This is the case, for example, for alginate or collagen gels, which flow *in situ* in a liquid state before being gelled. These polymers are non-Newtonian [67, 68]. Another example is that of whole blood, in which the presence of deformable red blood cells (RBCs) confers a shear-thinning rheological behavior [68–71]. Blood plasma has a Newtonian behavior, but the deformations of RBCs depend on the

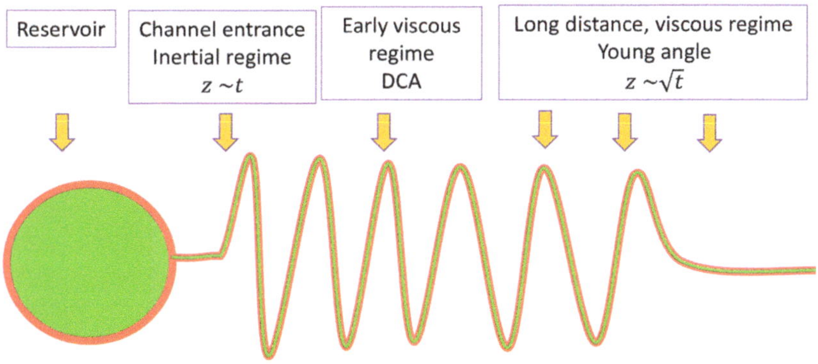

Figure 4.19. A schematic of the evolution of capillary flow showing the early inertial regime coupled to the viscous region, first with a DCA and then with the Young contact angle.

shear rate. The deformability of RBCs, which confers non-Newtonian behavior to whole blood, is shown in figure 4.20.

4.6.1 Shear-thinning fluids

The one-dimensional shear rate is defined by

$$\dot{\gamma} = \frac{\partial u}{\partial y}, \tag{4.82}$$

where u is the local velocity and y the unit vector perpendicular to the direction of u. The shear rate in a 1D system is represented in figure 4.21.

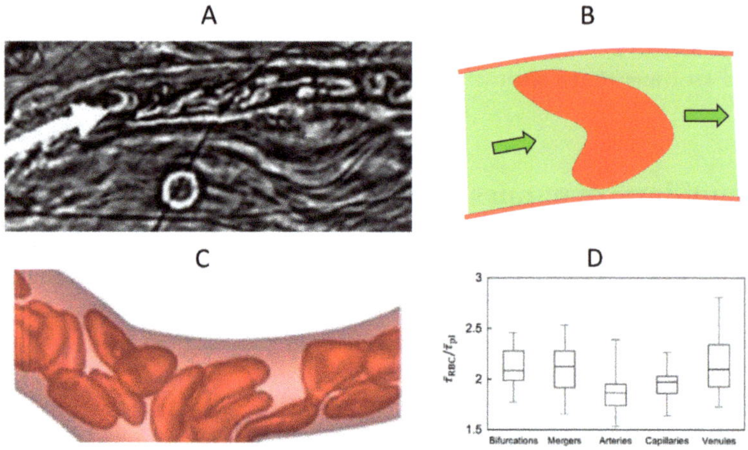

Figure 4.20. (A) A view of whole blood flow in a capillary, showing large deformations of the RBCs [71]; (B) a sketch of the deformation of an RBC; (C) the deformation of RBCs in a capillary; (D) the ratio between the friction of whole blood and the friction of the plasma component of blood in different channels of the human body [72]. The ratio falls roughly between 1.7 and 2.7. Figure (A) is reprinted from [71] Copyright (2006), with permission from Elsevier; figures (C) and (D) are reproduced from [72] CC BY 4.0.

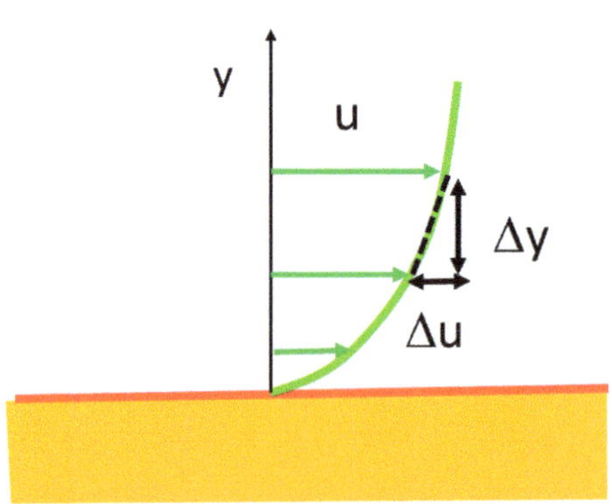

Figure 4.21. A sketch of the local shear rate $\dot{\gamma} = \frac{\partial u}{\partial y} \approx \frac{\Delta u}{\Delta y}$.

In a 2D case, the expression for the shear rate is

$$\dot{\gamma} = \frac{\partial u}{\partial y} + \frac{\partial v}{\partial x}. \tag{4.83}$$

where u,v are the two components of the velocity and x,y are the two coordinates. The most general expression of the shear rate is

$$\dot{\gamma} = \sqrt{\frac{1}{2}\left[4\frac{\partial u^2}{\partial x} + 4\frac{\partial v^2}{\partial y} + 4\frac{\partial w^2}{\partial z} + 2\left(\frac{\partial u}{\partial y} + \frac{\partial v}{\partial x}\right)^2 + 2\left(\frac{\partial v}{\partial z} + \frac{\partial w}{\partial x}\right)^2 + 2\left(\frac{\partial w}{\partial y} + \frac{\partial u}{\partial z}\right)^2\right]} \tag{4.84}$$

where u, v, and w are the three components of the velocity, and x, y, and z are the three coordinates. Expression (4.84) reduces to the one-dimensional expression (4.82) if we set $v = w = 0$ and the derivatives $\partial/\partial x = 0$ and $\partial/\partial z = 0$.

Depending on the liquid, different rheological laws describe the viscosity as a function of the shear rate. When the viscosity decreases with the shear rate, the fluid viscosity is called 'shear-thinning' (figure 4.21). The Ostwald or power law is one of the laws usually used to describe shear-thinning behavior—in particular, it describes the rheological behavior of alginates and xanthan [73] (figure 4.22)—and the viscosity expression is

$$\mu = K\,\dot{\gamma}^{n-1}. \tag{4.85}$$

where K and n are constants. Note that n is not usually an integer in (4.85). For example, for xanthan gum, n varies between 0.3 and 0.6, depending on the concentration. The friction term is then

$$\tau = \mu\,\dot{\gamma} = K\,\dot{\gamma}^n. \tag{4.86}$$

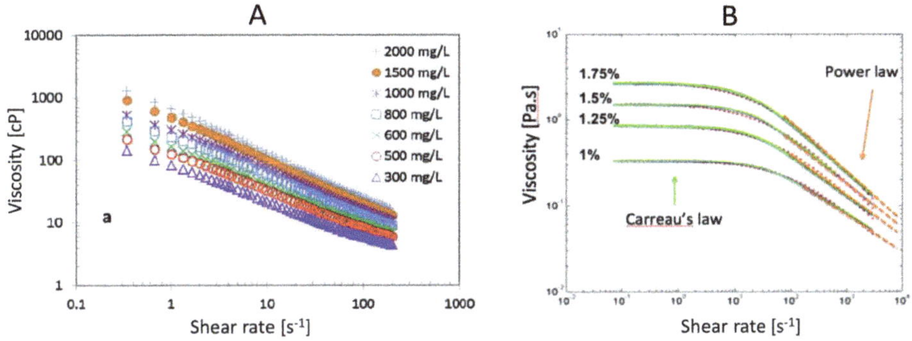

Figure 4.22. The viscosity of xanthan gum as a function of the shear rate and concentration, showing a power-law behavior: the viscosity is relatively high at low shear rates and becomes small at high shear rates; note that 1 cP = 10^{-3} Pa s; reprinted from [73], Copyright (2013), with permission from Elsevier; (B) the viscosity of alginate solutions as a function of the shear rate.

Note that $n = 1$ corresponds to a Newtonian fluid (μ = a constant). Carreau and Yasuda have proposed a rheological law that is more detailed than the Ostwald law [74, 75]:

$$\mu = \mu\,[1 + (\tau\dot{\gamma})^a]^{\frac{n-1}{a}}. \tag{4.87}$$

where τ is a relaxation time and a is a nondimensional coefficient. The exponent n is the same as for the Ostwald law.

Another rheological law of interest in biotechnology—in particular for the rheological behavior of human whole blood—is the Herschel–Bulkley (HB) rheological model [76, 77]:

$$\dot{\gamma} = 0 \text{ if } \tau < \tau_0$$

$$\tau = (K \mid \dot{\gamma} \mid^{n-1} + \tau_0 \mid \dot{\gamma} \mid^{-1})\,\dot{\gamma} \text{ if } \tau > \tau_0, \tag{4.88}$$

where τ_0 is the yield stress, i.e. the stress rate below which the liquid viscosity becomes infinite ($\mu = \tau/\dot{\gamma}$). K and n are two rheological parameters that depend on the fluid. Note that if $\tau_0 = 0$, this law reduces to the power law. Yield stress fluids are often denoted by the acronym YSF [76]. Note that the capillary flow of HB fluids must stop at some distance—contrary to Newtonian fluids, which in theory flow indefinitely—due to the yield stress. Whole blood is a YSF and its viscosity is shown in figure 4.23.

4.6.2 The case of whole blood

As we have seen above, whole blood has a Newtonian behavior at high shear rates and non-Newtonian behavior for low shear rates. The transition takes place at

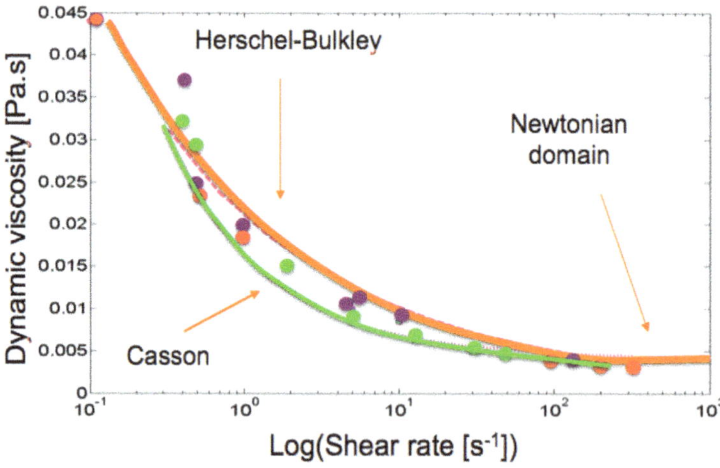

Figure 4.23. The viscosity of whole blood [78]: the shear-thinning domain and the Newtonian domain. The colored dots are experimental results from the literature. The HB model fit has been obtained using the values $\tau_0 = 0.013\,97$ Pa, $K = 0.0106$, and $n = 0.788$. Casson's law is similar to the HB law but uses the square root of the shear rate $\dot{\gamma}$.

approximately $\dot{\gamma} = 100\ \mathrm{s}^{-1}$. Above this threshold, blood has a Newtonian behavior [77]. The dispersion or aggregation mechanisms of RBCs explain this behavior (figure 4.24): at sufficiently high velocities, the shear rate is important and RBCs are randomly dispersed. Conversely, when the shear rate is smaller than 30–100 s^{-1}, RBCs form piles, called 'rouleaux' [79, 80].

These rouleaux may be viewed as polymeric chains—although they are very fragile—which confer non-Newtonian behavior on the flow. Note that the formation of 'rouleaux' locally densifies the blood. Pockets of plasma form in the bulk of the flow and/or in the vicinity of the solid walls.

The LWR law and its extension to non-cylindrical channels assume a Poiseuille quadratic profile of the flow for the wall friction. Newtonian fluids—i.e. fluids whose viscosity is independent of the shear stress—have such a profile, but non-Newtonian fluids do not show such a quadratic profile [79]. Hence, the LWR law does not strictly apply to such fluids. An expression modifying the Bosanquet equation for HB fluids has recently been derived [70]:

$$\frac{\partial^2 z}{\partial t^2} + \frac{p_\mathrm{w}}{\rho\, S_\mathrm{C}} \frac{K}{\bar{\lambda}^n} \left(\frac{\partial z}{\partial t}\right)^n - \frac{p_w}{\rho\, S_\mathrm{C}} \frac{\gamma \cos \theta^*}{z} = -\frac{p_\mathrm{w}}{\rho\, S_\mathrm{C}} \tau_0. \tag{4.89}$$

In the case of a Newtonian fluid ($n = 1$, $K = \mu$, and $\tau_0 = 0$) flowing in a cylindrical tube ($p_\mathrm{w} = 2\pi R$, $S_\mathrm{C} = \pi R^2$, and $\bar{\lambda} = R/4$), equation (4.89) reduces to the Bosanquet equation. The derivation of (4.89) is quite difficult and lengthy; the reader will find the whole derivation in reference [70]. In equation (4.89), the first term on the left-hand side is the inertial term. Assuming that we are outside the inertial regime—which is extremely short lived—this first term may be dropped and we are left with the transposition of the LWR law to non-Newtonian fluids.

In the case of a capillary flow of HB liquid in a circular duct, the travel distance is given by the equation

$$z\left(\frac{\partial z}{\partial t}\right)^n = \frac{1}{1 - C} \frac{1}{\left(\frac{n+1}{n R \alpha}\right)^n} (\gamma \cos \theta - z\, \tau_0). \tag{4.90}$$

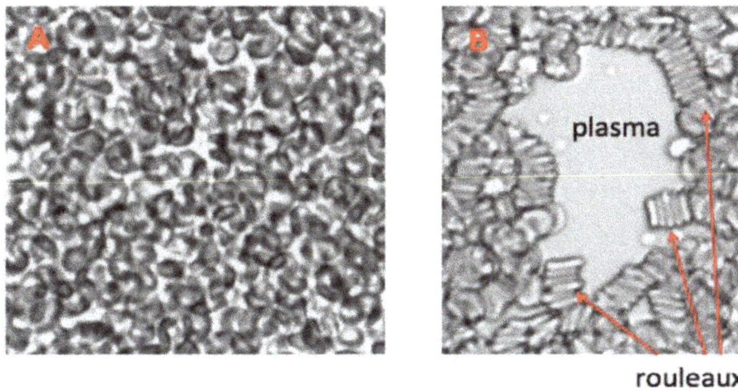

Figure 4.24. Two photographs of whole blood: (A) dispersed RBCs corresponding to Newtonian behavior; (B) the formation of RBC rouleaux, creating room for plasma pockets; this corresponds to non-Newtonian behavior. Photographs courtesy of J Berthier and M Cubizolles.

where C and α are constants. Equation (4.90) has no closed-form solution. A numerical solution must be set up. However, in the case in which $\tau_0 = 0$—which corresponds to the power law—the solution of (4.90) is the well-known Weissenberg–Rabinowitsch–Mooney (WRM) law [81, 82]:

$$z = \left[\frac{n+1}{3n+1}\frac{R^n}{K}\gamma\cos\theta\right]^{\frac{1}{n+1}} t^{\frac{n}{n+1}}. \tag{4.91}$$

Figure 4.25 shows the travel distance of whole blood in a tube as a function of time. At first, when the velocity is high, the flow is Newtonian and obeys the LWR law. At a distance of approximately 100 mm, rouleaux start to form and the experimental data do not fit the LWR anymore.

In chapter 11, we shall see how the whole-blood rheological behavior can be used to extract plasma.

4.7 Representation in 1/*V*

In chapter 3, we saw that the representation of the dynamics using the inverse of the velocity was of interest. A linear plot of the inverse of the velocity (*1/V*) as a function of distance (z) is associated with the LWR type of dynamics. Let us now investigate this representation for a DCA and for a non-Newtonian fluid.

First, consider the case of a water flow in a $w = 1$ mm, $h = 2$ mm rectangular open channel, which was analyzed in section 4.2 (figure 4.26). In the representation (*1/V*, z), the plot corresponding to the (real) dynamics that includes the DCA is linear and parallel to the plot corresponding to the LWR approach based on the static Young

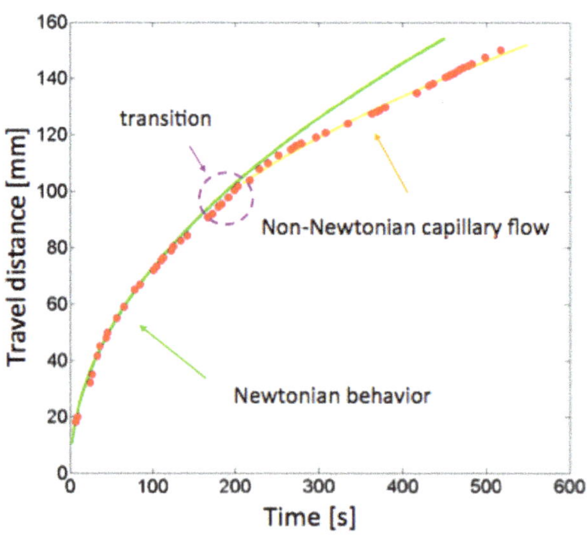

Figure 4.25. The capillary flow of whole blood in a cylindrical tube: the two flow regimes clearly appear; first, the Newtonian regime, then the non-Newtonian regime. In the non-Newtonian regime, the travel distance follows the extended LWR equation.

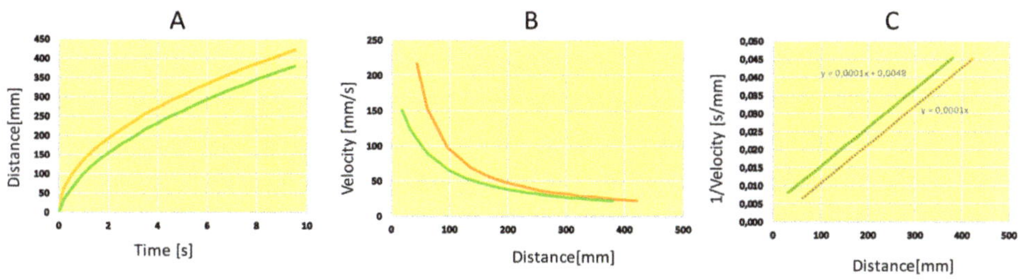

Figure 4.26. A water flow in a rectangular open channel. (A) Travel distance vs. time; (B) velocity vs. distance; (C) the inverse of velocity vs. distance. The orange line corresponds to the LW law, the green line to the DCA-corrected law.

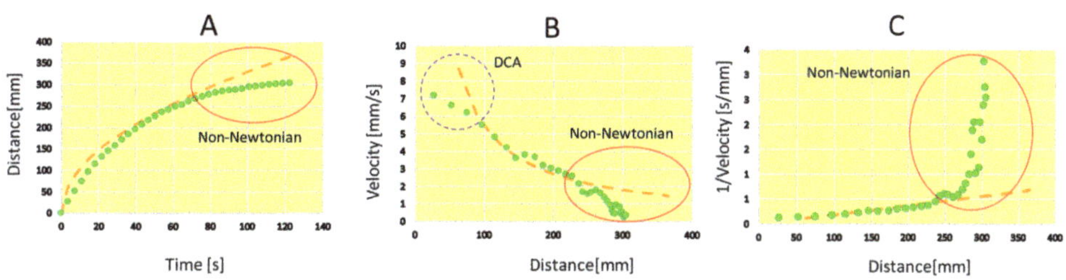

Figure 4.27. An alginate flow in a suspended channel. (A) Travel distance vs. time; (B) velocity vs. distance; (C) the inverse of velocity vs. distance. The orange line corresponds to the LW law, the green dotted line to the non-Newtonian behavior.

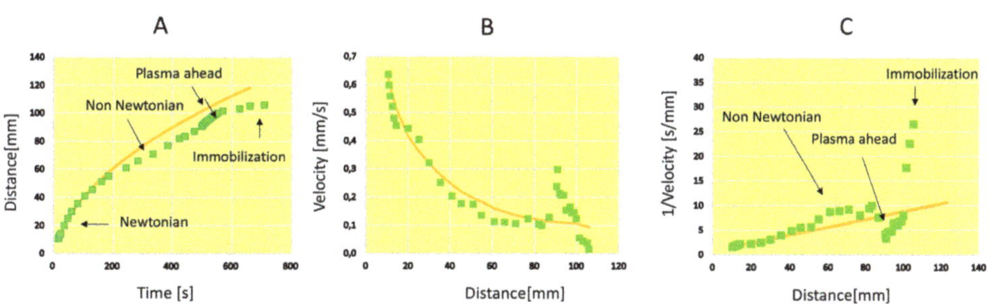

Figure 4.28. Whole blood flow in a rectangular open channel. (A) Travel distance vs. time; (B) velocity vs. distance; (C) the inverse of velocity vs. distance. The orange line corresponds to the LW law, the green dots to the observed blood flow.

contact angle. The difference is the intersection with the vertical axis ($z = 0$), which is nonzero in the case of the DCA.

Second, consider an alginate (1.5% concentration) flowing in a suspended channel ($w = 0.3$ mm, $h = 1$ mm). This case is detailed in reference [66]. Non-Newtonian behavior appears when the velocity and shear rate have sufficiently decreased. The dynamics plot departs from that of the LWR dynamics when the velocity is less than 2 mm s^{-1} (figure 4.27). The representation in the ($1/V,z$) system is very sensitive to the onset of non-Newtonian behavior.

Finally, let us examine the behavior of whole blood flowing in a rectangular open channel (figure 4.28). Four different phases can be observed: (1) the Newtonian

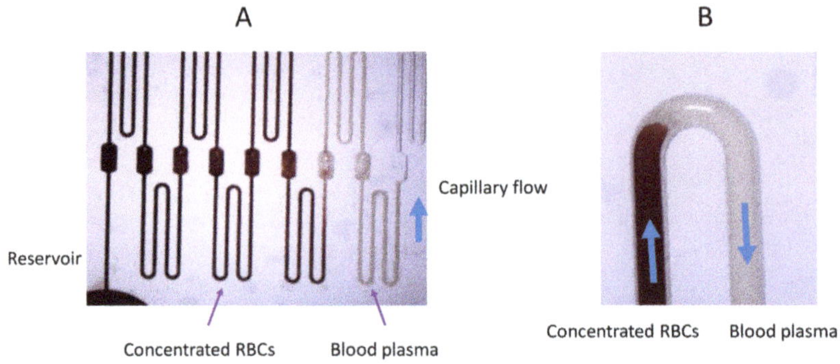

Figure 4.29. (A) Whole blood flows in a rectangular open channel; blood plasma flows ahead. (B) The sharp separation between RBCs and plasma. Photographs courtesy of Jean Berthier and Myriam Cubizolles.

phase, in which the RBCs are well mixed with blood plasma due to a sufficiently high velocity; (2) the non-Newtonian phase, in which the RBCs start aggregating; (3) plasma extraction: plasma flows ahead of the bulk due to its smaller viscosity, leaving a concentrated flow of RBCs behind; (4) the immobilization phase caused by the clogging of the channel by stacked RBCs (rouleaux).

RBC/plasma separation is shown in figure 4.29. Blood plasma flows first due to its smaller viscosity. It is interesting to note the rather brutal separation between RBCs and plasma.

References

[1] Bosanquet C H 1923 On the flow of liquids into capillary tubes *Philos. Mag.* **45** 525

[2] Berthier J, Gosselin D and Berthier E 2015 A generalization of the Lucas–Washburn–Rideal law to composite microchannels of arbitrary cross-section *Microfluid. Nanofluid.* **19** 497–507

[3] Berthier J, Brakke K A and Berthier E 2016 *Open Microfluidics* (New York: Wiley)

[4] Quéré D 1997 Inertial capillarity *Europhys. Lett.* **39** 533–8

[5] Dimitrov D I, Milchev A and Binder K 2007 Capillary rise in nanopores: Molecular dynamics evidence for the Lucas–Washburn equation *Phys. Rev. Lett.* **99** 054501

[6] Zhang S and Pei H 2020 Rate of capillary rise in quartz nanochannels considering the dynamic contact angle by using molecular dynamics *Powder Technol.* **372** 477–85

[7] Lucas R 1918 Ueber das Zeitgesetz des Kapillaren Aufstiegs von Flussigkeiten *Kolloid Z.* **23** 15

[8] Washburn E W 1921 The dynamics of capillary flow *Phys. Rev.* **17** 273–83

[9] Rideal E K 1922 On the flow of liquids under capillary pressure *Philos. Mag. Ser.* **6** 1152–9

[10] McDonough J M 2009 *Lectures in elementary fluid dynamics: physics, mathematics and applications (Lecture notes)* (Lexington, KY: University of Kentucky)

[11] Berthier J and Silberzan P 2012 *Microfluidics for Biotechnology* 2nd edn (Norwood, MA: Artech House Publishing)

[12] Shah R K and London A L 1978 *Laminar flow forced convection in ducts* (New York: Academic) Advances in Heat Transfer series Suppl. 1

[13] Sharp K V, Adrian R J, Santiago J G and Molho J I 2005 Liquid flows in microchannels *MEMS: Introduction and Fundamentals* ed M Gad-el-Hak (Hoboken, NJ: CRC Press)

[14] Bahrami M, Yovanovich M M and Culham J R 2007 A novel solution for pressure drop in singly connected microchannels of arbitrary cross-section *Int. J. Heat Mass Transfer* **50** 2492–502

[15] McHale J P and Garimella S V 2010 Heat transfer in trapezoidal microchannels of various aspect ratios *Int. J. Heat Mass Transfer* **53** 365–75

[16] Berthier J, Renaudot R, Dalle P, Blanco-Gomez G, Rivera F, Agache V and Caillat P 2010 COMSOL assistance for the determination of pressure drops in complex microfluidic channels *Proc. of the 2010 European COMSOL Conf. (Paris, France, Nov. 17–19)* (Stockholm: COMSOL AB) https://www.comsol.com/paper/comsol-assistance-for-the-determination-of-pressure-drops-in-complex-microfluidi-7789

[17] Baviere R, Ayela F, Le Person S and Favre-Marinet M 2005 Experimental characterization of water flow through smooth rectangular microchannels *Phys. Fluids* **17** 098105-1–4

[18] Sharp K V and Adrian R J 2004 Transition from laminar to turbulent flow in liquid filled microtubes *Exp. Fluids* **36** 741–7

[19] Judy J, Maynes D and Webb B W 2002 Characterization of frictional pressure drop for liquid flows through microchannels *Int. J. Heat Mass Transfer* **45** 3477–89

[20] Lee P S, Garimella S and Liu D 2005 Investigation of heat transfer in rectangular microchannels *Int. J. Heat Mass Transfer* **48** 1688–704

[21] Hirt C W and Nichols B D 1981 Volume of fluid (VOF) method for the dynamics of free boundaries *J. Comput. Phys.* **39** 201–25

[22] Dussan V 1979 On the spreading of liquids on solid surfaces: static and dynamic contact lines *Annu. Rev. Fluid Mech.* **11** 371

[23] Wu P, Nikolov A D and Wasan D T 2017 Capillary dynamics driven by molecular self-layering *Adv. Colloid Interface Sci.* **243** 114–20

[24] Wu P, Nikolov A D and Wasan D T 2017 Capillary rise: validity of the dynamic contact angle models *Langmuir* **33** 7862–72

[25] Burley R and Kennedy B S 1976 An experimental study of air entrainment at a solid/liquid/gas interface *Chem. Eng. Sci.* **31** 901311

[26] Zhu Y and Petkovic K 2010 Capillary flow in microchannel *Microfluid. Nanofluid.* **8** 275–82

[27] Hoffman R L 1975 A study of the advancing interface. I. Interface shape in liquid–gas systems *J. Colloid Interface Sci.* **50** 228–41

[28] Voinov O 1976 Hydrodynamics of wetting *Fluid Dyn.* **11** 714–21

[29] Dussan E B 1976 On the difference between a bounding surface and a material surface *J. Fluid Mech.* **75** 609–23

[30] Tanner L H 1979 The spreading of silicone oil drops on horizontal surfaces *Phys. D: Appl. Phys.* **12** 1473

[31] Eggers J and Stone H A 2004 Characteristic lengths at moving contact lines for a perfectly wetting fluid: the influence of speed on the dynamic contact angle *J. Fluid Mech.* **505** 309–21

[32] Bracke M, De Voeght E and Joos P 1989 The kinetics of wetting: the dynamic contact angle *Trends in Colloid and Interface Science III* (Heidelberg: Steinkopff) 79 142–9

[33] Seebergh J E and Berg J C 1992 Dynamic wetting in the low capillary number regime *Chem. Eng. Sci.* **47** 4455

[34] Jiang T S, Oh S G and Slattery J C 1979 Corrections for dynamic contact angle *J. Colloid Interface Sci.* **69** 74

[35] Shi Z, Zhang Y, Liu M, Hanaor D A H and Gan Y 2018 Dynamic contact angle hysteresis in liquid bridges *Colloids Surf.* A **555** 365–71

[36] Li X, Fan X, Askounis A, Wu K, Sefiane K and Koutsos V 2013 An experimental study on dynamic pore wettabilit *Chem. Eng. Sci.* **104** 988–97

[37] Blake T D and De Coninck J I 2002 The influence of solid–liquid interactions on dynamic wetting *Adv. Colloid Interface Sci.* **96** 21–36

[38] Blake T D 2006 The physics of moving wetting lines *J. Colloid Interface Sci.* **299** 1–13

[39] Duvivier D, Blake T D and De Coninck J I 2013 Toward a predictive theory of wetting dynamics *Langmuir* **29** 10132–40
[40] Hamraoui A, Thuresson K, Nylander T and Yaminsky V 2000 Can a dynamic contact angle be understood in terms of a friction coefficient? *J. Colloid Interface Sci.* **226** 199–204
[41] Berthier J, Gosselin D and Berthier E 2014 A general condition for spontaneous capillary flow in uniform cross-section microchannels *Microfluid. Nanofluid.* **16** 779–85
[42] Concus P and Finn R 1969 On the behavior of a capillary surface in a wedge *Proc. Natl Acad. Sci.* **63** 292–9
[43] Berthier J, Brakke K A, Gosselin D, Huet M and Berthier E 2014 Metastable capillary filaments *AIMS Biophys.* **1** 31–48
[44] Lade R K, Hippchen E J, Macosko C W and Francis L F 2017 Dynamics of capillary-driven flow in 3D printed open microchannels *Langmuir* **33** 2949–64
[45] Mohammad A A, Alkhaldi K H A E, Al Tuwaim M S and Al-Jimaz A S 2014 Effect of temperature and chain length on the viscosity and surface tension of binary bystems of *N*,*N*-dimethylformamide with 1-octanol, 1-nonanol and 1-decanol *J. Chem. Thermodyn.* **74** 7–15
[46] Kim S *et al* 2016 *PubChem substance and compound databases Nucleic Acids Res.* **44** D1202
[47] Lee J J, Berthier J, Theberge A B and Berthier E 2019 Capillary flow in open microgrooves: bifurcations and networks *Langmuir* **35** 10667–75
[48] Tokihiro J C *et al* 2024 On the Dynamic Contact Angle of Capillary-Driven Microflows in Open Channels *Langmuir* (accepted) https://doi.org/10.1021/acs.langmuir.4c00391
[49] Abadi G B and Bahrami M 2022 The effect of surface roughness on capillary rise in micro-grooves *Sci. Rep.* **12** 14867
[50] Abadi G B and Bahrami M 2020 A general form of capillary rise equation in micro-grooves *Sci. Rep.* **10** 19709
[51] Boussinesq J 1868 Mémoire sur l'influence des Frottements dans les Mouvements Réguliers des Fluides *J. Math. Pures Appl.* **13** 377–424
[52] Bejan A 2013 *Convection Heat Transfer* (New York: Wiley) 4th edn
[53] Gamrat G, Favre-Marinet M, Le Person S, Baviere R and Ayela F 2008 An experimental study and modelling of roughness effects on laminar flow in microchannels *J. Fluid Mech.* **594** 399–423
[54] Gloss D and Herwig H 2012 Wall roughness in laminar flows: an often ignored though significant issue *Exp. Fluids* **49** 461–70
[55] Liu Y, Li J and Smits A J 2019 Roughness effects in laminar channel flow *J. Fluid Mech.* **876** 1129–45
[56] Xing Y, Zhi T, Haiwang L and Yitu T 2016 Experimental investigation of surface roughness effects on flow behavior and heat transfer characteristics for circular microchannels *Chin. J. Aeronaut.* **29** 1575–81
[57] Eijo P M G, Cabaleiro J M and Artana G 2022 Capillary flow dynamics in composite rectangular microchannels with rough walls *Langmuir* **38** 13296–304
[58] Mann J J A, Romero L, Rye R R and Yost F G 1995 Flow of simple liquids down narrow V-grooves *Phys. Rev.* E **52** 3967
[59] Rye R R and Yost F G 1996 Wetting kinetics in surface capillary grooves *Langmuir* **12** 4625–2627
[60] Yost F G, Rye R R and Mann J A 1997 Solder wetting kinetics in narrow V-grooves *Acta Mater.* **45** 5337–45
[61] Romero L A and Yost F G 1996 Flow in an open channel capillary *J. Fluid Mech.* **322** 109–29
[62] Kitron-Belinkov M, Marmur A, Trabold T and Dadheech G V 2007 Groovy-drops: effect of groove curvature on spontaneous capillary flow *Langmuir* **23** 8406–10

[63] Baret J-C, Decré M M J, Herminghaus S and Seemann S 2007 Transport dynamics in open microfluidic grooves *Langmuir* **23** 5200–4
[64] Yang D, Krasowska M, Priest C, Popescu M N and Ralston J 2011 Dynamics of capillary-driven flow in open microchannels *J. Phys. Chem.* C **115** 18761–9
[65] Ouali F F, McHale G, Javed H, Trabi C, Shirtcliffe N J and Newton M I 2013 Wetting considerations in capillary rise and imbibition in closed square tubes and open rectangular cross-section channels *Microfluid. Nanofluid.* **15** 309–26
[66] Berthier J, Brakke K A, Gosselin D, Bourdat A-G, Nonglaton G, Villard N, Laffite G, Boizot F, Costa G and Delapierre G 2015 Suspended microflows between vertical parallel walls *Microfluid. Nanofluid.* **18** 919–29
[67] Le Vot S, Berthier J, Tiquet P, David N, Blanc R, Benhamou P-Y and Rivera F 2008 Non-Newtonian fluids in flow-focusing devices: encapsulation with alginates *Proc. of the 1st European Conf. on Microfluidics—Microfluidics 2008 (Bologna, December 10–12)*
[68] Berthier J, Le Vot S, Tiquet P, David N, Lauro D, Benhamou P Y and Rivera F 2010 Highly viscous fluids in pressure actuated flow focusing devices *Sensors Actuators* A **158** 140–8
[69] Fedosov D A, Pan W, Caswell B, Gompper G and Karniadakis G E 2011 Predicting human blood viscosity in silico *Proc. Natl Acad. Sci.* **108** 11772–7
[70] Gosselin D, Huet M, Cubizolles M, Rabaud D, Belgacem N, Chaussy D and Berthier J 2016 Viscoelastic capillary flow: the case of whole blood *AIMS Biophys.* **3** 340–57
[71] Jeong J H, Sugii Y, Minamiyama M and Okamoto K 2006 Measurement of RBC deformation and velocity in capillaries *in vivo Microvasc. Res.* **71** 212–7
[72] Balogh P and Bagchi P 2019 Three-dimensional distribution of wall shear stress and its gradient in red cell-resolved computational modeling of blood flow in *in vivo*-like microvascular networks *Phys. Rep.* **7** e14067
[73] Zhong L, Oostrom M, Truex M J, Vermeul V R and Szecsody J E 2013 Rheological behavior of xanthan gum solution related to shear thinning fluid delivery for subsurface remediation *J. Hazard. Mater.* **244–245** 160–70
[74] Shahsavari S and McKinley G H 2015 Mobility of power-law and Carreau fluids through fibrous media *Phys. Rev.* E **92** 063012
[75] Steffe J F 1982 *Rheological Methods in Food Process Engineering* 2nd edn (New York: Freeman Press)
[76] Herschel W H and Bulkley R 1926 Konsistenz-messungen von Gummi-Benzollösunge *Kolloid-Z* **39** 291–300
[77] Geraud B, Jørgensen L, Petit L, Delanoe-Ayari H, Jop P and Barentin C 2014 Capillary rise of yield-stress fluids *EPL* **107** 58002
[78] Berthier J, Brakke K A, Furlani E P, Karampelas I H, Poher V, Gosselin D, Cubizolles M and Pouteau P 2015 Whole blood spontaneous capillary flow in narrow V-groove microchannels *Sensors Actuators* B **206** 258–67
[79] Fu T, Carrier O, Funfschilling D, Ma Y, Li H and Chen G 2016 Newtonian and non-Newtonian flows in microchannels: in line rheological characterization *Chem. Eng. Technol.* **39** 987–92
[80] Morhell N and Pastoriza H 2016 Power law fluid viscometry through capillary filling in a closed microchannel *Sensors Actuators* B **227** 24–8
[81] Rabinowitsch B 1929 Uber die Viskosität *u*nd Elastizität von Solen *Phys. Chem.* A **145** 1–26
[82] Mooney M 1931 Explicit formulas for slip and fluidity *J. Rheol.* **2** 210–21

IOP Publishing

Open-Channel Microfluidics (Second Edition)
Fundamentals and applications
Jean Berthier, Ashleigh B Theberge and Erwin Berthier

Chapter 5

Common open-channel geometries

This chapter presents the behavior of open-capillary flows in different open-channel geometries, such as rectangular, rounded, suspended, and embossed channels. Approximate values of the average friction length are presented. An analysis of the spontaneous capillary flow (SCF) theory for thread-based microfluidics is also developed.

5.1 Introduction

This chapter is devoted to the analysis of capillary flows in the many different configurations found in microsystems. Some typical channel morphologies are shown in figure 5.1 [1]. We present the SCF condition in such open channels and determine their approximate friction length and Cassie angle for the dynamic analysis.

We also show that the SCF theory developed in the preceding chapters applies to threads and yarns (thread-based microfluidics).

Let us start with the simplest cases, those of suspended channels (vertical walls) and rails (horizontal walls).

5.2 Suspended channels

A frequently used open microchannel configuration is that of a channel partially or totally devoid of a bottom and a ceiling. Such a configuration falls within the category called 'suspended microfluidics' [2, 3] (figure 5.2). The advantage of such a configuration is that it increases access to the fluid from the top and/or the bottom. Different configurations exist for the bottom floor, such as holes or even the total absence of a floor [3, 4]. In this section, we investigate the suspended capillary flow between two vertical walls.

If w denotes the distance between the two walls and h the height of the walls, the SCF condition is

doi:10.1088/978-0-7503-5507-0ch5

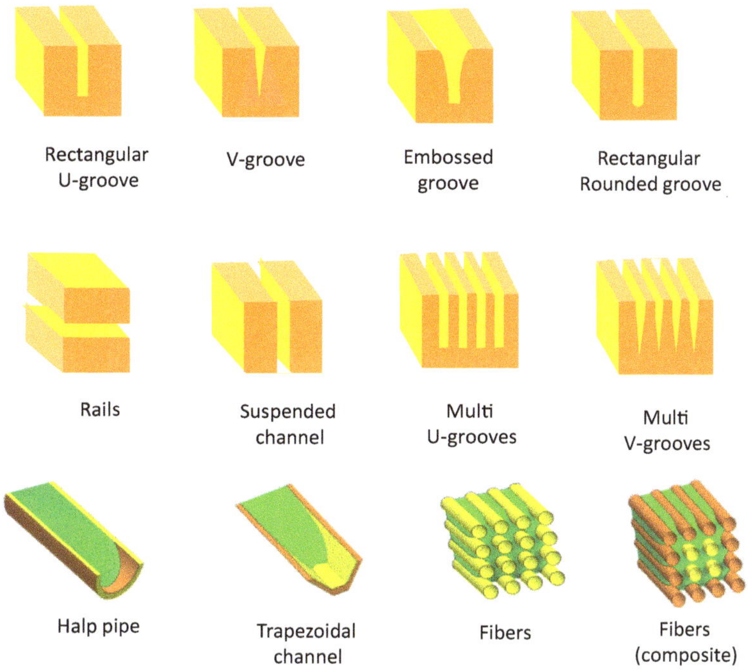

Figure 5.1. Some of the most common open-channel geometries.

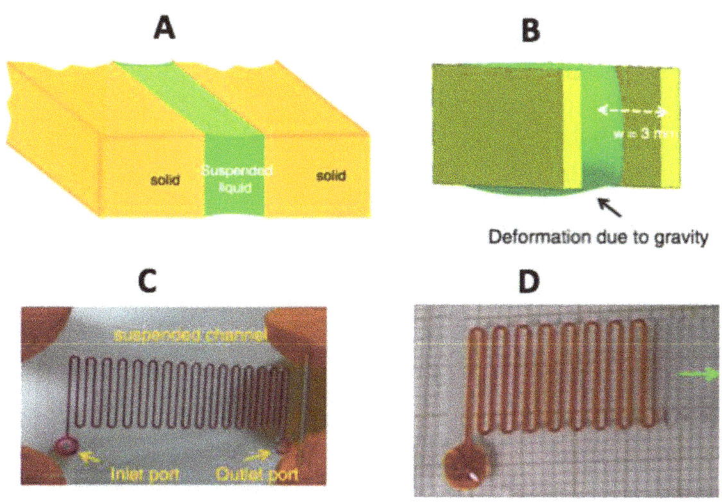

Figure 5.2. (A) A sketch of a capillary flow between two vertical walls. (B) When the distance between the two walls increases towards the capillary length, gravity distorts the horizontal free surfaces. (C) A photograph of a winding suspended channel milled in a PMMA plate. (D) Whole blood flowing in a suspended channel (photo courtesy of Myriam Cubizolles and Jean Berthier, CEA/Leti).

$$\cos\theta > \frac{p_F}{p_w} = \frac{2w}{2h} = \varepsilon, \tag{5.1}$$

where ε is the aspect ratio $\varepsilon = w/h$. Note that the width w must be sufficiently small, otherwise gravitational effects deform the free surface at the liquid bottom (figure 5.2(B)) up to the point at which the liquid free-falls.

The capillary length indicates this limit:

$$w < \lambda = \sqrt{\frac{\gamma}{\rho\, g}}. \tag{5.2}$$

If the width w is less than the capillary length λ, then gravity does not notably deform the interface. The Bond number (Bo) provides another equivalent approach. The Bond number in the case of a suspended channel is

$$Bo = \max\left(\frac{\rho\, g\, w^2}{\gamma}, \frac{\rho\, g\, h^2}{\gamma}\right) = \frac{\max\,(w^2, h^2)}{\lambda^2} < 1. \tag{5.3}$$

If Bo is smaller than unity, then gravity can be neglected. The SCF condition can also be determined using the software Evolver (figure 5.3).

The dynamics of the flow requires the generalized Cassie angle and the average friction length to be determined [1]. The following relation determines the Cassie angle:

$$\cos\theta^* = \frac{h\cos\theta - w}{w + h} = \frac{\cos\theta - \varepsilon}{\varepsilon + 1}. \tag{5.4}$$

The average friction length ($\bar{\lambda}$) may be approximated by simplifying the velocity field. An approximation relies on the assumption of a Poiseuille flow between the two plates (see chapter 3). If y denotes the axis perpendicular to the two vertical walls (figure 5.4), the velocity profile far from the top and bottom edges is approximately given by

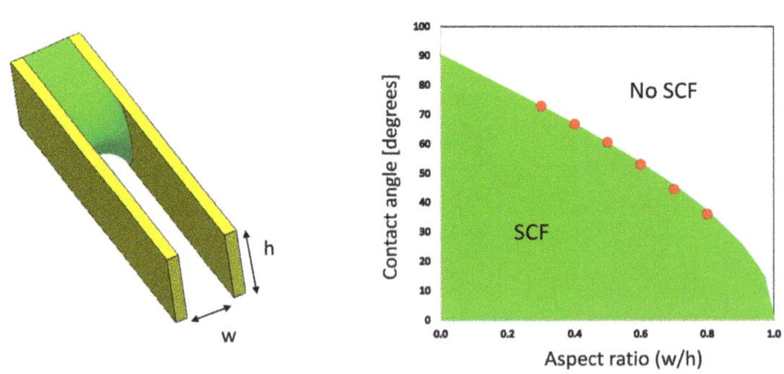

Figure 5.3. Left: a sketch of a suspended capillary flow; right: a comparison between the theoretical prediction and an Evolver simulation for the SCF domain.

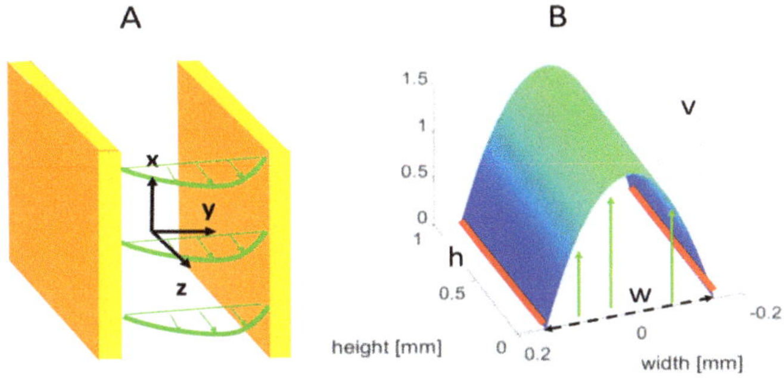

Figure 5.4. (A) A schematic of the velocity profile for a suspended capillary flow. (B) Approximate velocity profile; the two walls are symbolized by the red lines.

$$v(y) = V_{\max}\left[1 - \left(\frac{y}{w/2}\right)^2\right] = \frac{3}{2}V\left[1 - \left(\frac{y}{w/2}\right)^2\right], \tag{5.5}$$

where w is the channel width and x the horizontal cross-dimension. The shear rate at the wall is then given by

$$\dot{\gamma} = \left|\frac{\partial v}{\partial y}\right|(y = \pm w/2) = 6\frac{V}{w}. \tag{5.6}$$

Locally, at the wall, $\lambda = w/6$. Using the definition of the average friction length, we obtain

$$\frac{1}{\bar{\lambda}} \cong \frac{1}{2(w+h)}\int_{\Gamma_{\text{wall}}}\frac{1}{w/6}\,\mathrm{d}l = \frac{1}{2(w+h)}2\int_0^h\frac{1}{w/6}\,\mathrm{d}l = \frac{6h}{w\,(w+h)}. \tag{5.7}$$

Hence,

$$\bar{\lambda} = \frac{w}{6}\frac{w+h}{h} = \frac{w}{6}(\varepsilon + 1). \tag{5.8}$$

Using (5.4) and (5.8), the expression for the travel distance as a function of time is then

$$z = \sqrt{\frac{\gamma}{\mu}2\bar{\lambda}\cos\theta^*}\,\sqrt{t} = \sqrt{\frac{\gamma}{\mu}\left(\cos\theta - \frac{w}{h}\right)\frac{w}{3}}\,\sqrt{t}. \tag{5.9}$$

The other equivalent approach [5] consists of equating the capillary force and the wall friction:

$$F_{\text{cap}} = \gamma(2h)\cos\theta - \gamma(2w) = F_{\text{drag}} \cong \frac{6\mu\,V}{w}z(2h). \tag{5.10}$$

Then

$$V = \frac{\gamma}{\mu}\frac{w}{6}\frac{\left(\cos\theta - \frac{w}{h}\right)}{z}, \tag{5.11}$$

and, using $V = \mathrm{d}z/\mathrm{d}t$, relation (5.9) for the travel distance is retrieved:

$$z = \sqrt{\frac{\gamma}{\mu}\left(\cos\theta - \frac{w}{h}\right)\frac{w}{3}t}\,. \tag{5.12}$$

Figure 5.5 shows the travel distances of different solutions of IPA (isopropyl alcohol) and alginate (1%) in a suspended channel ($w = 0.3$ mm, $h = 1$ mm) milled in PMMA and treated with oxygen plasma [6]. The physical properties of the different liquids are listed in table 5.1.

The contact angles are measured by optical means (Krüss, Germany) and are shown in figure 5.5.

Figure 5.6 shows the progression of the tinted IPA solution in a 40 cm long suspended channel.

A comparison between the model and experimental results is shown in figure 5.7 for IPA 25%. The agreement between theory and experiment is remarkable if the

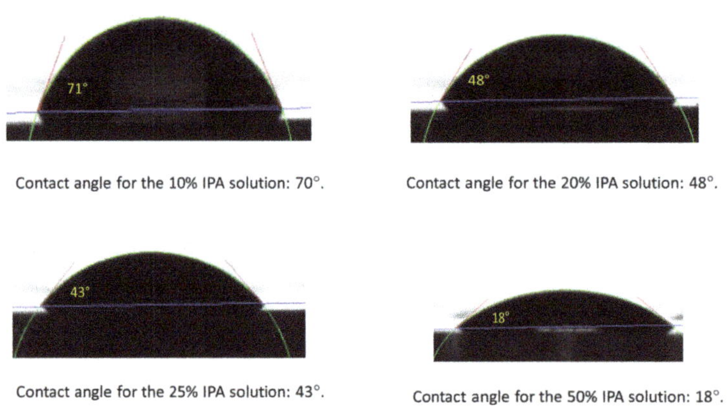

Figure 5.5. Contact angles of IPA solutions with O_2-plasma-treated PMMA.

Table 5.1. Physical properties of IPA solutions and alginate.

Liquids	Surface tension (mN m^{-1})	Viscosity (cP or mPa s)	Contact angle (°)
IPA 10%	40	1.5	71
IPA 20%	31	2.1	48
IPA 25%	29	2.4	43
IPA 50%	24	3.1	18
Alginate 1%	71	5.0	10

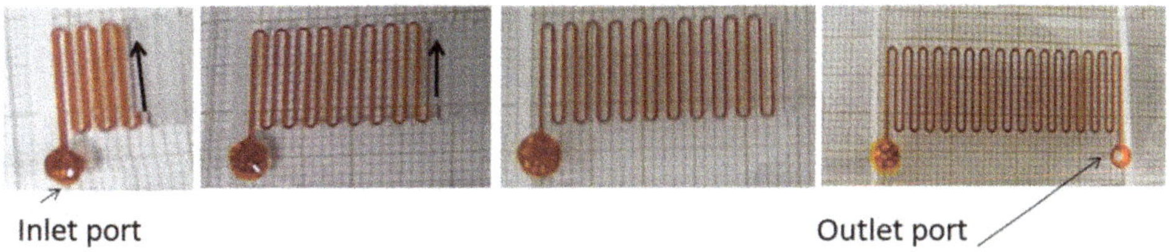

Figure 5.6. An IPA solution colored red progresses in a suspended winding channel (photo by David Gosselin and Jean Berthier, CEA/Leti).

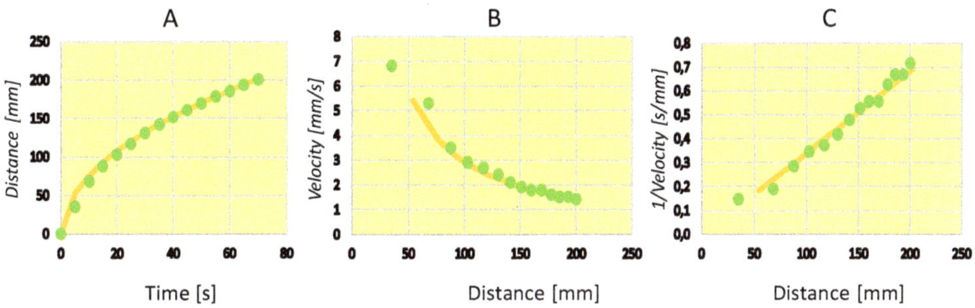

Figure 5.7. (A) Travel distance as a function of time for a 25% solution of IPA; green dots correspond to experimental results and orange continuous lines to theory. (B) Velocity as a function of distance in the channel. (C) The inverse of velocity as a function of distance in the channel.

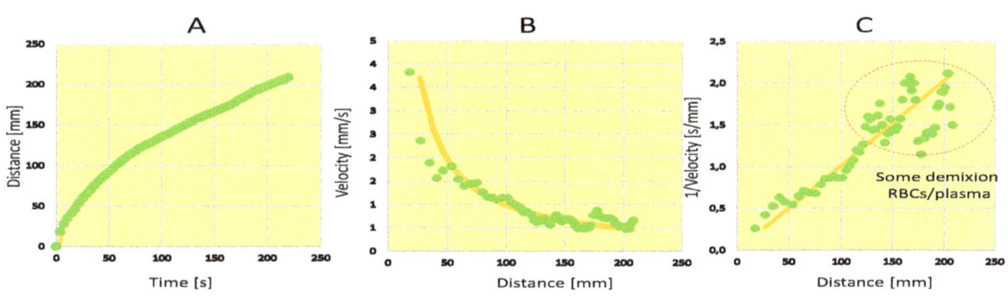

Figure 5.8. (A) Travel distance as a function of time for whole blood (with ethylenediaminetetraacetic acid (EDTA)); green dots correspond to experimental results and orange continuous lines to theory. (B) Velocity as a function of distance in the channel. (C) The inverse of velocity as a function of distance in the channel.

friction length $\bar{\lambda}$ is slightly adapted: relation (5.8) produces the theoretical value $\bar{\lambda} \cong 65\mu m$, while the fit indicates $\bar{\lambda} \cong 80\mu m$. Note that there is no appearance of a dynamic contact angle due to the relatively small velocities.

The same approach was used for human whole blood (stabilized by ethylenediaminetetraacetic acid (EDTA)) and the results are shown in figure 5.8. The difference—as was already mentioned in chapter 3 for non-Newtonian fluids—is the appearance of a complex regime after a travel distance of approximately 150 mm.

It is suspected that this regime is linked to some demixing of the blood plasma from hematocrits.

5.3 Rails

Rails may be defined as parallel horizontal solid structures guiding the capillary flow, as sketched in figure 5.9. There is an obvious similarity between suspended and rail microflows. Basically, rails are similar to suspended channels. However, the difference will appear later (chapter 6), when we consider enlargements of the fluidic path. Gravity acts differently in the two cases, and horizontal rails do not suffer from a gravitational effect when the fluidic path is enlarged. Besides, in contrast to suspended channels, rails can form networks.

One of the first rail-microfluidic devices reported in the literature was that of Satoh and coworkers [7], who used electrowetting to move aqueous fluid from one electrode to the next along a network of rails (figure 5.10). A similar approach was taken by Baret and coworkers [8]. A similar idea was exploited by Sung Hoon Lee and coworkers for cellular co-culture [9]. Horizontal rails—comprising hydrophilic pathways and hydrophobic boundaries—have also been used by Bouaidat and coworkers [10]. One of the advantages of this approach is that no alignment is needed.

The SCF condition for rails is similar to that for suspended channels, except that the axes are switched, i.e.

$$\cos\theta > \frac{p_F}{p_w} = \frac{2h}{2w} = \varepsilon, \tag{5.13}$$

where the aspect ratio ε is now defined as $\varepsilon = h/w$. The generalized Cassie angle is also similar to that of suspended channels:

$$\cos\theta^* = \frac{w\cos\theta - h}{w + h} = \frac{\cos\theta - \varepsilon}{\varepsilon + 1}, \tag{5.14}$$

as is the average friction length:

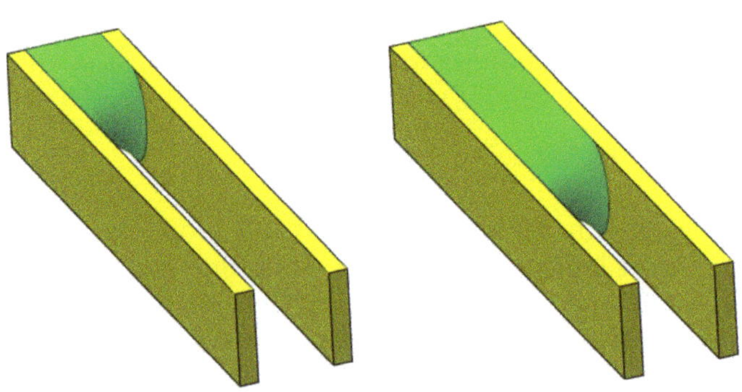

Figure 5.9. A capillary flow of liquid between two horizontal rails calculated using Evolver (w = 40 μm, h = 30 μm, θ = 30°).

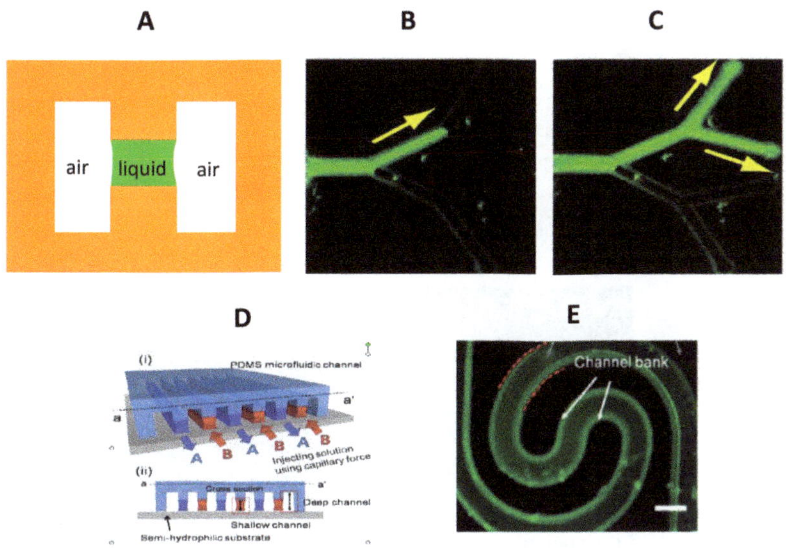

Figure 5.10. Top, electrowetting rails: (A) schematic of the cross-section. (B) and (C) liquid progression guided by electrowetting, reprinted with permission from [7] Copyright (2005) American Chemical Society. (D) A schematic of the device imagined by Lee and coworkers and (E) a view of the winding rails, reprinted with permission from [9] Copyright (2010) American Chemical Society.

$$\bar{\lambda} = \frac{h}{6}\frac{w+h}{w} = \frac{h}{6}(\varepsilon + 1). \tag{5.15}$$

To date, no report of the dynamics of the capillary flow in rails has been completed. Behavior similar to that of suspended channels is expected.

5.4 Rectangular channels

Channels that have rectangular cross-sections are amongst the most used [11, 12] (figure 5.11(A)). They can be fabricated using a milling machine with a rectangular drill bit (figure 5.11(B)).

5.4.1 The SCF condition

The SCF condition for a rectangular open channel of width w and depth h can be obtained using the generic inequality

$$\cos\theta > \frac{p_F}{p_w} = \frac{w}{w+2h} = \frac{\varepsilon}{\varepsilon+1}, \tag{5.16}$$

where ε is the aspect ratio defined by $\varepsilon = w/2h$. If condition (5.16) is satisfied, a capillary flow is established in the channel. Relation (5.16) can be cast in the form

$$\varepsilon = \frac{w}{2h} < \frac{\cos\theta}{1-\cos\theta}. \tag{5.17}$$

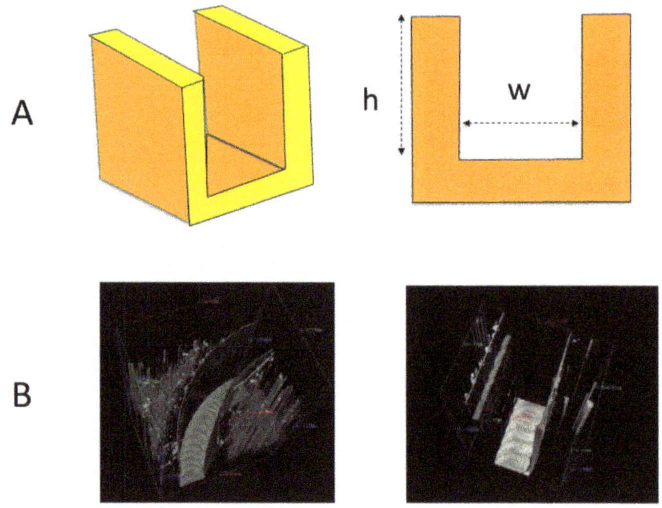

Figure 5.11. (A) A sketch of a rectangular open groove; (B) photographs of rectangular channels milled in PMMA (courtesy of David Gosselin and François Boizot).

However, another condition must be taken into account, depending on whether precursor fingers or filaments are desired. The condition for the formation of filaments is presented in chapter 7. We just indicate here the Concus–Finn (CF) condition for filaments in a rectangular sharp corner [13, 14]:

$$\theta < \frac{\pi}{4}, \tag{5.18}$$

If this condition is met, two capillary filaments form in the bottom corners. In the following, it is assumed that there are no capillary filaments, i.e. $\theta > \pi/4$. The theoretical SCF condition (5.16) has been compared to that obtained by the software Surface Evolver [15], which iteratively moves an interface to its position of minimal energy. This comparison is shown in figure 5.12. Note that Evolver does not account for the inertia of friction. It performs a static calculation, but in some cases—as here for the onset of SCF—it can be accurate enough for the determination of a slow SCF.

5.4.2 The generalized Cassie angle

Let us next consider the flow dynamics. As already shown in this book, the first parameter is the generalized Cassie angle, denoted here by θ^*. The Cassie angle is easily determined from

$$\cos\theta^* = \frac{(w + 2h)\cos\theta - w}{2(w + h)} = \frac{(\varepsilon + 1)\cos\theta - \varepsilon}{(2\varepsilon + 1)}. \tag{5.19}$$

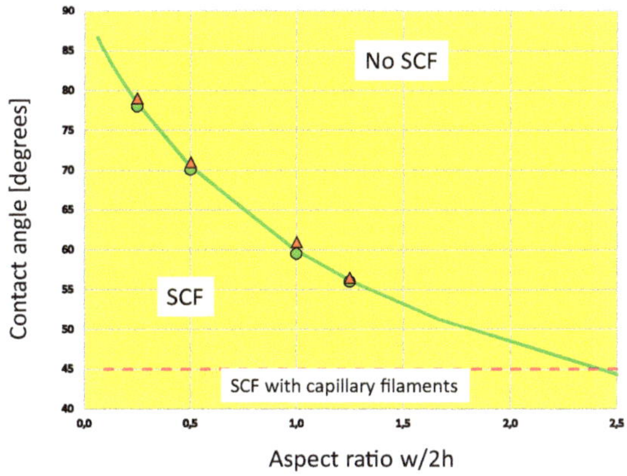

Figure 5.12. The SCF domain as a function of the aspect ratio and contact angle. The green line represents condition (5.16); the red triangles represent Evolver results corresponding to the recession of the liquid in the channel and the green circles correspond to advancing liquid (SCF).

5.4.3 Average friction length

The second parameter is the average friction length $\bar{\lambda}$. In the case of suspended channels, a simple approximation has been found. The problem is much more difficult in the case of a rectangular open channel because the flow field is not described by an analytical function. A very coarse approach is to estimate the shear rate along the two vertical walls using $\dot{\gamma}_{\text{vertical}} \approx 6\frac{V}{w}$ and on the floor using $\dot{\gamma}_{\text{floor}} \approx 3\frac{V}{h}$. These values are estimated using quadratic profiles, which is a coarse approximation. Using these approximations, the wall friction can be deduced to be

$$F_{\text{drag}} = z\ \mu\ V\left[2\int_0^h \frac{6}{w}\,\mathrm{d}l + \int_{-w/2}^{w/2} \frac{3}{h}\,\mathrm{d}l\right] = 6z\ \mu\ V\left(\frac{2h}{w} + \frac{w}{2h}\right). \tag{5.20}$$

A first expression for the average length can be deduced from (5.20):

$$\bar{\lambda} = \frac{1}{3}\frac{w + h}{\left(\frac{2h}{w} + \frac{w}{2h}\right)}. \tag{5.21}$$

However, expression (5.21) has two weaknesses: it does not account for the bottom corner effect and the velocity along the vertical is not parabolic when the aspect ratio *w*/*2h* is small or moderate. Figure 5.13 sketches the velocity field for two different open-channel aspect ratios, showing the estimated velocity fields in the corners I and J (in red on the figure) and along the vertical axis.

Obviously, expression (5.21) overestimates the average friction distance. Note that the term *w*/*2h* in relation (5.21) is linked to a parabolic vertical profile that only appears for high aspect ratios *w*/*2h*. So, we next investigate different expressions, built similarly to (5.21):

$$\bar{\lambda}_1 = \frac{1}{3}\frac{w + h}{\left(\frac{2h}{w} + \frac{w}{h}\right)}, \tag{5.22}$$

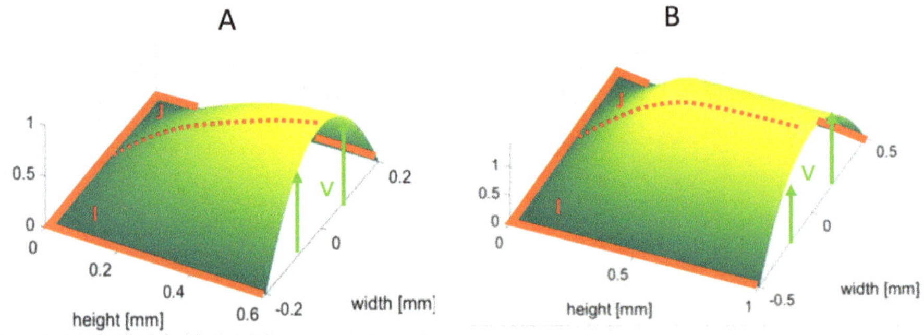

Figure 5.13. A sketch of the velocity field in a rectangular open channel: (A) a moderate aspect ratio, (B) a small aspect ratio. The red dotted line represents the velocity profile along the vertical axis.

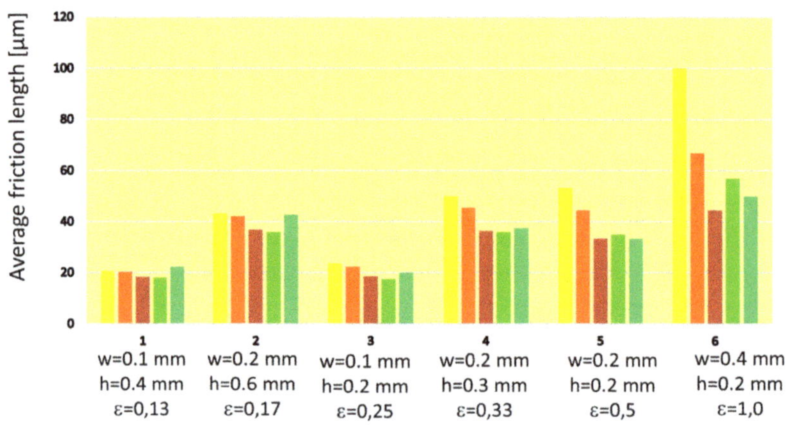

Figure 5.14. A comparison between friction lengths obtained using different formulas and a numerical approach (COMSOL). The yellow bars correspond to relation (5.21), orange bars to relation (5.22), red bars to relation (5.23), green bars to COMSOL, and dark green bars to relation (5.24).

and

$$\overline{\lambda_2} = \frac{1}{6}\frac{w + 2h}{\left(\frac{2h}{w} + \frac{w}{h}\right)}. \tag{5.23}$$

Finally, we consider an expression based on a modified hydraulic diameter

$$\overline{\lambda_3} = \frac{1}{2}\frac{w\,h}{w + 2h}. \tag{5.24}$$

Comparisons with COMSOL results are shown in figure 5.14 for six different channel dimensions.

Expressions (5.23) and (5.24) are the closest to the COMSOL calculations. These friction lengths are detailed in table 5.2.

Table 5.2. The comparison between the $\bar{\lambda}$ values obtained using analytical formulations (5.23) and (5.24) and those obtained via numerical calculation (COMSOL).

w (μm)	h (μm)		$\bar{\lambda}$ (5.23) (μm)	$\bar{\lambda}$ (5.24) (μm)	$\bar{\lambda}$ (COMSOL) (μm)
100	400	0.125	18.2	22.2	18.0
200	600	0.17	36.8	42.8	36.0
100	200	0.25	18.5	20.0	17.5
200	300	0.33	36.4	37.5	36.0
200	200	0.50	33.3	33.3	38.0
400	200	1.00	44.4	50.0	57.0

5.4.4 The homothetical rule

Expression (5.23) is convenient as a first approximation of the friction length, which is helpful when performing a fit with experimental results. Expression (5.23) also suggests that the friction lengths of two rectangular open channels of same aspect ratio are related. Consider two such channels, denoted by c1 and c2:

$$h_{c2}/w_{c2} = \frac{h_{c1}}{w_{c1}}. \tag{5.25}$$

The two dimensions w and h are thus linked by a proportionality ratio α:

$$h_{c2}/h_{c1} = \frac{w_{c2}}{w_{c1} = \alpha}, \tag{5.26}$$

and substituting this into (5.23), we find

$$\bar{\lambda}_{c2} = \alpha\, \bar{\lambda}_{c1}. \tag{5.27}$$

Hence, we have the property that two homothetic channels have friction lengths in the same homothetic ratio. This is an important property, since the generalized Cassie angles are the same—for the same wall material—in the two cases, as we shall see in the next section.

5.4.5 Other approaches

We report here the approach followed by Ouali and coworkers [16] for rectangular open channels. For a horizontal channel, they derived the following expression of the Lucas–Washburn law:

$$z = \sqrt{\frac{2b}{a}\, t}, \tag{5.28}$$

where $a = 3\mu/\rho h^2 \zeta_0(h/w)$ and $b = \gamma[(1 + 2h/w)\cos\theta - 1]\rho h$. For h/w smaller than two, they found the approximation $\zeta^{-1}(h/w) \approx 1 + 0.671h/w + 4.17h^2/w^2$. In terms of average friction length, relation (5.28) yields

$$\bar{\lambda} = \frac{2}{3}\frac{h}{w}\, \zeta(h/w)\,(w + h). \tag{5.29}$$

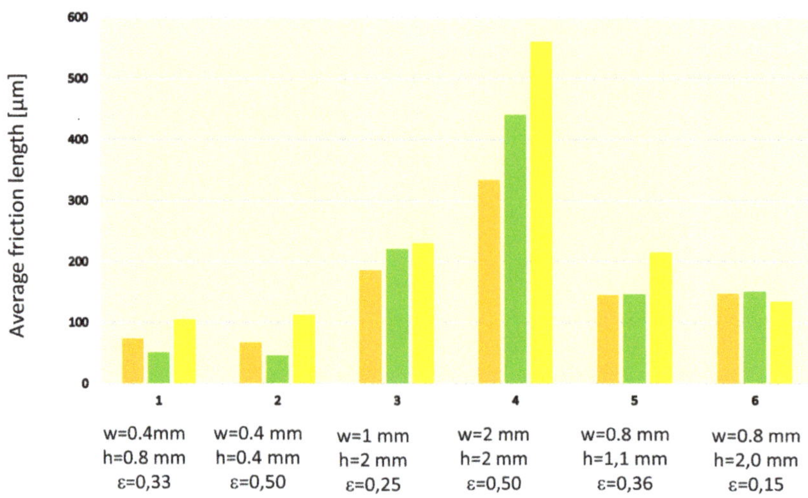

Figure 5.15. A comparison of the friction lengths obtained from formula (5.23), formula (5.29), and experiments. The orange bars correspond to relation (5.23), yellow bars to relation (5.29), and green bars to the experimental results.

Figure 5.15 compares the average friction lengths obtained from relation (5.23), relation (5.29), and experimental results.

5.4.6 Dynamics

The inertial regime—in which dynamics effects are non-negligible—is very short-lived. In the following viscous regime, inertia is negligible and, if the flow velocities are small, it has been shown that the Surface Evolver software can be used to predict the flow morphology [15]. Figure 5.16 shows the advancing capillary flow in a channel that has an aspect ratio of $\varepsilon = w/2h{=}1/4$ and a Young contact angle of 50°.

It has been shown in the preceding chapter that the velocity V is linked to the travel distance z by the relation

$$V = \frac{\gamma}{\mu} \frac{\bar{\lambda} \cos\theta^*}{z}. \tag{5.30}$$

A first remark concerns homothetic channels. As we have seen in the preceding section, homothetic channels whose homothetic ratio is α have friction lengths in the same ratio. In addition, their generalized Cassie angles are the same

$$\begin{aligned}\cos\theta^*_{c2} &= \frac{(w_{c2} + 2h_{c2})\cos\theta - w_{c2}}{2(w_{c2} + h_{c2})} = \frac{(\varepsilon + 1)\cos\theta - \varepsilon}{2\left(\varepsilon + \frac{1}{2}\right)} \\ &= \frac{(w_{c1} + 2h_{c1})\cos\theta - w_{c1}}{2(w_{c1} + h_{c1})} = \cos\theta^*_{c1},\end{aligned} \tag{5.31}$$

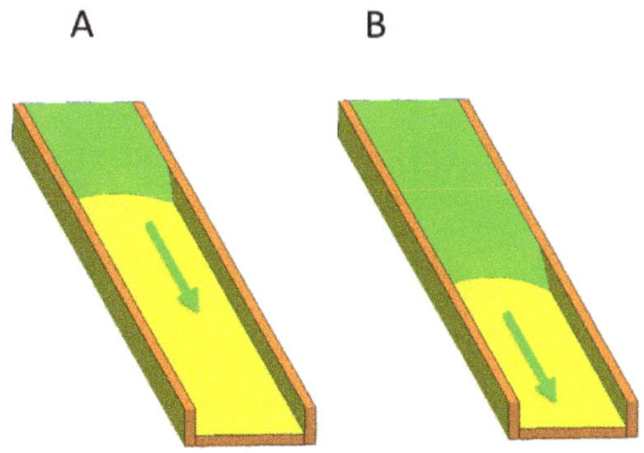

Figure 5.16. SCF in a U-groove (quasi-static Evolver calculation) showing the curved triple line (w = 1 mm, h = 200 μm, θ = 50°).

with $\varepsilon = w/2h$. Hence, if α denotes the homothetic ratio, then at the same distance z in the channels, the velocities are linked by

$$V_{c2} = \frac{\gamma}{\mu}\frac{\bar{\lambda}_{c2}\cos\theta_{c2}^*}{z} = \alpha\frac{\gamma}{\mu}\frac{\bar{\lambda}_{c1}\cos\theta_{c1}^*}{z} = \alpha V_{c1}. \tag{5.32}$$

Homothetically increasing the dimensions of a rectangular open channel increases the velocity in the same ratio. This result is valid as long as the contact angle is the static contact angle, i.e. there is no dynamic contact angle (DCA) effect. As we have seen in the preceding chapters, a DCA may appear in large channels.

A second remark pertains to the aspect ratio. It is often advantageous to obtain the highest possible velocities. The question is then: for a given channel depth h, what is the optimal value of w (or ε) that produces the maximum velocity V? Substituting the approximate value of the friction length (5.23) into (5.30), we find

$$\begin{aligned} V &= \frac{\gamma}{\mu}\frac{\bar{\lambda}\cos\theta^*}{z(t)} \cong \frac{\gamma}{\mu}\frac{1}{6}\frac{w+2h}{\left(\frac{2h}{w}+\frac{w}{h}\right)}\frac{[(w+2h)\cos\theta - w]}{(w+h)}\frac{h}{z} \\ &= \frac{\gamma}{\mu}\frac{1}{6}\frac{1+\varepsilon}{\left(\frac{1}{\varepsilon}+2\varepsilon\right)}\frac{[(1+\varepsilon)\cos\theta-\varepsilon]}{\left(\varepsilon+\frac{1}{2}\right)}\frac{h}{z}. \end{aligned} \tag{5.33}$$

In other words, we have to find the maximum of the nondimensional function

$$f = \frac{(1+\varepsilon)}{\left(\frac{1}{\varepsilon}+2\varepsilon\right)}\frac{[(1+\varepsilon)\cos\theta-\varepsilon]}{\left(\varepsilon+\frac{1}{2}\right)}. \tag{5.34}$$

The function f is plotted in figure 5.17 for five values of the contact angle (20°, 30°, 40°, 60°, and 70°). Note that the maximum velocity results from a competition between the Cassie contact angle and the friction length. On the one hand, the Cassie angle decreases with a reduction in the channel width because of the associated decrease in the free surface area—and the cosine of the Cassie angle increases.

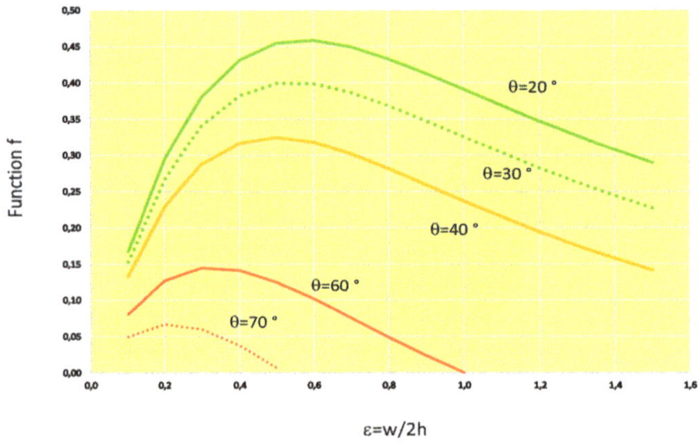

Figure 5.17. A plot of the function f as a function of the aspect ratio for five values of the contact angle (20°, 30°, 40°, 60°, and 70°). For each contact angle, the function shows a maximum. SCF occurs only when f is positive.

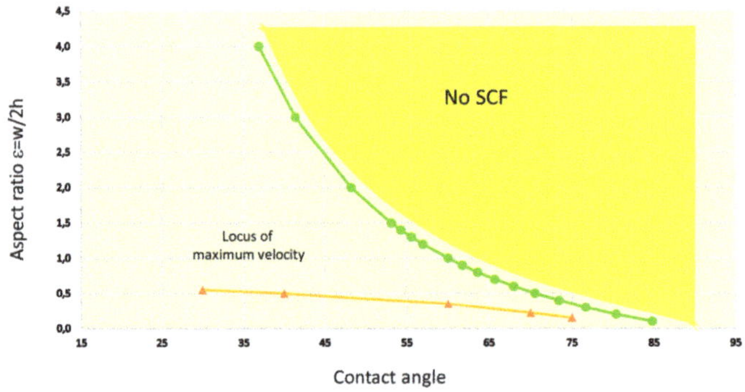

Figure 5.18. The SCF limit (green) and optimal aspect ratio (red) in rectangular U-grooves: the optimal aspect ratio corresponds to the maximum velocity.

On the other hand, the friction length decreases with the reduction of the channel width. The two parameters vary oppositely. The product of these two parameters has a maximum, as shown in figure 5.17.

We can plot the locus of the maximum of the function f—relation (5.34)—in a coordinate system defined by the aspect ratio versus the contact angle (figure 5.18). The maximum velocity occurs for aspect ratios that lie between 0.25 and 0.5, which corresponds to $h/2 \lesssim w \lesssim h$.

Let us now present experiments performed in a PMMA rectangular open channel 300 μm wide and 1 mm deep. This winding channel is shown in figure 5.19. The device was treated with oxygen plasma prior to the experiments to make it hydrophilic.

The velocity vs. distance is given by relation (5.30), and the travel distance vs. time is

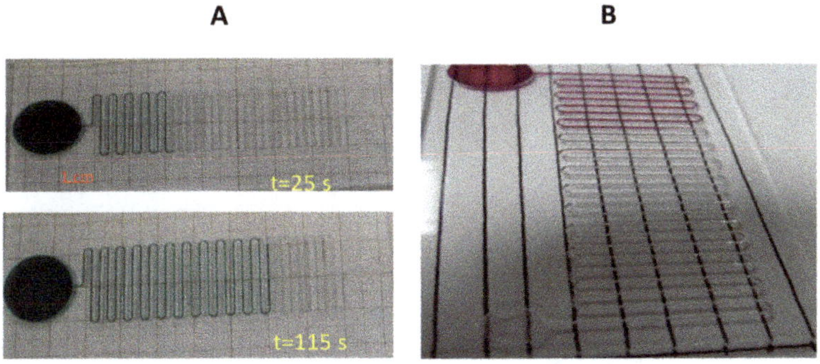

Figure 5.19. SCF in a winding, rectangular, open channel (w = 300 μm, h = 1 mm; ε = 0.15); the channel was milled in PMMA and oxygen plasma treated in order to decrease the static contact angle from 85° to 55°. The fluid is water tinted with food coloring. Photographs courtesy of Jean Berthier and David Gosselin.

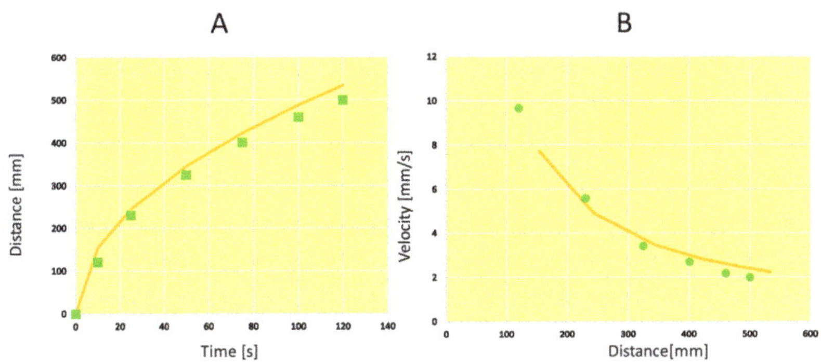

Figure 5.20. (A) Travel distance vs. time: a comparison between experimental results (green dots) and theoretical relation (5.33) (orange line); (B) velocity vs. distance (shown using the same color code).

$$z = \sqrt{\frac{\gamma}{\mu} 2\bar{\lambda} \cos\theta^* \, t} = \sqrt{\frac{\gamma}{\mu} \frac{f(\varepsilon)}{3} h \, t}. \tag{5.35}$$

Experimental results and relation (5.35) have been plotted in figure 5.20. A good fit is obtained using a value of 49 μm for the average friction length, instead of the value of 55 μm predicted by (5.23).

5.5 Rounded channels

Due to the fabrication process, rectangular channels are sometimes rounded at the bottom. Sometimes they are also specially designed to avoid capillary filaments that would appear in the bottom sharp corners of a rectangular groove [13, 14] (such filaments are presented in chapter 7). Rounded drill bits that can mill such devices are commercially available. Figure 5.21 shows a typical cross-section of a rounded channel.

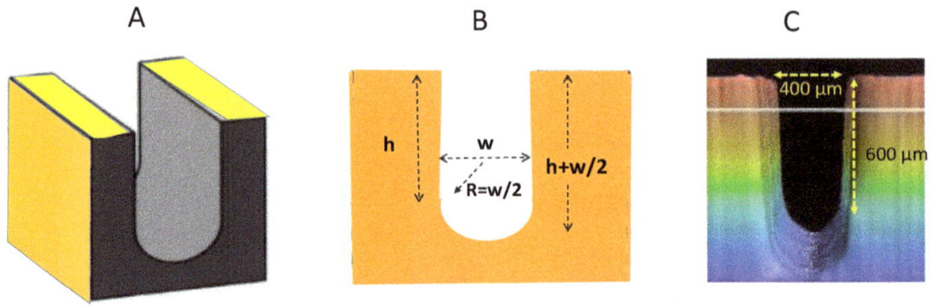

Figure 5.21. Rounded rectangular open channels: (A) and (B) a sketch of a rounded U-groove; (C) a profilometric view of a milled rounded channel. (C) Profilometry courtesy of Jodie Tokihiro.

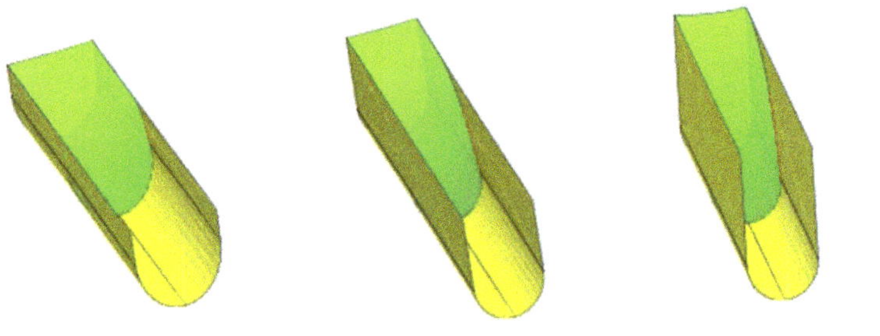

Figure 5.22. An Evolver simulation of SCF in rounded grooves of different aspect ratios w/h.

In such channels, the SCF condition is

$$\cos\theta > \frac{p_\mathrm{F}}{p_\mathrm{w}} = \frac{w}{2h + \frac{\pi w}{2}} = \frac{\varepsilon}{1 + \varepsilon\frac{\pi}{2}}, \tag{5.36}$$

where ε is the aspect ratio defined by $\varepsilon = w/2h$.

Relation (5.36) can be recast in the form

$$\varepsilon = \frac{w}{2h} < \frac{\cos\theta}{1 - \frac{\pi}{2}\cos\theta}. \tag{5.37}$$

The SCF condition can also be determined using the software Evolver (figure 5.22). Using energy minimization, the shape of the interface is calculated under a quasi-static assumption, namely that dynamics effects are negligible, which is valid at the onset of the motion. The condition for SCF can then be determined.

A plot can be made under the SCF and non-SCF conditions. The SCF condition is the curve at the limit between the two domains (figure 5.23). One observes that there is complete agreement between the Evolver results and the theoretical approach.

Note that when $w \to 0$, the channel becomes very thin and the open surface decreases to zero. Hence, the SCF contact angle in this narrow slot converges

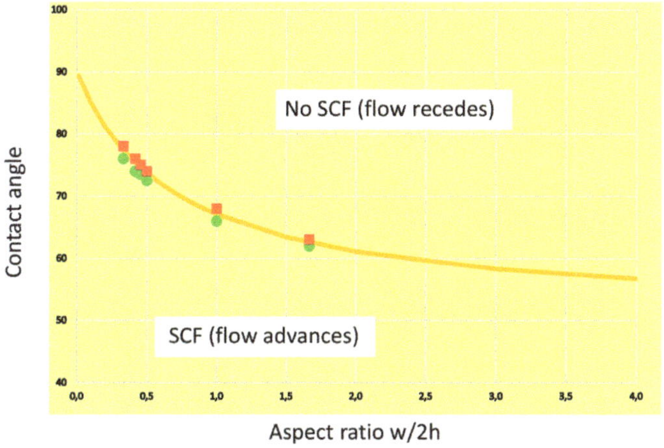

Figure 5.23. The SCF domain as a function of the aspect ratio: a comparison between theory (continuous orange line) and Evolver (red squares for receding flow and green dots for advancing SCF).

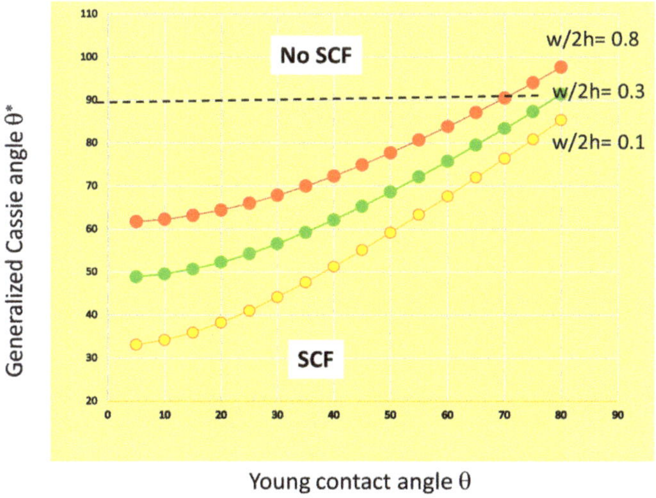

Figure 5.24. The Cassie angle vs. the Young contact angle for three aspect ratios.

towards 90°. On the other hand, when h decreases and reaches the zero value ($h = 0$), the channel shape becomes semicylindrical, and a contact angle of less than 50.5° is required to observe SCF.

The dynamics of the flow first requires the determination of the generalized Cassie angle. The generalized Cassie angle is given by

$$\cos\theta^* = \frac{\left(2h + \frac{\pi w}{2}\right)\cos\theta - w}{w + 2h + \frac{\pi w}{2}} = \frac{\left(1 + \frac{\pi}{2}\varepsilon\right)\cos\theta - \varepsilon}{\varepsilon + \left(1 + \frac{\pi}{2}\varepsilon\right)}, \tag{5.38}$$

where $\varepsilon = w/2h$. In figure 5.24, the Cassie angle defined by (5.38) is plotted as a function of the liquid–solid contact angle for three different values of the aspect ratio. The Cassie angle is much larger than the Young angle, especially for large aspect ratios $w/2h$, which correspond to large open surface areas.

The second parameter required to determine the dynamics of the capillary flow in a rounded channel is the average friction length, denoted by $\bar{\lambda}$. In the preceding section, we have approximated the average friction length for rectangular channels. For more complicated geometries—such as rounded channels—the determination of the friction length requires a numerical calculation or an experimental approach. A numerical approach has already been detailed in chapter 4. We have run COMSOL for four different channel widths and depths and obtained the data shown in table 5.3. In the range of dimensions considered in table 5.3, $\bar{\lambda} \approx 0.23\, w$.

In the following, we present experimental results for the capillary flow of different oils in a rounded U-groove ($w = 0.8$ mm, $h = 1.1$ mm, $\varepsilon = 0.36$) milled in PMMA. The organic liquids used for the experiments are dodecane, pentanol, and nonanol. Figure 5.25 shows the capillary flow of pentanol in the device.

Using the universal relation $z = \sqrt{\frac{\gamma}{\mu} 2\bar{\lambda} \cos\theta^*\, t}$, the physical data for the three liquids listed in table 5.4, and a friction length of 190 μm, we obtain the plots shown in figure 5.26.

Table 5.3. Friction lengths for different rounded open channels, calculated using COMSOL.

w (μm)	h (μm)	eps = $w/2h$	V (μm s^{-1})	Wall shear rate COMSOL (1 s^{-1})	Average shear rate (1 s^{-1})	Average friction length $\bar{\lambda}$ (μm)	$\bar{\lambda}/w$
800	1600	0.25	1000	6.73	5.45	183.4	0.229
800	1100	0.36	1000	6.68	5.41	184.8	0.231
1200	1100	0.55	1000	4.45	3.60	277.4	0.231
800	600	0.67	1000	6.67	5.40	185.1	0.231

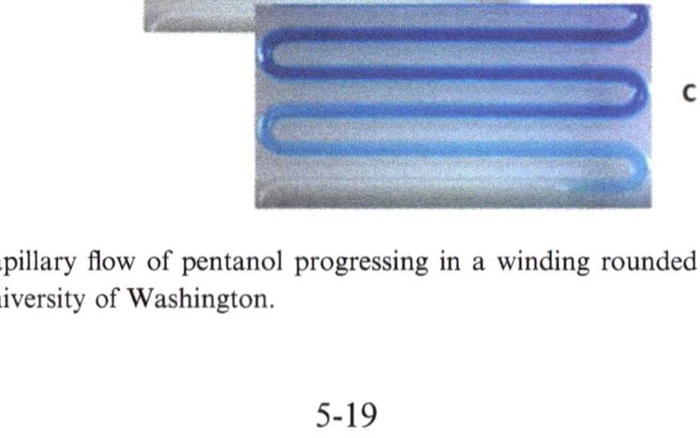

Figure 5.25. An open-capillary flow of pentanol progressing in a winding rounded U-groove. Photographs courtesy of Jing Lee, University of Washington.

Figure 5.27 shows the travel distance as a function of time for nonanol, for two different rounded channels of the same width $w = 0.8$ mm; for the first, $h = 1.1$ mm (and $\varepsilon = 0.36$) and for the second, $h = 2.6$ mm (and $\varepsilon = 0.15$). Because the friction

Table 5.4. Some physical properties of pentanol, dodecane, and nonanol.

Liquid	Viscosity (mPa s)	Surface tension *w*. air (mN m^{-1})	γ/μ (m s^{-1})	$\sqrt{\frac{\gamma}{\mu}}$ ((mm s)$^{1/2}$)	Contact angle θ (°)	Cosine Cassie angle (°)
Pentanol	3.50	26.0	7.4	86	12.5°	0.60
Dodecane	1.25	25.3	21.1	142	26.7°	0.54
Nonanol	11.7	28.7	1.56	25	13.0°	0.60

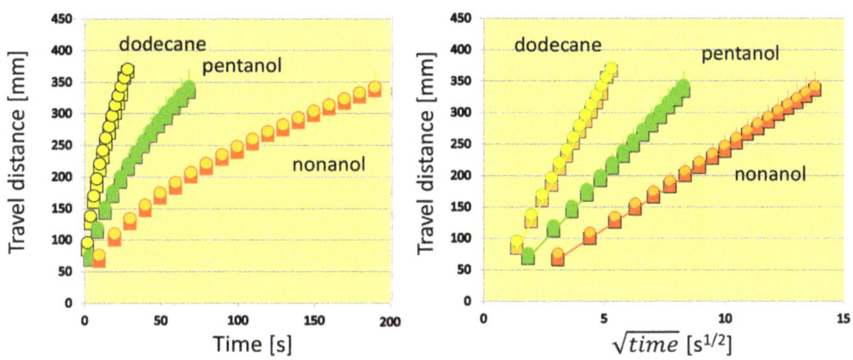

Figure 5.26. Capillary flows of dodecane, pentanol, and nonanol in a PMMA rounded U-groove. Left: travel distance vs. time; dots correspond to experiments and squares to theory. Right: travel distance vs. the square root of time, showing a clear linearity between the travel distance and the square root of time.

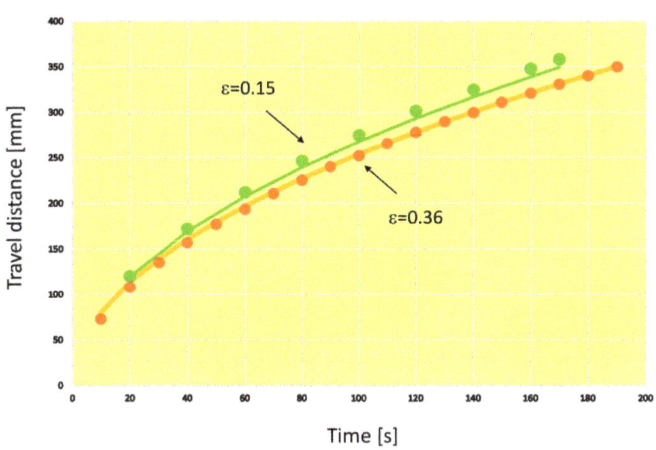

Figure 5.27. Capillary flows of nonanol in two different rounded channels: $h = 2.6$ mm (green curve) and $h = 1.1$ mm (orange curve). The dots correspond to experiments, while the continuous lines represent the theoretical expression.

length is nearly the same in the two cases—as shown in table 5.3—the difference is due to the Cassie angle, which is less for the small-aspect-ratio geometry.

5.6 Semicylindrical channels

A rounded channel becomes a semicylindrical channel when $h = 0$ in relation (5.36). The SCF condition in such channels is then given by

$$\cos\theta > \frac{p_F}{p_W} = \frac{w}{\pi\, w/2} = \frac{2}{\pi}. \tag{5.39}$$

This condition yields

$$\theta < 50.46°. \tag{5.40}$$

This condition has been checked with the software Evolver (figure 5.28). The fluid advances in the half-cylinder when the contact angle is 50° and recedes when the contact angle is 51°.

Note that the shape of the advancing interface is largely unaffected by the contact angle in a semicylindrical open channel, as shown in figure 5.29. This is due to the curvature of the solid surface.

No experimental approach has been reported so far for such geometries. The reason for this is probably the difficulty of fabricating such semicircular channels.

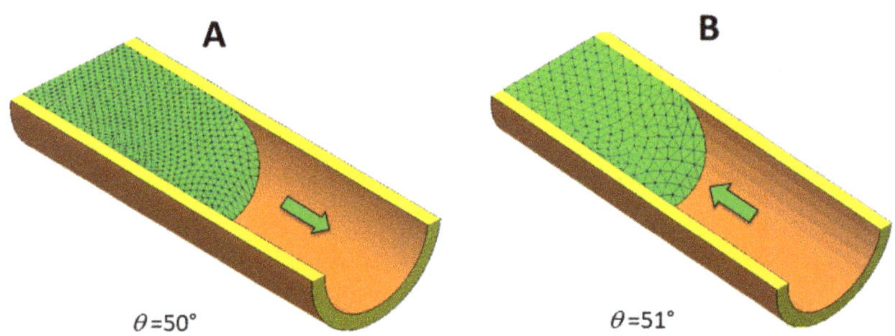

Figure 5.28. Capillary flow in a semicylindrical channel calculated using Evolver: (A) fluid advances when $\theta = 50°$; (B) fluid recedes when $\theta = 51°$.

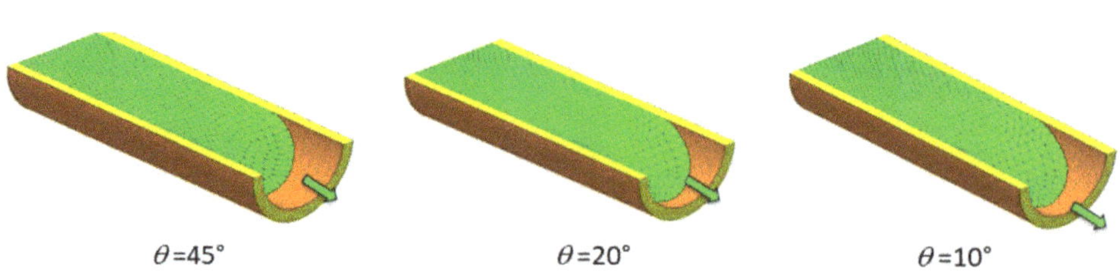

Figure 5.29. The shape of the advancing interface is largely unaffected by the value of the contact angle in a semicylindrical open channel.

However, recent developments in embossing techniques may be used to achieve such semicircular channels [17].

5.7 Embossed channels

Embossing consists of imprinting a shape in a substrate using a master mold and sometimes its counter-shape at a pressure of a few kilo-bars (10–40 kbars) and a temperature of 70 ° C –100 ° C. It is a rapid and low-cost fabrication process that is highly compatible with manufacturing activities [17]. Plastic and paper are typical substrates for embossing. For example, embossed paper channels coated with impermeable styrene-butadiene (SB) and hydrophilic reticulated poly(vinyl alcohol) (PVA) have been used to perform glucose tests [18]. Figure 5.30 shows an example of an embossed channel.

Embossing cannot produce vertical channel sides and smears out the side boundaries of the channel. If the relief created by embossing is sufficient and the contact angle is sufficiently small, SCF can occur, as demonstrated in figure 5.31.

If the channel cross-section is approximated by a flat bottom with rounded sides, as sketched in figure 5.32(A), the SCF condition is

Figure 5.30. Aqueous liquid flowing in an open channel embossed in paper and coated with styrene-butadiene. Photograph courtesy of David Gosselin and Jean Berthier.

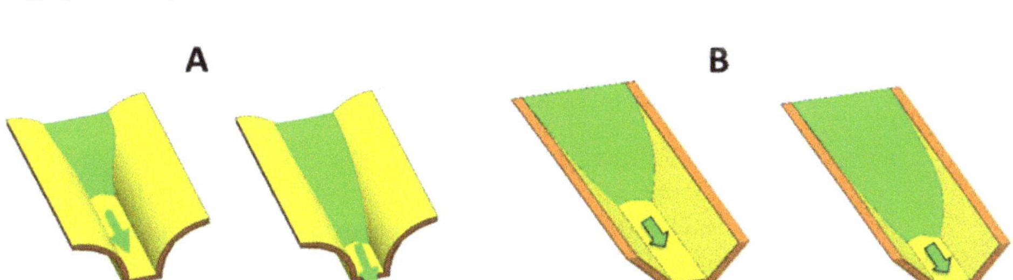

Figure 5.31. A sketch of an open-capillary flow in an embossed channel: (A) rounded sides; (B) a trapezoidal channel.

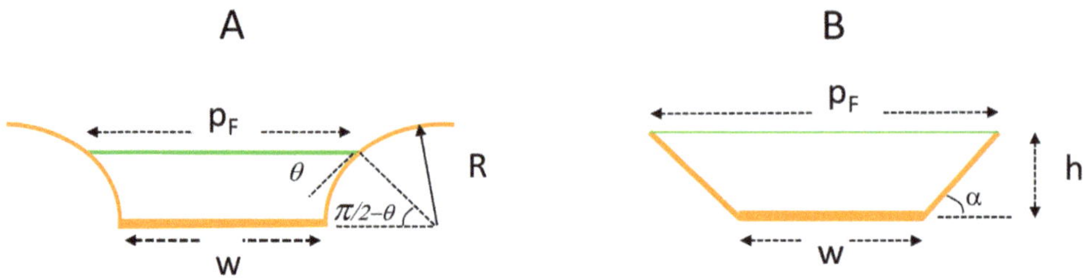

Figure 5.32. (A) A schematic of the cross-section of the embossed channel; (B) a trapezoidal approximation of an embossed channel.

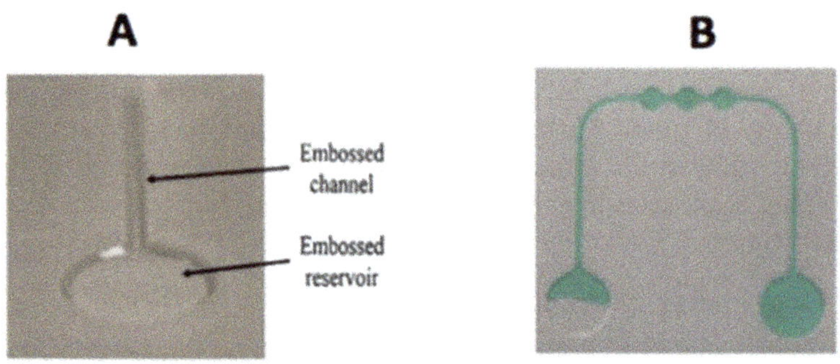

Figure 5.33. (A) An embossed channel in paper with the inlet port (courtesy of David Gosselin); (B) an embossed paper channel with secondary chambers for chemical reactions. Reprinted from [18], Copyright (2017), with permission from Elsevier.

$$\cos\theta > \frac{p_F}{p_w} = \frac{w + 2R\,(1 - \sin\theta)}{w + R\,(\pi - 2\theta)}, \tag{5.41}$$

or if $\varepsilon = w/R$,

$$\cos\theta > \frac{\varepsilon + 2(1 - \sin\theta)}{\varepsilon + (\pi - 2\theta)}. \tag{5.42}$$

An SCF condition (5.39) is satisfied if the relief created by embossing has a sufficiently small aspect ratio. In the case of a channel morphology closer to a trapezoidal shape (figure 5.32(B)), the SCF condition is

$$\cos\theta > \frac{p_F}{p_w} = \frac{w + 2h/\tan\alpha}{w + 2h/\sin\alpha} = \frac{\varepsilon + \cos\alpha/\sin\alpha}{\varepsilon + 1/\sin\alpha}, \tag{5.43}$$

where $\varepsilon = w/2h$. Figure 5.33 shows an open channel embossed in a paper matrix (coated successively by a layer of styrene-butadiene rubber (SBR) and a layer of PVA). In this example, the contact angle is of the order of 20°, due to the high wettability of the PVA layer.

Figure 5.34 shows the capillary flow in an embossed channel in a paper matrix (coated successively by a layer of SBR and a layer of PVA).

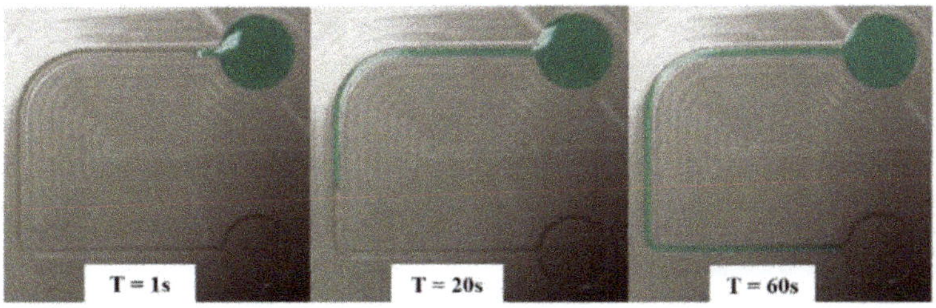

Figure 5.34. Open capillary flow in an embossed paper channel. Reprinted from [18], Copyright (2017), with permission from Elsevier.

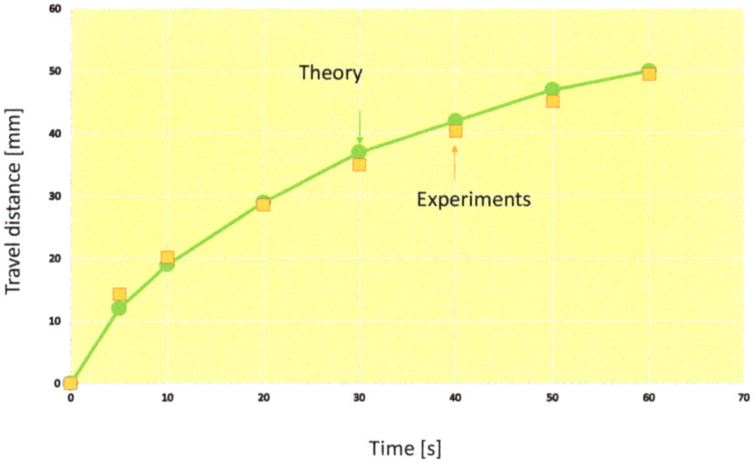

Figure 5.35. A comparison between experimental results and theoretical results for the capillary flow of tinted water in a channel embossed in paper [18].

Using the general expression for the travel distance as a function of time

$$z = \sqrt{\frac{\gamma}{\mu} 2\bar{\lambda} \cos \theta^* t}, \tag{5.44}$$

with $\theta^* = 40°$, $\bar{\lambda} = 300$ μm, and $\gamma/\mu = 72$ mN s^{-1}, we obtain the plots shown in figure 5.35, where the theoretical expression (5.44) is compared to experimental results.

5.8 Fiber bundles and flow caging

Threads and yarns can be considered to be open channels because of their outer boundary with the surrounding air. In chapter 11, we shall see that the idea of guiding microflows by wireframe structures is now considered for advanced open channel devices. Thread bundles were the first step towards such systems.

We can consider threads and yarns as a special case of microporous media and use the developments of chapter 8 to predict the capillary flow in such geometries. On the other hand, a simplified approach can be taken by making use of the general SCF condition for open microfluidics.

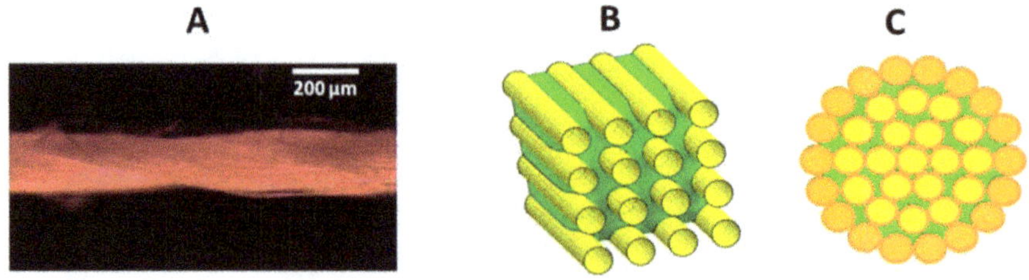

Figure 5.36. (A) A thread used for capillary flow. Reprinted from [20], Copyright (2013), with permission from Springer. (B) and (C) Two schematizations of the thread bundle.

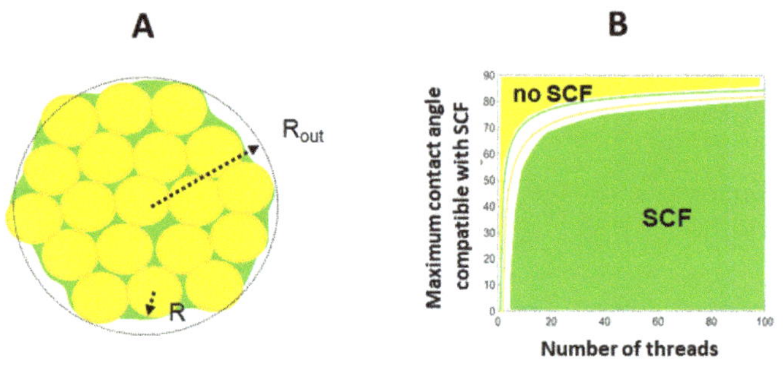

Figure 5.37. (A) A cross-section of a compact homogeneous fiber bundle. (B) The SCF domain. The two curves correspond to $\tau = 1$ (green curve) and $\tau = 2$ (orange curve).

Let us consider an ideal fiber bundle in which the fibers are parallel, as sketched in figure 5.36 [19].

As we assimilate the thread into a parallel fiber bundle, the cross-section is uniform, and the SCF condition applies. Let us consider the schematic figure 5.37, where the fibers are closely packed, i.e. each fiber is in contact with its neighbors.

Assuming that the liquid/fiber contact angle is small, i.e. the fibers have high wettability, the free perimeter p_F can be approximated by

$$p_F \approx 2\pi\, R_{out}. \tag{5.45}$$

On the other hand, assuming that all the fibers are totally wetted—which is true except for some parts of the exterior fibers—the wetted perimeter is simply approximated by

$$p_W \approx n 2\pi\, R, \tag{5.46}$$

where n is the number of fibers in the bundle. We derive the following condition for SCF in the fiber bundle:

$$\cos\theta > \frac{p_F}{p_W} \approx \frac{2\pi\, R_{out}}{n 2\pi\, R} = \frac{R_{out}}{n\, R}. \tag{5.47}$$

This inequality shows that a single thread cannot induce SCF (unless the contact angle is zero): in the case of a single thread, $n = 1$, and if it is the case that $R_{out} > R$, condition (5.47) can never be satisfied.

5.8.1 Two parallel rods

At least two parallel threads are required to obtain an SCF, as shown in figure 5.38 (calculated by Surface Evolver [15]). In such a case, the SCF condition is

$$\cos\theta > \frac{p_F}{p_W} = \frac{p_{F1} + p_{F2}}{p_{W1} + p_{W2}}. \tag{5.48}$$

If e is the interval between two rods of radius R, the 'free' perimeter $p_{F1} + p_{F2}$ is given by

$$p_{F1} = p_{F2} = (2R + e) - 2R\sin\left(\frac{\pi}{2} - \theta\right) = R\left[\left(2 + \frac{e}{R}\right) - 2\cos\theta\right], \tag{5.49}$$

and

$$p_{W1} = p_{W2} = 2\left(\frac{\pi}{2} - \theta\right)R. \tag{5.50}$$

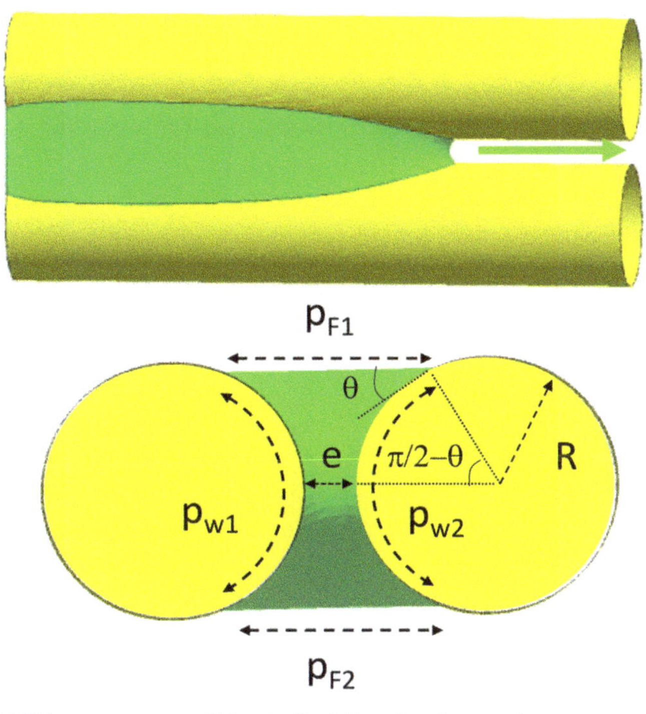

Figure 5.38. SCF between two parallel rods: liquid is colored green; the contact angle is $\theta = 30°$.

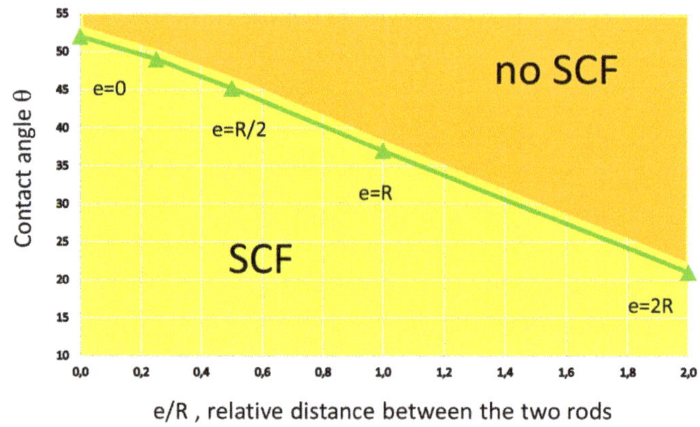

Figure 5.39. SCF between two parallel rods.

Equation (5.48) then becomes

$$\cos\theta > \frac{p_{\mathrm{F}}}{p_{\mathrm{W}}} = \frac{R\left[\left(2+\frac{e}{R}\right) - 2\cos\theta\right]}{2\left(\frac{\pi}{2}-\theta\right)R} = \frac{\left(1+\frac{e}{2R}\right) - \cos\theta}{\frac{\pi}{2}-\theta}. \tag{5.51}$$

Figure 5.39 shows the maximum contact angle that can lead to SCF as a function of the relative distance between the two rods, obtained using (5.49):

$$\left(\frac{\pi}{2} - \theta + 1\right)\cos\theta - 1 + \frac{e}{2R} = 0. \tag{5.52}$$

When the two rods are in contact, SCF is obtained for a contact angle $\theta < 52°$.

5.8.2 More than two parallel rods

Consider now the case in which $n>1$. In this case, referring to figure 5.37, the area of the cross-section is

$$A \approx \pi\, R_{\mathrm{out}}^2, \tag{5.53}$$

and the area of the fibers

$$A_{\mathrm{fi}} \approx n\,\pi\, R^2, \tag{5.54}$$

The fraction $\tau = A/A_{\mathrm{fi}}$ in the compact bundle depends on the arrangement of the fibers, but in any case, this fraction lies between

$$1 < \tau = \frac{A}{A_{\mathrm{fi}}} < 2. \tag{5.55}$$

The lower bound stems from the definitions of A and A_F. The upper bound $\tau = 2$ is linked to the compaction of the fiber bundle to the point at which each thread touches its neighbors. This upper limit is clearly seen in figure 5.37(A), where the free

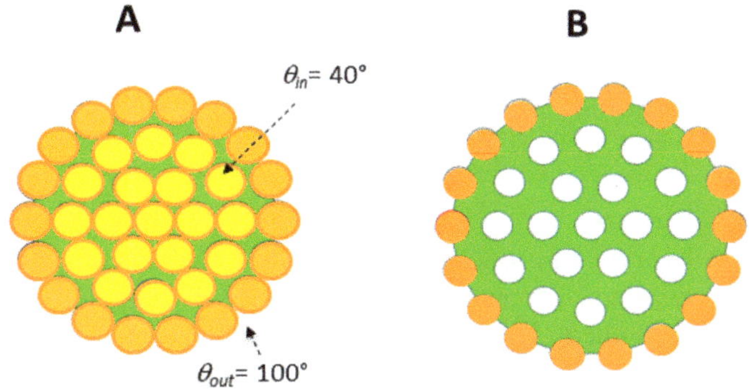

Figure 5.40. Caging a liquid using external hydrophobic threads [19]; (A) a compact bundle, (B) a spaced bundle.

(green) area is smaller than the (yellow) area occupied by the solid. Substituting (5.53) and (5.54) into (5.55) yields

$$\frac{R_{\text{out}}}{R} = \sqrt{n\,\tau}\,. \tag{5.56}$$

Substituting (5.56) into (5.47) yields the SCF condition

$$\cos\theta > \sqrt{\frac{\tau}{n}}, \tag{5.57}$$

with $1<\tau<2$. Relation (5.57) is plotted in figure 5.37(B). The figure shows an asymptote just above 80°. This shows that it is unnecessary to twist together more than 30 fibers in a bundle: Young contact angles of up to 75° can produce SCF in such bundles. This observation related to a number of approximately 25–30 threads is in agreement with the threads used by Ballerini and Wei in their experiments [21]. Moreover, this latter publication showed that the Lucas–Washburn–Rideal (LWR) law is valid as long as an adequate capillary radius is used and evaporation is negligible (see chapter 8).

It has been shown [19] that a capillary flow can be 'imprisoned' by an outer layer of hydrophobic threads while hydrophilic inner threads promote a capillary flow. This property can be called the 'caging' of a capillary flow, and is schematized in figure 5.40.

5.9 Capillary rise and uphill open-capillary flows

So far in this book, we have mostly considered horizontal capillary flows. Capillary rise occurs when a hydrophilic (lyophilic in the general case) solid structure—such as a plate, a plate with reliefs, paper, or thread—is placed vertically (or at a slanting angle) in contact with a liquid surface. The physics of capillary rise has been widely investigated [22–25]. The height reached by capillarity is determined by Jurin's law [26].

5.9.1 Jurin's law for capillary rise

Jurin's law was established for vertical cylindrical tubes, but it also applies to different geometries. In the following sections, Jurin's law is developed for the cases of cylindrical tubes and vertical plates.

5.9.1.1 Cylindrical tubes

Let us briefly recall Jurin's approach: when a capillary tube is plunged into a volume of wetting liquid, capillary forces make the liquid rise against gravitational forces until equilibrium is reached. It is observed that the height reached by the liquid is inversely proportional to the inner radius of the tube (figure 5.41). This observation is called Jurin's law.

An easy method used to derive Jurin's law is to balance the hydrostatic pressure that pulls the fluid down, as was done, for example, by Princen [22], with the Laplace pressure that pushes the fluid up. If ρ denotes the liquid density and g the gravitational constant, the hydrostatic pressure is

$$P_{\text{hydro}} = \rho\, g\, h, \tag{5.58}$$

where h is the equilibrium height of the fluid. On the other hand, the capillary pressure is

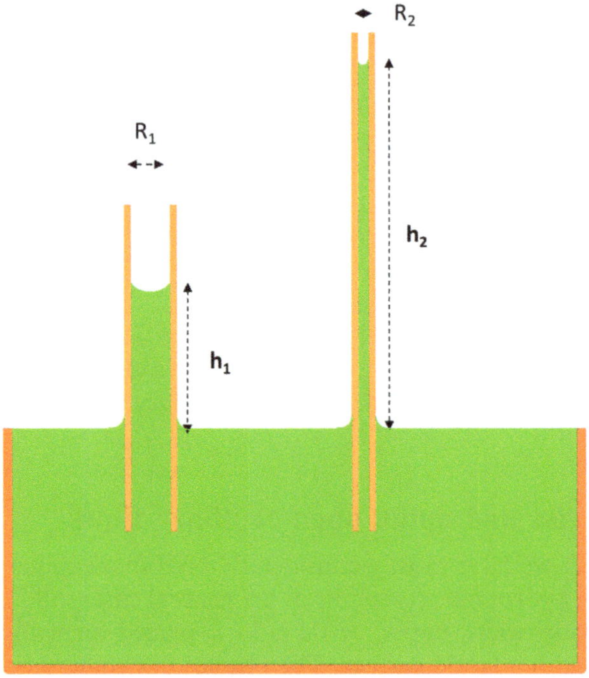

Figure 5.41. Jurin's law: the capillary rise is inversely proportional to the inner radius of the tube: $h_2/h_1 = R_1/R_2$.

$$P_{cap} = \frac{2\pi\ R\,\gamma\cos\theta}{\pi\ R^2} = \frac{2\gamma\cos\theta}{R}, \tag{5.59}$$

where R is the cylinder radius, γ the surface tension, and θ the contact angle. When the liquid reaches the equilibrium height, the two pressures balance each other, and

$$h_{eq} = \frac{2\gamma\cos\theta}{\rho\, gR}. \tag{5.60}$$

The capillary length $\kappa^{-1} = \sqrt{\gamma/\rho g}$ can be introduced into (5.60) as follows:

$$h_{eq} = 2\kappa^{-2}\frac{\cos\theta}{R}. \tag{5.61}$$

5.9.1.2 The general case: a vertical channel of arbitrary cross-section

In a channel—open or closed—of arbitrary cross-section, the capillary pressure is

$$P_{cap} = \frac{F_{cap}}{S_C} = \frac{\gamma\, p\cos\theta^*}{S_C}, \tag{5.62}$$

where S_C is the cross-sectional area of the channel and p the (total) perimeter. On the other hand, the hydrostatic pressure is

$$P_{hydro} = \rho\, g\, h. \tag{5.63}$$

Balance between the two pressures yields the equilibrium height

$$h_{eq} = \frac{\gamma\, p\cos\theta^*}{\rho\, g\, S_C}. \tag{5.64}$$

It is straightforward to check that (5.69) reduces to (5.60)—Jurin's law—if $p_F = 0$, $p_W = 2\pi R$, and $\theta^* = \theta$, $S_C = \pi R^2$. Another expression for the equilibrium height can be obtained by defining an equivalent hydraulic diameter

$$D_H = 2R_H = \frac{4S_C}{\sum_i p_{w,\,i} + p_F} = \frac{4S_C}{p}, \tag{5.65}$$

where the index i refers to the different walls—in case of a composite channel—and p_F to the free perimeter—in the case of an open channel. Equation (5.64) can then be written as

$$h_{eq} = \frac{2\gamma\cos\theta^*}{\rho\, g\,(D_H/2)} = \frac{2\gamma\cos\theta^*}{\rho\, g\, R_H} = 2\kappa^{-2}\frac{\cos\theta^*}{R_H}. \tag{5.66}$$

Relation (5.66) can be cast into a nondimensional form by normalizing h_{eq} and R_H using $h^*_{eq} = h_{eq}/\kappa^{-1}$ and $R^*_H = R_H/\kappa^{-1}$. One obtains

$$h_{\text{eq}}^* = \frac{h_{\text{eq}}}{\kappa^{-1}} = 2\frac{\cos\theta^*}{R_{\text{H}}^*}. \tag{5.67}$$

Relation (5.67) is theoretical. To be practical, the value of $\cos\theta^*$ must be determined. This determination is easy for closed channels, but difficult for open channels, because the exterior boundaries of the flow are unknown. In the following, some examples of capillary rise in closed and open geometries are given.

5.9.1.3 Capillary rise in a vertical square tube

Relations (5.66) or (5.67) can easily be used to determine the equilibrium height inside vertical square tubes. If a denotes the length of the side of the square tube, $R_{\text{H}} = a/2$, and $a^* = a/\kappa^{-1}$, relation (5.67) becomes

$$h_{\text{eq}}^* = 4\frac{\cos\theta}{a^*}. \tag{5.68}$$

Figure 5.42 shows a comparison between formula (5.68) and the experimental results of Wu *et al* [27].

5.9.1.4 Capillary rise between two parallel, vertical plates

a. Analytical approach

Let us consider two vertical parallel plates plunged into a liquid at rest (figure 5.43(A)). This configuration has been theoretically studied by Princen and Wu and coworkers [22, 27]. In order to obtain a tractable problem, Princen simplified the configuration by assuming a 2D geometry—neglecting the curvature of the interface in the plane parallel to the plates (figure 5.43(B)).

In such a case, referring to figure 5.43(B), the local curvature radius of the meniscus r is

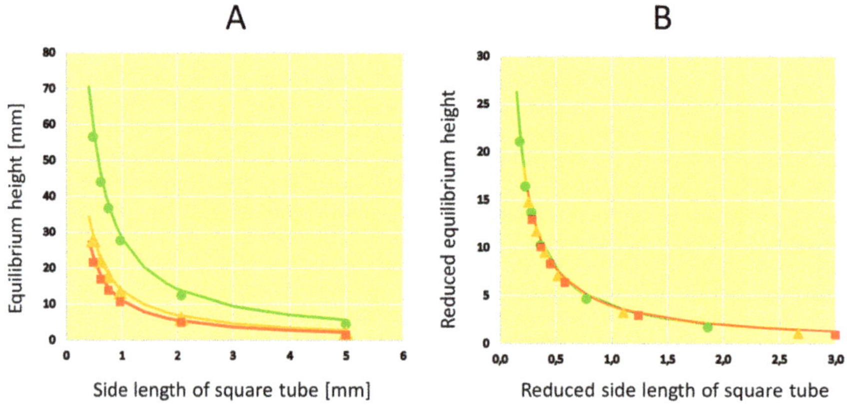

Figure 5.42. The equilibrium height in square tubes: a comparison between relations (5.66) and (5.67) and Wu *et al*'s experimental results for water (green), hexadecane (orange) and ethanol (red). (A) The height vs. the side length of the square tube; (B) the reduced height vs. the reduced length of the side tube.

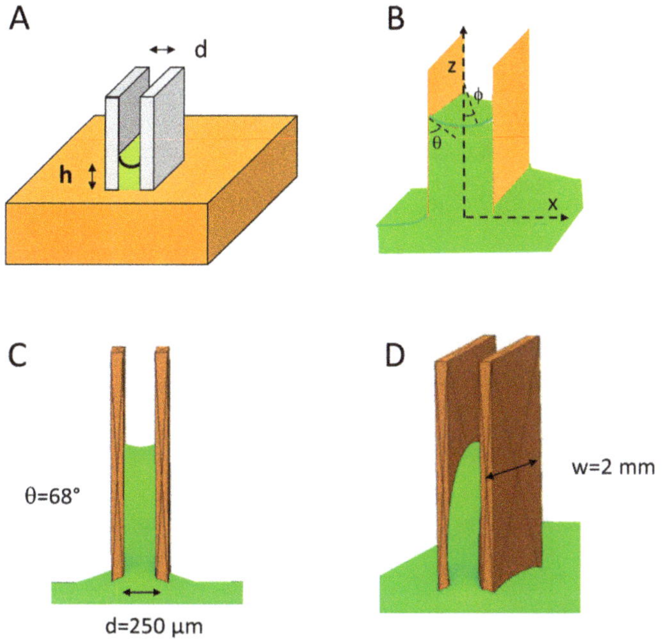

Figure 5.43. (A) A sketch of the device with two parallel vertical plates; (B) the sketch used by Princen for the analysis; (C) and (D) the capillary rise between the two plates obtained from an Evolver calculation.

$$\frac{1}{r} = \frac{\mathrm{d}\sin\phi}{\mathrm{d}x}, \tag{5.69}$$

and—assuming a noncircular interface due to the effect of the open boundaries—the local Laplace pressure of the interface is

$$P = \gamma\frac{\mathrm{d}\sin\phi}{\mathrm{d}x}. \tag{5.70}$$

The balance between the hydrostatic pressure and the Laplace pressure of the meniscus is then

$$\gamma\frac{\mathrm{d}\sin\phi}{\mathrm{d}x} = \rho g h. \tag{5.71}$$

The integration of (5.71) is complicated, as it requires elliptic integrals. The form found by Princen is

$$\frac{d}{\kappa^{-1}} = 2k\ G(k,\,\theta), \tag{5.72}$$

where $k = 4/\left(\frac{h}{\kappa^{-1}}\right)^2 + 4$, and G is a combination of elliptic integrals.

b. Approximate approach

An approximate approach can be taken by balancing the capillary forces and gravitational forces for a meniscus. The capillary force on the plates' surfaces is

$$F_{\text{cap}} \cong 2w\,\gamma\cos\theta - 2\gamma d. \tag{5.73}$$

On the other hand, at equilibrium, the gravitational force is

$$F_{\text{grav}} \cong \rho g\; wd\; h_{\text{eq}}. \tag{5.74}$$

Note that relations (5.74) and (5.75) are only approximate because of the meniscal curvature in the direction parallel to the plates. At equilibrium, the two forces balance each other, yielding a capillary rise of

$$h_{\text{eq}} \cong 2\frac{\gamma}{\rho g}\left(\frac{\cos\theta}{d} - \frac{1}{w}\right). \tag{5.75}$$

Upon introducing the capillary length $\kappa^{-1} = \sqrt{\gamma/\rho g}$, the generalized Cassie angle $\cos\theta^{*} = (2w\cos\theta - 2d)/2(w + d)$, and the hydraulic radius $R_{\text{H}} = 2S/p = \frac{wd}{(w+d)}$, relation (5.75) becomes

$$h_{\text{eq}} \sim 2\kappa^{-2}\frac{\cos\theta^{*}}{R_{\text{H}}}. \tag{5.76}$$

c. Numerical approach

The most accurate approach is numerical. Figures 5.43(C) and (D) show the capillary rise between two identical, vertical parallel plates calculated with the software Evolver. The software Surface Evolver relocates the interface until it finds a minimum energy location, taking into account constraints such as gravity [15]. Using this software allows the capillary rise in more complex vertical structures to be determined, as illustrated in figure 5.44.

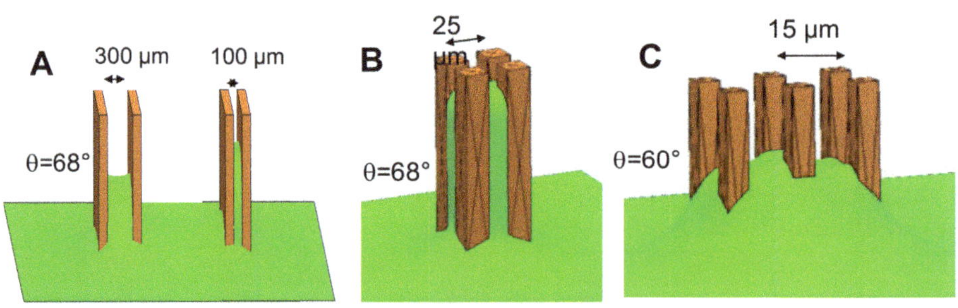

Figure 5.44. Surface Evolver results for the calculation of capillary rise in complex structures: (A) parallel vertical plates; (B) four square pillars arranged in a square pattern; and (C) six square pillars arranged in a rectangular pattern.

5.9.2 Uphill open-capillary flow

Capillary forces can also move droplets upwards. This takes place, for example, in the case of a droplet moving up a step or an inclined plane from a hydrophobic region to a hydrophilic region (figure 5.45) [28].

Open microflows can also move upwards, at least for some distance, in a similar way to capillary rise. Consider a rectangular open channel tilted upwards (figure 5.46). Once again, as in the case of capillary rise, the approach consists in

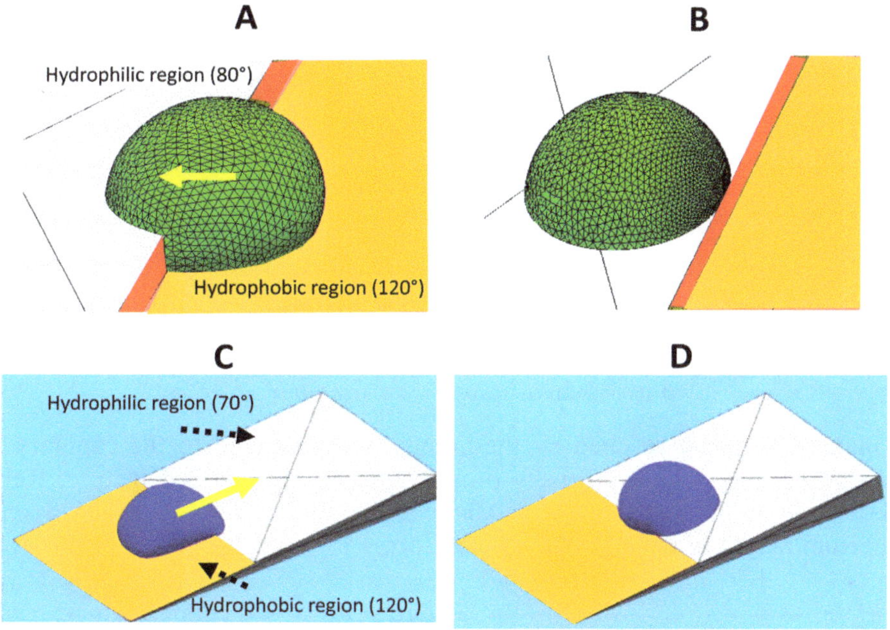

Figure 5.45. (A) and (B) A droplet moving up a step from a hydrophobic region to a hydrophilic region; (C) and (D) a droplet moving up a slanting plane from a hydrophobic region to a hydrophilic region (Evolver calculation).

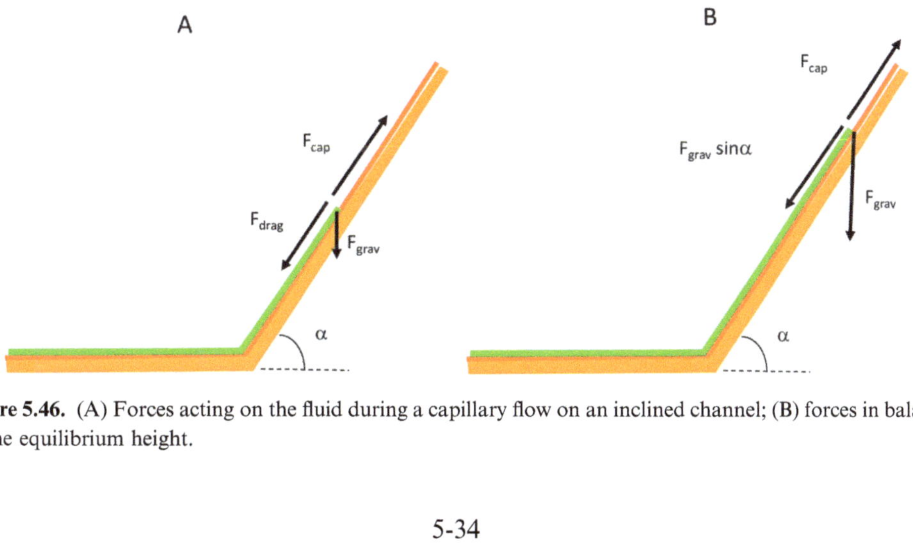

Figure 5.46. (A) Forces acting on the fluid during a capillary flow on an inclined channel; (B) forces in balance at the equilibrium height.

balancing the hydrostatic pressure with the capillary pressure. The hydrostatic pressure is

$$P_{\text{hydro}} = \rho\, g\, h, \tag{5.77}$$

and the hydrostatic gravity force—in the direction of the slanting plane—is

$$F_{\text{hydro}} = \rho\, g\, h\, w\, d \sin\alpha. \tag{5.78}$$

The capillary force must take account of the boundary with air and the wetting properties of the channel walls. The generalized capillary force is based on the generalized Cassie angle

$$F_{\text{cap}} = \gamma\, p \cos\theta^* = \gamma\,(2d + w)\cos\theta - \gamma w. \tag{5.79}$$

We recall that the generalized Cassie contact angle is defined by

$$\cos\theta^* = \sum_i \frac{p_{\text{W, i}}}{p}\cos\theta_i - \frac{p_{\text{F}}}{p} = \frac{(2d + w)\cos\theta - w}{2(d + w)}. \tag{5.80}$$

The balance between the two forces yields the equilibrium height

$$h_{\text{eq}} = \frac{\gamma}{\rho\, g}\frac{p\cos\theta^*}{wd\sin\alpha} = \frac{\gamma}{\rho\, g}\frac{(2d + w)\cos\theta - w}{wd\sin\alpha}. \tag{5.81}$$

The maximum height reached by the liquid is proportional to the capillary force $\gamma[(2d + w)\cos\theta - w]$ and inversely proportional to the liquid density ρ. Relation (5.81) is similar to that found by Abadi and Bahrami [29, 30], who wrote the equilibrium height in the nondimensional form

$$h^*_{\text{eq}} = \frac{h_{\text{eq}}}{\frac{\gamma}{\rho\, g\sin\alpha}\frac{p_W}{S_C}} = \frac{h_{\text{eq}}}{\frac{\gamma}{\rho\, g\sin\alpha}\frac{(2d + w)}{wd}} = \cos\theta - \frac{w}{(2d + w)} = \cos\theta - \frac{w}{p_W}. \tag{5.82}$$

An example of capillary uphill flow was given by Ghosh and coworkers [31] (figure 5.47). In this example, the slope of the channel was 8°. The channel walls were a composite of fluoroacrylic polymer coated with TiO_2 particles which became superhydrophilic upon exposure to UV light (390 nm).

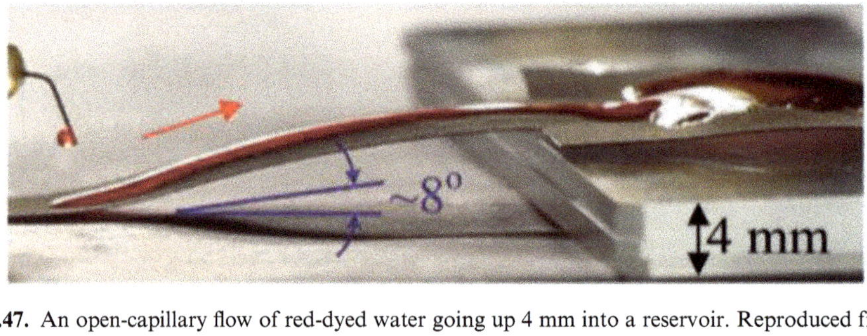

Figure 5.47. An open-capillary flow of red-dyed water going up 4 mm into a reservoir. Reproduced from [31] with permission of The Royal Society of Chemistry.

5.9.3 The dynamics of capillary rise

The dynamics of capillary rise has been widely investigated for vertical cylindrical tubes [32–37]. Studies of the capillary rise in arbitrary shaped closed channels and open channels are more recent [38–42].

5.9.3.1 The dynamics of capillary rise in cylindrical tubes

The dynamics of capillary rise inside cylindrical tubes was studied by Lucas and Washburn [32, 33]. By balancing the capillary forces, wall friction, and gravity—neglecting the evanescent inertial regime—and integrating the resulting equation, they obtained the relation linking time t to the meniscal height h [34]:

$$t(h) = -T\left[\frac{h}{L} + \cos\theta \ln\left(1 - \frac{h}{L\cos\theta}\right)\right], \tag{5.83}$$

with $T = \frac{16\gamma\mu}{\rho^2 g^2 R^3}$ and $L = \frac{2\gamma}{\rho g R}$. The dimension of T is time and that of L is length. For small times, using a Taylor expansion of the logarithm $\ln\left(1 - \frac{h}{L\cos\theta}\right) \cong -\frac{h}{L\cos\theta} - \frac{1}{2}\left(\frac{h}{L\cos\theta}\right)^2$, relation (5.83) becomes

$$h = \sqrt{2\frac{L^2}{T}\cos\theta\, t} = \sqrt{\frac{\gamma R\cos\theta}{2\mu}t}. \tag{5.84}$$

Over relatively small times, the effect of gravity does not contribute to the flow dynamics. But later in time, gravity limits the capillary rise and an asymptotic height is reached by the liquid (equilibrium height h_{eq}), as shown in figure 5.47. An approximate transition time t_c is determined by the intersection of relation (5.84) with the asymptote corresponding to the equilibrium height $h_{eq} = \frac{2\gamma\cos\theta}{\rho g R}$. Hence

$$t_c = \frac{8\gamma\mu\cos\theta}{\rho^2 g^2 R^3}. \tag{5.85}$$

Figure 5.48 graphically represents the determination of the transition time t_c and the dynamics of the capillary rise in the tube.

5.9.3.2 The dynamics of capillary rise in open channels

The dynamics of the vertical rise of liquids in open geometries is more complicated. We briefly indicate here some theoretical investigations. In the case of arbitrary open channels, the use of numerical methods is the most tractable approach.

A precursor study dating from 1980 was made by Levine [38], who managed to integrate the Navier–Stokes equation for the vertical flow and find a solution under the assumptions of inertialess motion, translation of the meniscus during the rise, and constant Laplace pressure of the meniscus. After long theoretical developments, Levine found a relation of the type: $dh/dt = \frac{A}{[h(t) + B]}$. However, tables are needed to obtain the value of B, which reduces the practicality of the solution. Besides, an additional integration is needed to obtain the expression for $h(t)$.

Ouali and coworkers and Ponce de Leon investigated the dynamics of capillary rise in rectangular open channels [16] and between two parallel plates with reliefs (figure 5.49) [42] by adding a gravity term to the Bosanquet equation.

The approach is very similar to that used for the cylindrical tube but with different coefficients. We recall the most general Bosanquet equation

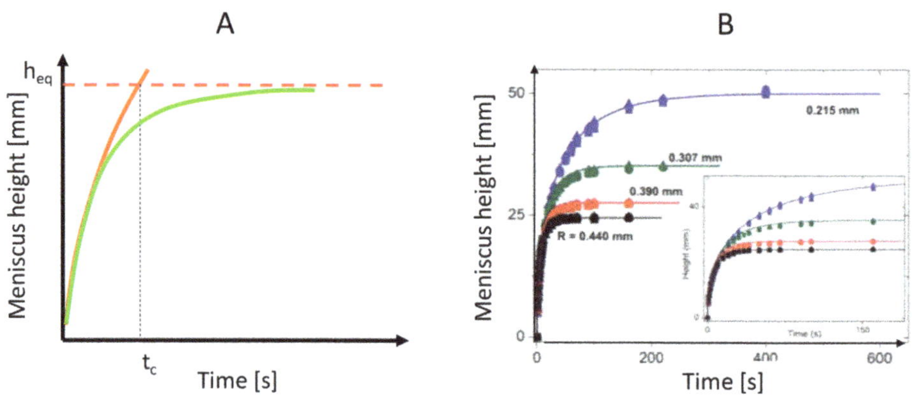

Figure 5.48. (A) Meniscal height vs. time. The orange line corresponds to relation (5.84), the green line to the real height, and the red dotted line to the asymptote. (B) Experimental results from O'Loughlin (85% glycerol) in vertical tubes of different radii. Reprinted from [36], Copyright (2013), with permission from Elsevier.

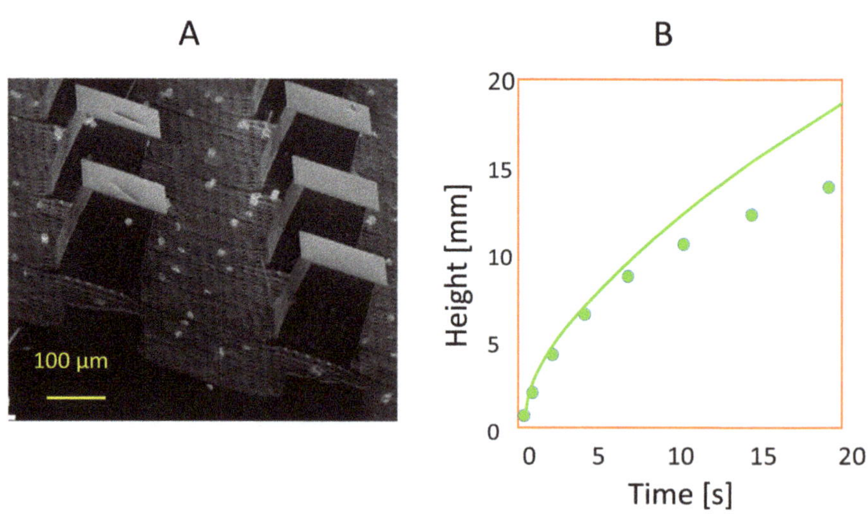

Figure 5.49. (A) Details of the reliefs on the plate. Reproduced from [42]. © IOP Publishing Ltd. CC BY 3.0. (B) Meniscal height vs. time: a comparison between the height calculated using the model without gravity (continuous line) and the experimentally observed height (dots).

$$\frac{1}{2}\frac{\mathrm{d}}{\mathrm{d}t}\left(\frac{\mathrm{d}z^2}{\mathrm{d}t}\right) = b - g\sin\alpha z - az\frac{\mathrm{d}z}{\mathrm{d}t}, \tag{5.86}$$

where z is the liquid height, b the capillary force, α the tilt angle, and a the wall friction coefficient. For a rectangular open channel, the coefficient a is semi-empirical. The first term on the left is inertia, the first term on the right-hand side is the capillary force, the second the gravity force, and the third the wall friction force. A nondimensional form of relation (5.86) is

$$\frac{\mathrm{d}^2 Z^2}{\mathrm{d}T^2} = 1 - G\sin\alpha\, Z - \frac{\mathrm{d}Z^2}{\mathrm{d}T}, \tag{5.87}$$

with $T = at$, $Z = az/\sqrt{2b}$, and $G = g\sqrt{2b}/a$. Ouali and coworkers found that there is first an inertial regime for which friction can be neglected, followed by a viscous–gravitational regime. The most accurate solution is numerical.

5.10 Conclusions

This chapter was devoted to the analysis of capillary flow in many different open-channel geometries. From a static point of view, it has been shown that the general SCF condition determines whether or not a capillary flow occurs in these geometries. From a dynamic point of view, the generalized/modified LWR law governs the velocity of the capillary flow. In particular, the notions of the generalized Cassie angle and the average friction length are required to access the dynamics of the flow.

The particular case of triangular channels has not been treated in this chapter because it is associated with capillary filament microflows and will be treated in chapter 7.

So far, uniform cross-section open channels have been investigated. In the following chapter, we present channels that have nonuniform cross-sections.

References

[1] Berthier J, Brakke K A and Berthier E 2016 *Open Microfluidics* (New York: Wiley)

[2] Casavant B P *et al* 2013 Suspended microfluidics *Proc. Natl Acad. Sci.* **110** 10111–1011

[3] Berthier J, Brakke K A, Gosselin D, Bourdat A-G, Nonglaton G, Villard N, Laffite G, Boizot F, Costa G and Delapierre G 2015 Suspended microflows between vertical parallel walls *Microfluid. Nanofluid.* **18** 919–29

[4] Humayun M, Chow C-W and Young E W K 2018 Microfluidic lung airway-on-a-chip with arrayable suspended gels for studying epithelial and smooth muscle cell interactions *Lab Chip* **18** 1298–130

[5] Berthier J, Theberge A B and Berthier E 2018 *Open-Channel Microfluidics (IOP Series)* (San Rafael, CA: Morgan & Claypool) 1st edn

[6] Tan S H, Nguyen N T, Chua Y C and Kang T G 2010 Oxygen plasma treatment for reducing hydrophobicity of a sealed polydimethylsiloxane microchannel *Biomicrofluidics* **4** 32204

[7] Satoh W, Hosono H and Suzuki H 2005 On-chip microfluidic transport and mixing using electrowetting and incorporation of sensing function *Anal. Chem.* **77** 6857–63

[8] Baret J-C, Decré M M J, Herminghaus S and Seemann R 2007 Transport dynamics in open microfluidic grooves *Langmuir* **23** 5200–4

[9] Lee S H, Heinz A J, Shin S, Jung Y-G, Choi S-E, Park W, Roe J-H and Kwon S 2010 Capillary based patterning of cellular communities in laterally open channels *Anal. Chem.* **82** 2900–6

[10] Bouaidat S, Hansen O, Bruus H, Berendsen C, Bau-Madsen N K, Thomsen P, Wolff A and Jonsmann J 2005 Surface-directed capillary system; theory, experiments and applications *Lab Chip* **5** 827–36

[11] Ouali F F, McHale G, Javed H, Trabi C, Shirtcliffe N J and Newton M I 2013 Wetting considerations in capillary rise and imbibition in closed square tubes and open rectangular cross-section channels *Microfluid. Nanofluid.* **15** 309–26

[12] Yang D, Krasowska M, Priest C, Popescu M N and Ralston J 2011 Dynamics of capillary-driven flow in open microchannels *J. Phys. Chem.* C **115** 18761–9

[13] Concus P and Finn R 1969 On the behavior of a capillary surface in a wedge *Proc. Natl Acad. Sci.* **63** 2.92299292E8–9

[14] Concus P and Finn R 1994 Capillary surfaces in a wedge—differing contact angles *Microgravity Sci. Technol.* **7** 1.52155152E8–5

[15] Brakke K A 1992 The surface evolver *Exp. Math.* **1** 141–65

[16] Ouali F F, McHale G, Javed H, Trabi C, Shirtcliffe N J and Newton M I 2013 Wetting considerations in capillary rise and imbibition in closed square tubes and open rectangular cross-section channels *Microfluid. Nanofluidics* **15** 309–26

[17] Liu L, Jacobi A M and Chvedov D 2009 A surface embossing technique to create micro-grooves on an aluminum fin stock for drainage enhancement *J. Micromech. Microeng.* **19** 035026

[18] Gosselin D, Belgacem M N, Joyard-Pitiot B, Baumlin J M, Navarro F, Chaussy D and Berthier J 2017 Low-cost embossed-paper micro-channels for spontaneous capillary flow *Sensors Actuators* B **248** 395–401

[19] Berthier J, Brakke K A, Gosselin D, Berthier E and Navarro F 2017 Thread-based microfluidics: flow patterns in homogeneous and heterogeneous microfiber bundles *Med. Eng. Phys.* **48** 55–61

[20] Wei Y-C, Fu L-M and Lin C-H 2013 Electrophoresis separation and electrochemical detection on a novel thread-based microfluidic device *Microfluid. Nanofluid.* **14** 583–90

[21] Ballerini D R, Li X and Shen W 2011 Flow control concepts for thread-based microfluidic devices *Biomicrofluidics* **5** 014105

[22] Princen H M 1969 Capillary phenomena in assemblies of parallel cylinders: I. Capillary rise between two cylinders *J. Colloid Interface Sci.* **30** 697569–75

[23] Bico J and Quéré D 2002 Rise of liquids and bubbles in angular capillary tubes *J. Colloid Interface Sci.* **247** 1.62166162E8–6

[24] Liu T, Choi K F and Li Y 2007 Capillary rise between cylinders *J. Phys. D: Appl. Phys.* **40** 5. 00650125006E11–12

[25] Trabi C L, Ouali F F, McHale G, Javed H, Morris R H and Newton M I 2016 Capillary penetration into inclined circular glass tubes *Langmuir* **32** 1.28912981289E11–98

[26] Jurin J 1719 An account of some experiments shown before the Royal Society; with an enquiry into the cause of the ascent and suspension of water in capillary tubes *Philos. Trans. R. Soc.* **30** 7.39747739E8–47

[27] Wu P, Zhang H, Nikolov A and Wasan D 2016 Rise of the main meniscus in rectangular capillaries: experiments and modeling *J. Colloid Interface Sci.* **461** 195–202

[28] Berthier J and Brakke K A 2014 *The Physics of Microdroplets* (Beverly, MA: Scrivener Publishing)

[29] Abadi G B and Bahrami M 2020 A general form of capillary rise equation in micro-grooves *Sci. Rep.* **10** 19709

[30] Abadi G B and Bahrami M 2022 The effect of surface roughness on capillary rise in micro-grooves *Sci. Rep.* **12** 14867

[31] Ghosh A, Ganguly R, Shutzius T M and Megaridis C M 2014 Wettability patterning for high-rate, pumpless fluid transport on open, non-planar microfluidic platforms *Lab Chip* **14** 1538–50

[32] Lucas R 1918 Ueber das Zeitgesetz des Kapillaren Aufstiegs von Flussigkeiten *Kolloid Z.* **23** 15

[33] Washburn E W 1921 The dynamics of capillary flow *Phys. Rev.* **17** 273–83

[34] Extrand C W and Moon S I 2013 Experimental measurement of forces and energies associated with capillary rise in a vertical tube *J. Colloid Interface Sci.* **407** 488–92

[35] Barry D A, Wissmeier L, Parlange J-Y, Sander G C and Lockington D A 2009 Comment on 'An analytic solution of capillary rise restrained by gravity' by N. Fries and M. Dreyer *J. Colloid Interface Sci.* **338** 293–5

[36] O'Loughlin M, Wilk K, Priest C, Ralston J and Popescu M N 2013 Capillary rise dynamics of aqueous glycerol solutions in glass capillaries: a critical examination of the Washburn equation *J. Colloid Interface Sci.* **411** 257–64

[37] Masoodi R, Languri E and Ostadhossein A 2013 Dynamics of liquid rise in a vertical capillary tube *J. Colloid Interface Sci.* **389** 268–72

[38] Levine S, Lowndes J, Watson E J and Neale G 1980 A theory of capillary rise of a liquid in a vertical cylindrical tube and in a parallel-plate channel: Washburn equation modified to account for the meniscus with slippage at the contact line *J. Colloid Interface Sci.* **73** 136–51

[39] Wang Z, Chang C-C, Hong S-J, Sheng Y-J and Tsao H-K 2012 Capillary rise in a microchannel of arbitrary shape and wettability: hysteresis loop *Langmuir* **28** 16917–26

[40] Gurumurthy V T, Roisman I V, Tropea C and Garoff S 2018 Spontaneous rise in open rectangular channels under gravity *J. Colloid Interface Sci.* **527** 151–8

[41] Wang H J, Tsai H C, Chen H K and Shing T K 2005 Capillary of rectangular micro grooves and their application to heat pipes *J. Appl. Sci. Eng.* **8** 249–55

[42] Ponce de Leon P J and Velásquez-García L F 2016 Optimization of capillary flow through open-microchannel and open-micropillar arrays *J. Phys.* D **49** 055501

IOP Publishing

Open-Channel Microfluidics (Second Edition)
Fundamentals and applications
Jean Berthier, Ashleigh B Theberge and Erwin Berthier

Chapter 6

Capillary filaments

6.1 Introduction

Capillary filaments (CFs) are often observed at the front end of capillary flows, where they progress quickly in the sharp inner corners of the channels. They are sometimes called 'fingers' in the literature. In essence, these filaments are flowing 'openly' as they extend in the corners. Figure 6.1 shows capillary filaments of dodecane in a rectangular open channel milled in poly(methyl methacrylate) (PMMA) and whole blood in a rectangular channel treated with O_2 plasma to make it hydrophilic.

The conditions for the formation of capillary filaments were initially investigated by Concus and Finn in the 1950s [1, 2]. In this section, we first present the conditions under which these filaments form. Next, the behavior of CFs in loose and sharp channel turns is revealed. Then, the metastable behavior of CFs in nonuniform channels is investigated. Finally, first thoughts about the filament velocity are given.

6.2 Capillary filaments: the Concus–Finn condition

Liquid interfaces in contact with highly wetting solid walls that form a sharp wedge spread inside the corner. This behavior results from the fact that the interface curvature is strongly reduced in the corner (in figure 6.2, the vertical curvature radius is small and the horizontal curvature radius is imposed by the two walls). As a result, the Laplace pressure is low in the corner area and liquid spreads in the corner. Another way to look at this phenomenon is to observe that the tip of an equilibrium filament in a wedge must have a tangent plane that satisfies the contact angle requirement on both walls. Given this fact, it is immediately clear that if the wedge angle is more than twice the contact angle, there cannot be a tip tangent plane that differs from each wall normal by the contact angle.

Concus and Finn developed a theory for these capillary filaments and produced a very simple criterion for capillary motion in the corner of the wedge [1–3].

doi:10.1088/978-0-7503-5507-0ch6

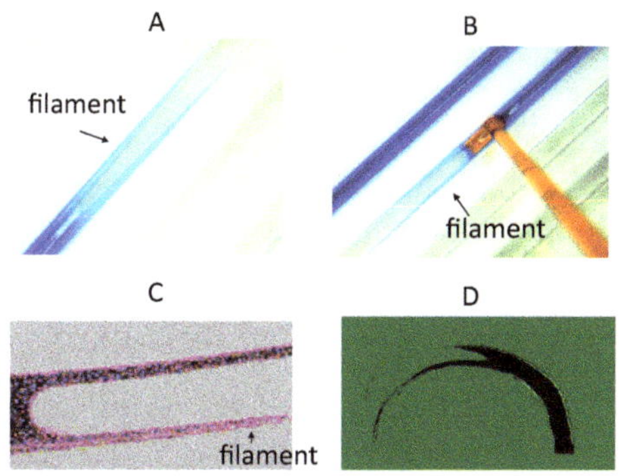

Figure 6.1. (A) Precursor capillary filaments of 'blue oil' in the bottom corner of a rectangular open channel milled in PMMA; (B) the same flow with the addition of an immiscible plug (the filaments stay ahead of the flow); (C) and (D) filaments of colored isopropyl alcohol (IPA) in PMMA open channels. Photos (A) and (B) courtesy of Jing Lee and Jean Berthier, University of Washington; photos (C) and (D) courtesy of David Gosselin and Jean Berthier.

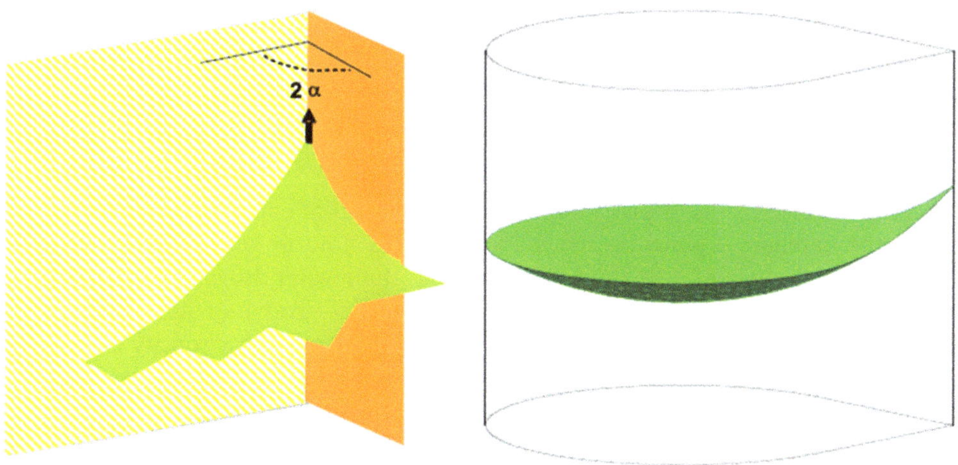

Figure 6.2. A liquid interface is lifted in the corner of a wedge made of two wetting plates. This phenomenon is due to the need for a common tangent plane that makes the proper contact angles on the two walls. The tube-with-wedge on the right has a wedge angle of $2\alpha = 90°$ and a contact angle of 60°.

If ï ± is the Young contact angle with the two planes and α the wedge half-angle, the condition for capillary self-motion leading to an endlessly growing filament—in the absence of gravity—is

$$\theta < \frac{\pi}{2} - \alpha. \tag{6.1}$$

The Concus–Finn condition has been extended to wedges and corners of different wettability of each plane, θ_1 and θ_2 [3]:

$$\theta_1 + \theta_2 < \frac{\pi}{2} - \alpha. \tag{6.2}$$

Such a configuration frequently occurs for closed channels for which the top cover is made of a different material than the channel itself, for example in the case of a glass cover on top of a silicon or plastic substrate.

6.3 The case of V-grooves

Bulk capillary flow and capillary filaments are found together in triangular open channels and open V-grooves. Let us write the spontaneous capillary flow (SCF) condition for a triangular channel using the notations of figure 6.3:

$$\frac{p_{\mathrm{F}}}{p_{\mathrm{W}}} = \frac{w}{2d} < \cos\theta \tag{6.3}$$

Using the half-angle of the wedge α, condition (6.3) becomes

$$\sin\alpha < \cos\theta, \tag{6.4}$$

which can be transformed into the Concus–Finn condition

$$\theta < \frac{\pi}{2} - \alpha. \tag{6.5}$$

Hence, in open V-grooves, the capillary flow condition is the same as the Concus–Finn condition.

V-groove geometry is of great interest for the design of capillary systems, because, due to the capillary filaments, the velocity of the flow is fast compared to other possible geometries, and even fluids more viscous than water move at a relatively high velocity. This is the case, for example, for the sharp V-groove of figures 6.4(A) and (B), which is used to move whole blood (whose non-Newtonian viscosity is four times

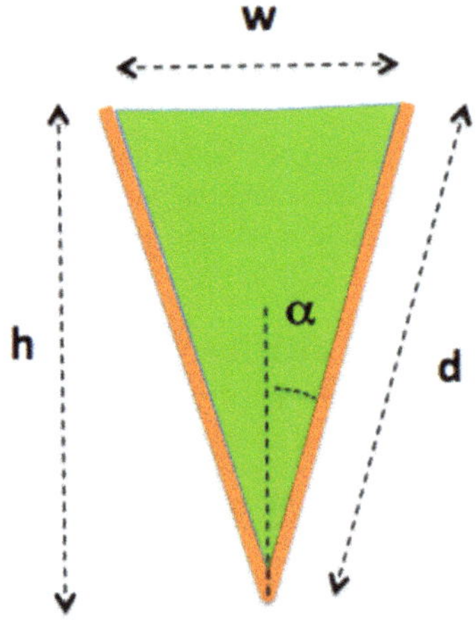

Figure 6.3. A sketch of a V-groove.

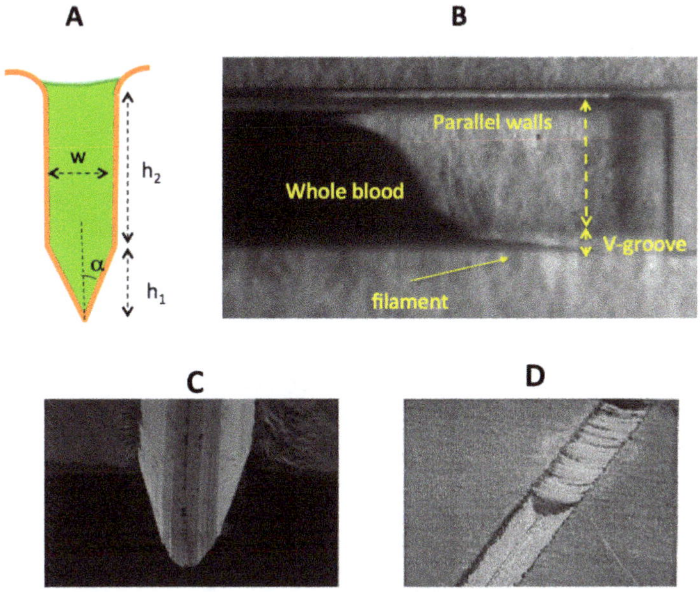

Figure 6.4. (A) A schematic of the cross-section of a V-groove [4]. (B) Whole blood capillary filling of the device, showing the capillary filament at the bottom in the triangular region and the bulk between the parallel plates. Courtesy of P Pouteau, Avalun SA. (C) A cross-section of one of the first V-grooves used by Rye and coworkers, who remarked that this V-groove does not have a sharp interior corner. Reprinted with permission from [6]. Copyright (1998) American Chemical Society. (D) Liquid solder flowing in the V-groove. Reprinted from [7], Copyright (1997), with permission from Elsevier.

that of water when the flow is sufficiently fast) [4]. It is also the case for liquid metal flowing in V-grooves to make electrodes after cooling: this type of capillary flow in open triangular channels has been documented by Rye, Mann, Yost, and Romero for the flow of liquid solder (figures 6.4(C) and (D)) [5–7]. They have shown that the Lucas–Washburn–Rideal (LWR) law for the travel distance vs. time is valid under the condition that a geometrical coefficient is introduced in the formulation [8–10]. Kitron-Belinkov and coworkers studied the free surface of the capillary filament [11].

6.4 Capillary filaments in open-channel turns

Straight channels are not frequently used for biotechnical devices, because their length is not compatible with the format of the device. Most of the time, the channels form zigzags (figure 6.5). If an SCF without capillary filaments is unaffected by the turns, it is not the case of capillary filaments: the interior filament precedes the exterior filament [12]. If turns in an open channel do not affect the spontaneous capillary flow, the situation changes when capillary filaments are present; the inner filament comes before the outer filament.

Indeed, in a channel where no filament exists—as is the case for rounded rectangular channels or suspended channels—the front meniscus keeps its shape all along the turn, as if it were a rotation. Figure 6.6 shows such behaviors for rectangular and suspended channels.

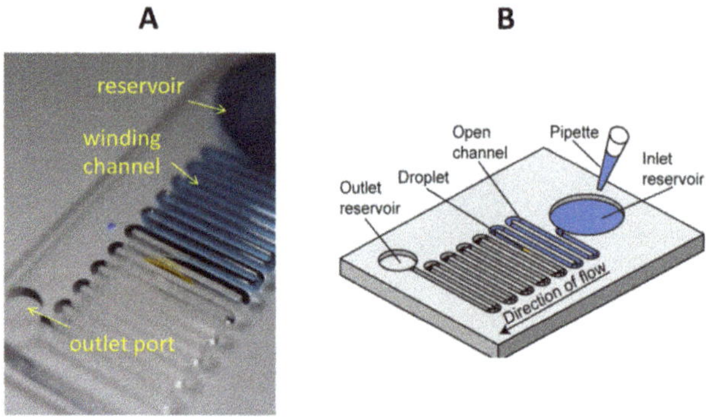

Figure 6.5. (A) A typical open-channel device with zigzags; (B) details of the device. Photograph courtesy of Jing Lee, University of Washington.

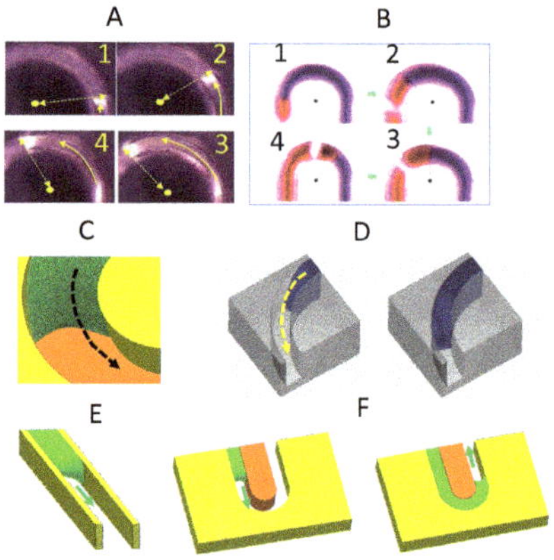

Figure 6.6. (A) and (B) Different locations of the front meniscus during the capillary flow of blood in a rectangular open channel (photograph courtesy of Myriam Cubizolles and Jean Berthier). (C) and (D) Details of the meniscus predicted by Surface Evolver software and Flow3D. (E) and (F) Locations of the front meniscus in a turning suspended channel predicted by Surface Evolver.

The picture is quite different in the case of capillary filaments, as shown in figure 6.7. The interior filament seems to precede the exterior filament. In fact, the lengths of the filaments are similar. Due to the curvature of the channel, in a uniform cross-section channel, the interior filament appears to precede the exterior filament, but the exterior filament catches up after a turn.

Interestingly, in the case of non-homogeneous liquids—such as liquids carrying particulates or cells—the capillary filaments discriminate between the less viscous components and the most viscous components. For example, in the case of whole blood, the blood plasma—whose viscosity is a little higher than that of water viscosity—concentrates in the filaments, while the red blood cells (RBCs) mostly remain in the bulk of the capillary flow (figure 6.8).

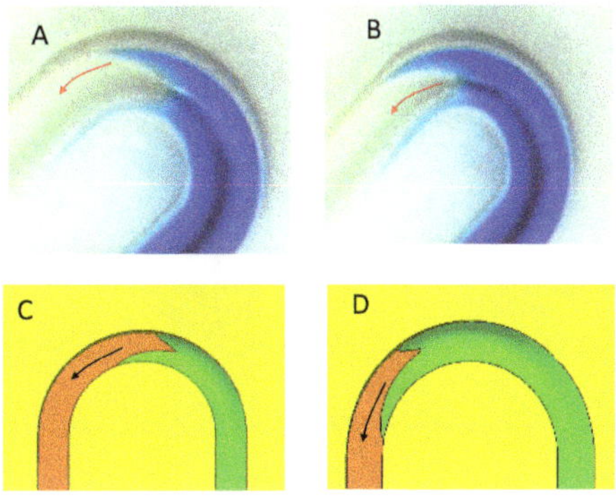

Figure 6.7. (A) and (B) Colored pentanol flowing in a turn of a rectangular open channel (photograph courtesy of Jing J Lee and Jean Berthier); (C) and (D) the shapes of the advancing meniscus predicted by the Surface Evolver software.

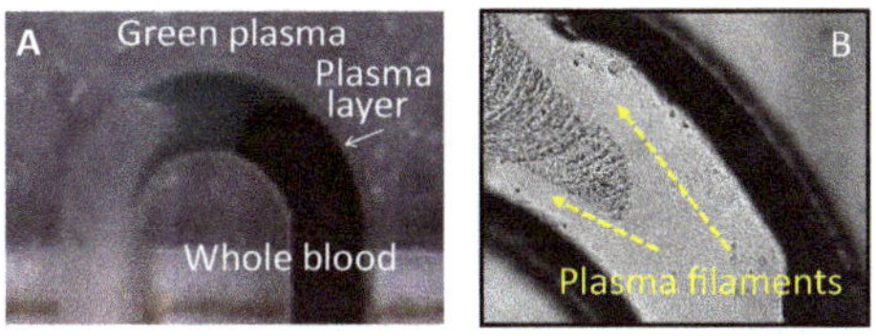

Figure 6.8. (A) Blood plasma—colored green—flows ahead of the whole blood in a turning rectangular open channel (photograph courtesy of Myriam Cubizolles and Jean Berthier). (B) Details of blood plasma filaments in a turning open channel. Milling marks can be seen on the channel bottom between the advancing filaments (photograph courtesy of David Gosselin and Myriam Cubizolles).

6.5 Capillary filaments in nonuniform channels

Capillary filaments are somewhat independent from the global geometry of the channel. They flow endlessly if the Concus–Finn condition is satisfied by the wedge in which they flow and if there is enough liquid available.

Let us consider a channel of nonuniform cross-section, as shown in figure 6.9. In the case of an asymmetric enlargement (figures 6.9(A) and (B)), the interior filament precedes the exterior filament. The exterior filament catches up later with the interior filament in the channel beyond the chamber. Figures 6.9(C) and (D) show the filaments progressing along the walls, while the interior of the chamber stays dry.

6.6 Detached capillary filaments

Capillary filaments may exist by themselves or may progress independently from the bulk of the liquid [3, 13–15]. Figure 6.10 illustrates the case of capillary filament flowing alone in interstices; it shows the spread of liquid glue—which is very viscous —in nanoscopic interstices.

Consider now the case of a straight, rectangular U-groove that has a large aspect ratio, i.e. $\varepsilon = w/h \gg 1$, where w is the channel width and h is the channel depth.

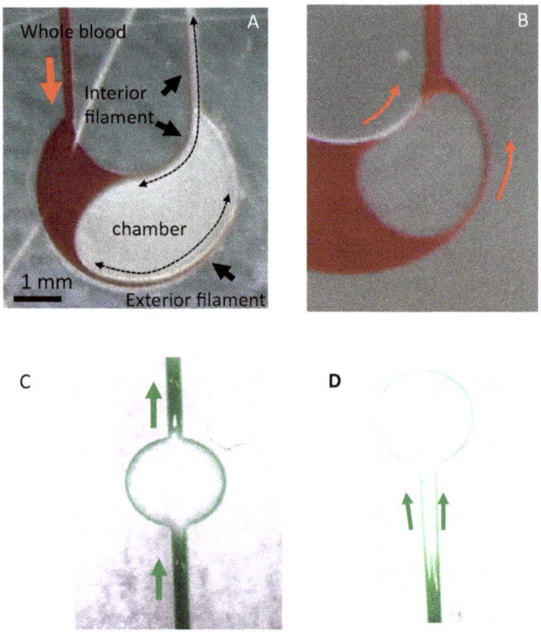

Figure 6.9. Capillary filaments in open channels of nonuniform cross-section. (A) and (B) Whole blood capillary filaments in an enlargement; due to the particular geometry, the interior filament is the first to penetrate the channel after the chamber. The exterior filament catches up later. (C) Blood plasma—colored green—separating into two capillary filaments in a large chamber, which merge later in a constricted channel. (D) Blood plasma filaments—colored green—separating in a diverging channel. Photographs courtesy of Jean Berthier and Myriam Cubizolles.

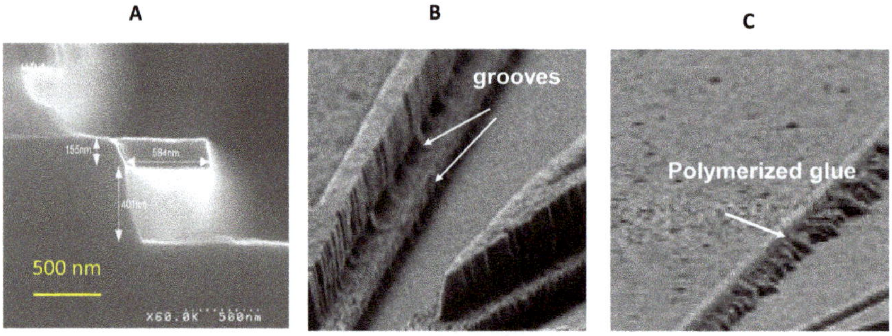

Figure 6.10. (A) A view of a nano-channel etched in silicon, open on its side; (B) liquid glue has spread in the grooves and gelled; (C) a detailed view of the polymerized glue alongside the nano-groove. Photographs courtesy of François Boizot, CEA-Leti.

This morphology has been experimentally investigated and reported in the literature [13]. In that publication, the rectangular cross-section channel had a height $h = 8.8$ µm and four different widths: w = 160, 180, 200, and 220 µm (figure 6.11). For all these, the aspect ratios were greater than 18, namely ε = 18.18, 20.45, 22.73, and 25. The inlet pressure was supposed to be zero (a large reservoir with a flat liquid surface).

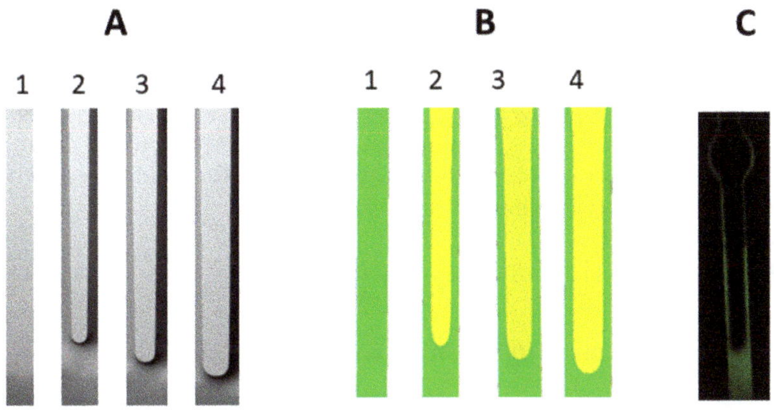

Figure 6.11. Capillary filaments progressing ahead of the bulk flow. (A) and (B) A comparison between experimental results [13] and Surface Evolver numerical results [15, 16]. Channel dimensions: the wall height is 8.8 μm, the channel widths are 1–160, 2–180, 3–200, and 4–220 μm. The contact angle is 14°. Figure (A) reprinted from [13], Copyright (2008), with permission from Springer. (C) Capillary filaments of fluorescein progressing ahead of the bulk flow in a silicon channel (photograph courtesy of D Gosselin and Jean Berthier, CEA-Leti).

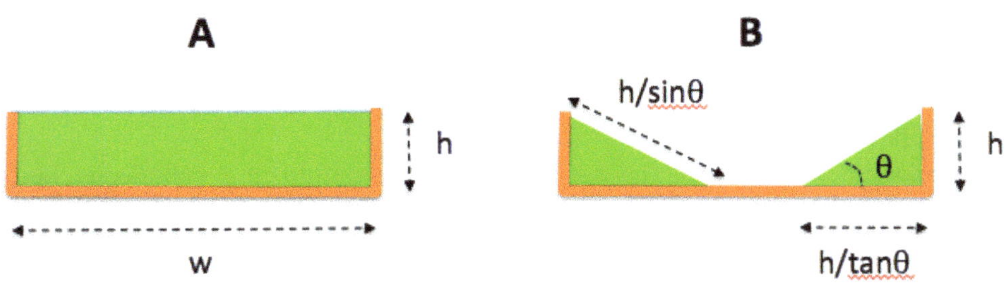

Figure 6.12. The two morphologies of capillary flow in a high-aspect-ratio channel: (A), completely filled channel; (B), partly filled channel. Note that because the pressure is zero (flow from an infinite flat reservoir), all the surfaces are flat.

Two morphologies of the flow are observed, depending on the value of the aspect ratio. In the first case (#1 in figure 6.11), which has the smallest aspect ratio, the channel is filled by the fluid and two filaments advance ahead of the bulk flow. In the second case (#2–4 in figure 6.11), the bulk flow is stopped but two filaments spread endlessly in the corners. Hence, filaments can advance endlessly (assuming there is enough liquid available), even if the front end of the bulk liquid stays immobile.

Let us consider the two following cases illustrated in figure 6.12: (1) a totally wetted channel, (2) a non-wetted channel center. In the first case, the surface energy variation over a distance Δx is

$$\Delta E_1 = \Delta x\,[(2h + w)\,\gamma_{SL} + w\,\gamma_{LG}], \tag{6.6}$$

while, in the second case, it is

$$\Delta E_2 = \Delta x\left[\left(w - 2\frac{h}{\tan\theta}\right)\gamma_{SG} + 2\left(h + \frac{h}{\tan\theta}\right)\gamma_{SL} + \left(2\frac{h}{\sin\theta}\right)\gamma_{LG}\right]. \quad (6.7)$$

The sign of $\Delta E_1 - \Delta E_2$ indicates whether the center of the channel is wetted or not. Subtracting (6.6) from (6.7) yields

$$\Delta E = \Delta E_2 - \Delta E_1 = \Delta x\left[\left(w - 2\frac{h}{\tan\theta}\right)\gamma_{SG} + \left(\frac{2h}{\tan\theta} - w\right)\gamma_{SL} + \left(2\frac{h}{\sin\theta} - w\right)\gamma_{LG}\right]. \quad (6.8)$$

Using Young's law, relation (6.8) can be written as

$$\Delta E = \Delta E_2 - \Delta E_1 = \Delta x\,\gamma_{LG}\left[\left(w - 2\frac{h}{\tan\theta}\right)\cos\theta + \left(2\frac{h}{\sin\theta} - w\right)\right]. \quad (6.9)$$

The center of the channel is not wetted if $\Delta E = \Delta E_2 - \Delta E_1 < 0$, i.e.

$$\varepsilon = \frac{w}{h} > \frac{2\sin\theta}{1 - \cos\theta}. \quad (6.10)$$

If relation (6.10) is satisfied, the smaller surface energy morphology is that of two separate CF filaments. In this case, the bulk flow is stopped just after the outlet of the reservoir, but two capillary filaments flow separately alongside the walls. Otherwise, if relation (6.10) is not satisfied, then the smaller surface energy morphology is that of the fully wetted channel. Note that in this case, there are two CF filaments ahead of the bulk flow.

An experimental view of the capillary flow of a liquid in a cylinder etched in silicon is shown in figure 6.13, using water marked with a fluorescent dye. The aspect ratio of the channel is $\varepsilon = w/h = 1/4$. In this particular case, the outlet channel aspect ratio does not satisfy (6.10) and the two filaments merge.

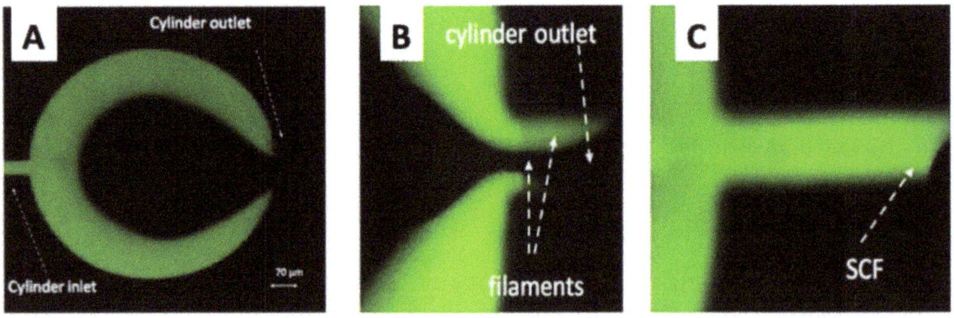

Figure 6.13. Experimental views of fluorescent water flow in a wide cylinder; (A) the flow separates into two filaments that advance alongside the curved corners (forming a catenoidal shape); (B) the two capillary filaments separately enter the cylinder outlet; (C) the filaments grow laterally to reach their nominal width; because relation (6.10) is not satisfied, the two filaments merge and a bulk SCF is observed. Photograph courtesy of Jean Berthier and David Gosselin, CEA-Leti.

6.7 Metastable capillary filaments

Now consider case B of figure 6.12: in the case in which the fluid inlet consists of two separated flows, two separated capillary filaments advance in parallel in the open channel. Using a trigonometric formulation for the free perimeter p_F and the wetted perimeter p_W (figure 6.12(B)), we find that the SCF for a filament is

$$\cos\theta > \frac{p_F}{p_W} = \frac{h}{\sin\theta}\frac{1}{\left(h + \frac{h}{\tan\theta}\right)} = \frac{1}{\sin\theta + \cos\theta}, \tag{6.11}$$

which—upon multiplying both sides of (6.11) by $\sin\theta + \cos\theta$ and using the trigonometric identity $\sin^2\theta = 1 - \cos^2\theta$—can be simplified to

$$\cos\theta > \sin\theta. \tag{6.12}$$

The condition for motion is then $\theta < \pi/4$, which is simply the Concus–Finn condition for perpendicular corners. On the other hand, if we again use the rectangular shape of the profile in a cross-section, the condition for the two filaments to stay separated, i.e. not to touch each other, is

$$\varepsilon = \frac{w}{h} > \frac{2}{\tan\theta}. \tag{6.13}$$

An interesting observation stems from the comparison of relations (6.10) and (6.13). Whatever the value of θ, the inequality

$$\frac{2}{\tan\theta} < \frac{2\sin\theta}{1 - \cos\theta} \tag{6.14}$$

is always true. If the value of the aspect ratio ε is between the two limits of inequality (6.14), i.e.

$$\frac{2}{\tan\theta} \leqslant \varepsilon \leqslant \frac{2\sin\theta}{1 - \cos\theta}, \tag{6.15}$$

the channel can sustain either a bulk SCF flow that fills the channel or two separate capillary filaments. The left-hand side of the double inequality (6.15) is the condition for the two filaments to stay separated, and the right-hand side is the condition for the wetting of the channel center.

The full and filament modes can change in a single channel. This property is illustrated in figure 6.14, where a large-aspect-ratio open channel ($\varepsilon = w/h$) has been enlarged by a cylindrical well of the same depth h as the channel. When the liquid exits the large reservoir, the channel completely fills, and the middle of the channel is covered with liquid because condition (6.10) is not satisfied. When the SCF enters a cylindrical enlargement, the flow separates if the cylinder radius is sufficiently large. The condition for observing a dry disc in the middle is given in [15], based on the catenoidal shape of the liquid surface. The flow is separated into two filament subflows, and if the cylinder outlet aspect ratio satisfies (6.15), two independent capillary filaments continue to flow in the straight outlet channel.

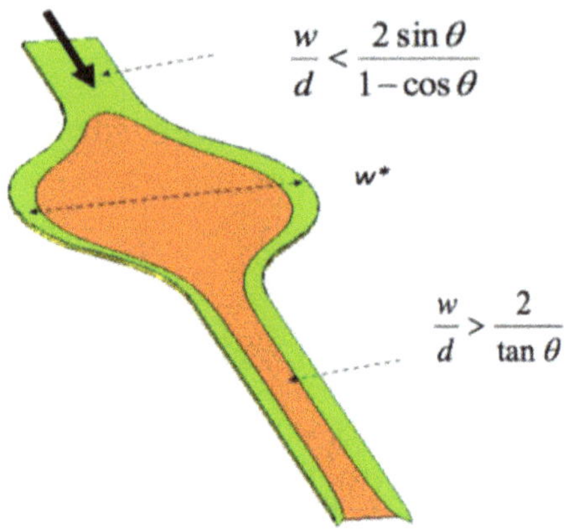

Figure 6.14. A capillary flow in a shallow open channel: at the top, the liquid enters the whole channel and the channel is progressively filled with liquid. The large cylindrical part acts as a separator for the flow, and two filaments continue in the channel past the cylinder.

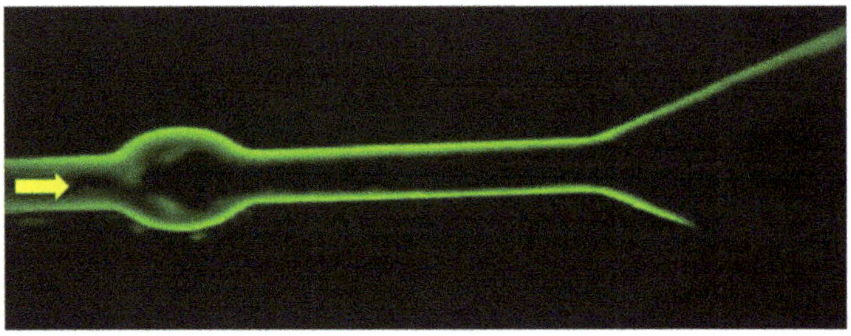

Figure 6.15. A capillary flow of fluorescein separates into two filaments after an enlargement. The rectangular open channel has an aspect ratio of five and is etched in silicon. Photograph courtesy of Jean Berthier and David Gosselin, CEA-Leti.

The Evolver calculation is illustrated by the experimental photograph shown in figure 6.15.

The two filaments are independent, they never merge, and the middle of the channel is never wetted. However, this morphology is metastable: if the two filaments are forced to contact each other, the fully wetted situation is restored, as shown in figure 6.16. Note that the restoration of the fully wetted morphology also progresses backward towards the large cylinder.

The corresponding experimental result is shown in figure 6.17: a neck in the channel forces the separated subflows to merge. The merged part then extends in both directions, upstream and downstream.

We have seen in chapter 3 that for contact angles greater than 45°, the SCF condition in a open U-channel of rectangular cross-section is

$$\varepsilon = \frac{w}{h} < \frac{2\cos\theta}{1 - \cos\theta}. \tag{6.16}$$

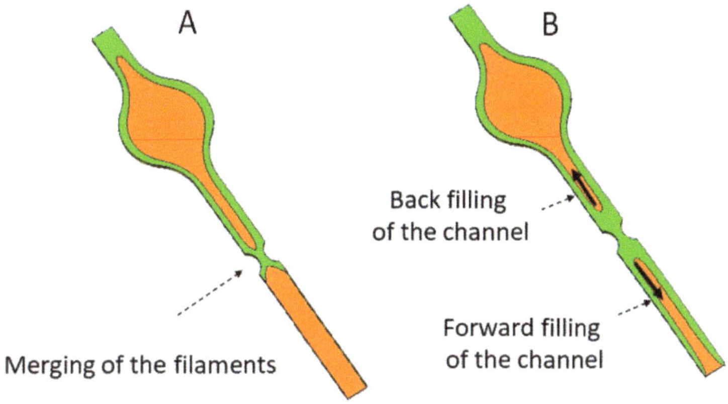

Figure 6.16. The same conditions as those shown in figure 6.14, but the two filaments are forced to contact downstream in a narrow section, and a fully wetted morphology is restored.

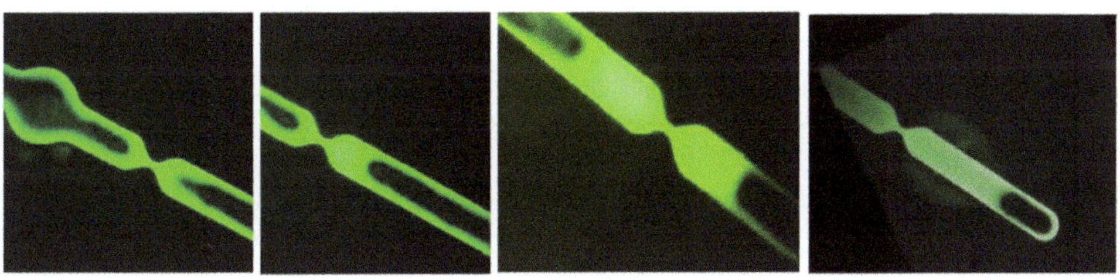

Figure 6.17. Capillary filaments of fluorescent water forced to merge at a neck in the channel. The substrate is made of PMMA, treated to become hydrophilic by exposition to oxygen plasma. The contact angle is approximately 30°, the height of the walls is 100 μm, and the width of the channel is 500 μm. The aspect ratio is 5 and satisfies the metastability condition (6.15). Photograph courtesy of Jean Berthier and David Gosselin, CEA-Leti.

In such a case, there is no formation of capillary filaments because the Concus–Finn limit is not respected. If an aspect ratio does not comply with condition (6.16), neither bulk SCF nor capillary filaments are observed.

When the contact angle is smaller than 45°, the situation is more complicated:

(1) Above the limit given by (6.10) $\varepsilon = w/h > 2\sin\theta/(1 - \cos\theta)$, no bulk SCF is observed, but two CF filaments spread in the corners.
(2) Below the limit given by (6.13) $\varepsilon = w/h < 2/\tan\theta$, bulk SCF is observed preceded by CF filaments.
(3) For an aspect ratio between $2/\tan\theta < w/h < 2\sin\theta/(1 - \cos\theta)$, the situation is metastable and depends on the history of the flow: either separated CF filaments progress in the channel, or full SCF occurs preceded by two CF filaments.

All the above conditions are shown in figure 6.18.

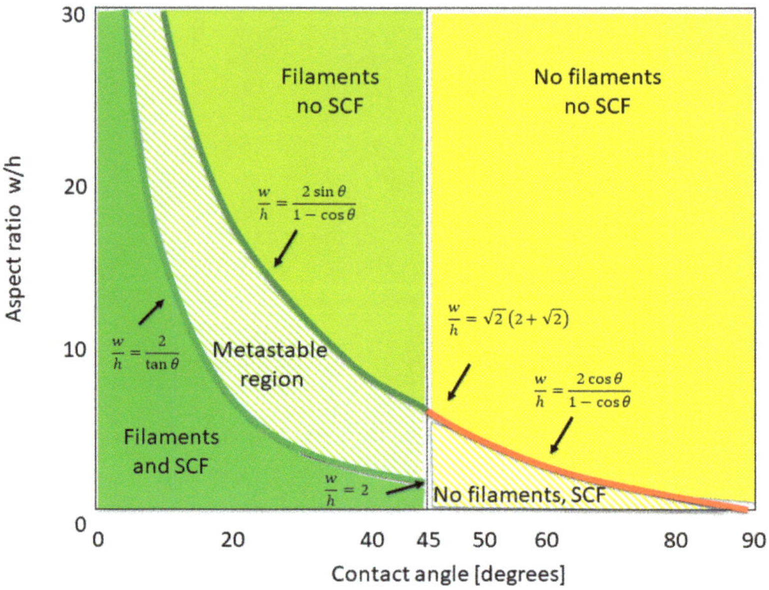

Figure 6.18. The different capillary flow domains for a rectangular, open channel: the notation 'SCF' means that the bulk is advancing in the channel, and the notation 'filaments' means that two CF filaments progress alongside the interior corners. Above a contact angle of 45°, there are no filaments; at smaller angles, there are always filaments.

6.8 Capillary filaments driving spontaneous capillary flow

We have seen that liquid pinning on sharp edges blocks the capillary flow if the triple line of the advancing meniscus is completely anchored (chapter 3). For open-surface capillary flow, this means that the contact line of the liquid with the solid is completely anchored on the sharp edges. An interesting observation is that capillary flow is only delayed if there is a geometrical possibility of the formation of capillary filaments, under the assumption that the enlargement aspect ratio w_e/h still satisfies the SCF condition, i.e.

$$\frac{p_F}{p_W} = \frac{w_e}{w_e + 2h} < \cos\theta. \tag{6.17}$$

Figure 6.19 shows a capillary flow of fluorescein-tinted water reaching a sudden enlargement. The bulk flow is first stopped while filaments form on both sides. The pinning is then lost, and the flow suddenly restarts.

6.9 The dynamics of capillary filaments

Little is known about the dynamics of capillary filaments. Two difficulties arise:

1. First, in the case of an autonomous capillary filament, i.e. a capillary filament progressing alone in a corner, the cross-sectional area occupied by the liquid decreases from the inlet (the beginning of the filament) towards the

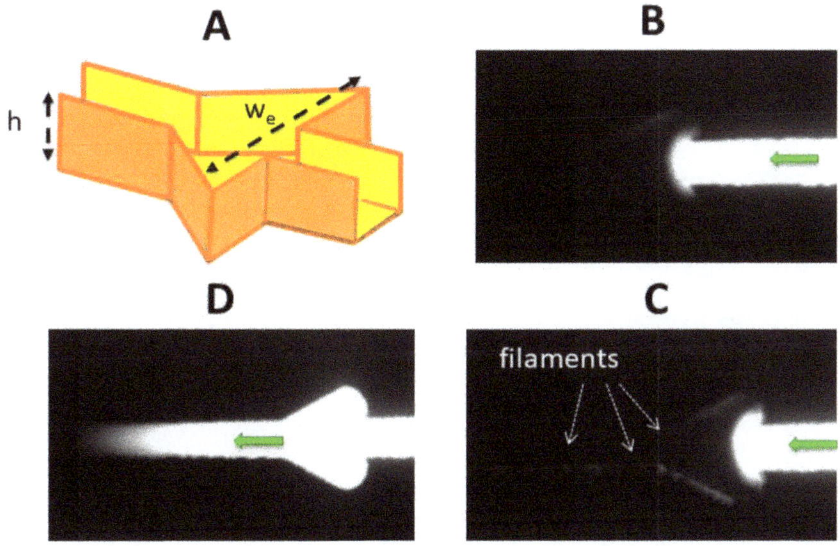

Figure 6.19. (A) A schematic of the channel; (B) the capillary flow of fluorescein is first stopped at the sudden enlargement; (C) capillary filaments progress along the corners at each side of the channel; (D) SCF suddenly restarts. Photographs courtesy of Jean Berthier and Noémie Villard, CEA-Leti.

tip. Because the filament is not in equilibrium or even in quasi-equilibrium, the cross-sectional shape of the filament is difficult to determine and depends on the dynamics. Hence, the frictional and capillary forces that govern the motion are difficult to determine.

2. Second, in the case of co-flowing filaments and bulk liquid, the interaction between the bulk velocity and the filament velocity is an unknown parameter of the problem.

However, it seems that the relation between the travel distance and time is still a square-root law, as in the case of the LWR law. We relate here some experimental results for an SCF preceded by filaments. We first consider the capillary flow of toluene in a rectangular open channel that has a width of 0.8 mm and a depth of 1.0 mm milled in a PMMA substrate and monitor the locations of the filament tip and the bulk liquid. The results are displayed in figure 6.20. The two travel distances follow the square-root law $z_b \approx A_b\sqrt{t}$ and $z_f \approx A_f\sqrt{t}$—where the indices b and f denote the bulk and the filament, respectively—so that the length of the growing filament is $z_f - z_b \approx \left(A_f - A_b\right)\sqrt{t}$ and the velocity of the growing filaments decreases as $1/\sqrt{t}$. Note that in the case of figure 6.20, the bulk and the filaments advance together. The velocity of detached filaments—filaments that flow while the bulk is not flowing—is still to be investigated. However, similar behavior has been observed for the capillary flow in a V-groove, on a smaller timescale [4].

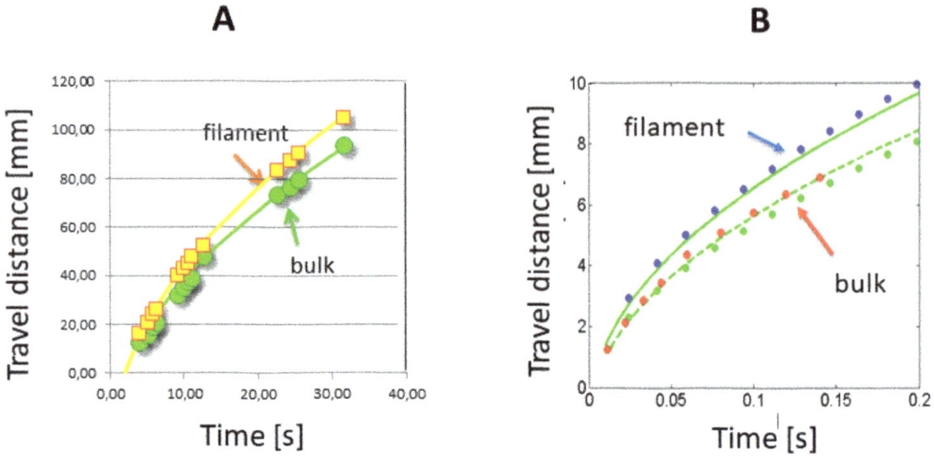

Figure 6.20. (A) A comparison of the travel distances of the filament tip and the bulk flow in a rectangular open channel for the capillary flow of pentanol in a PMMA channel (case of pentanol flowing in a rectangular open channel of width 0.8 mm and depth 1.0 mm). (B) A comparison of the travel distances of the filament tip and the bulk flow for whole blood in a very narrow V-groove [4].

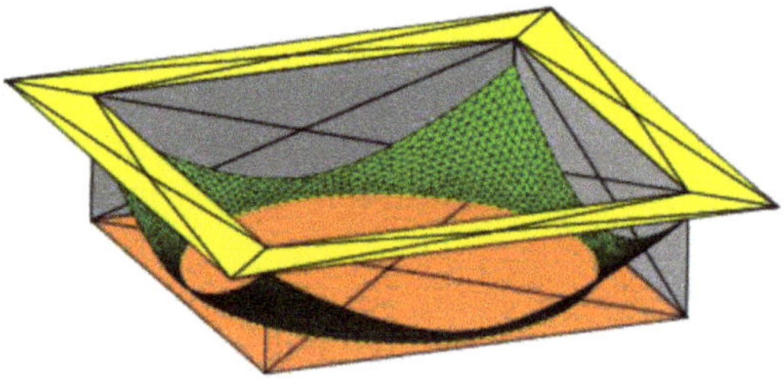

Figure 6.21. A model of evaporating liquid in a rectangular open well: the liquid retreats into the corners [3].

6.10 The drying of capillary filaments

When a liquid flowing—or at rest—in an open channel with sharp angles is left to evaporate, the evaporating liquid concentrates progressively in the corners, leaving only highly concentrated filaments [17]. This property has been checked with the numerical software Surface Evolver, as shown in figure 6.21.

This behavior is experimentally illustrated in figure 6.22, where the evaporating capillary flow in the V-groove has left a deposit of trehalose in the bottom corner [18]. Note that trehalose is an interesting sugar used in medicine, especially for dry eye treatment [19].

Similarly, when an open capillary flow of whole blood in a rectangular open channel dries—in fact the blood plasma evaporates—it concentrates the RBCs in the corners, as shown in figure 6.23.

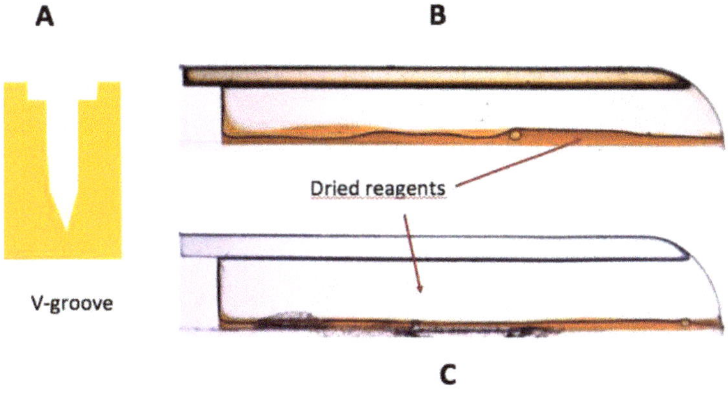

Figure 6.22. (A) A schematic of the cross-section of the channel; (B) a deposit of dried trehalose (brown) in the bottom corner produced by a highly concentrated solution; (C) a deposit of dried trehalose (brown) in the bottom corner produced by a low-concentration solution. Photographs courtesy of Maxime Huet, CEA-Leti.

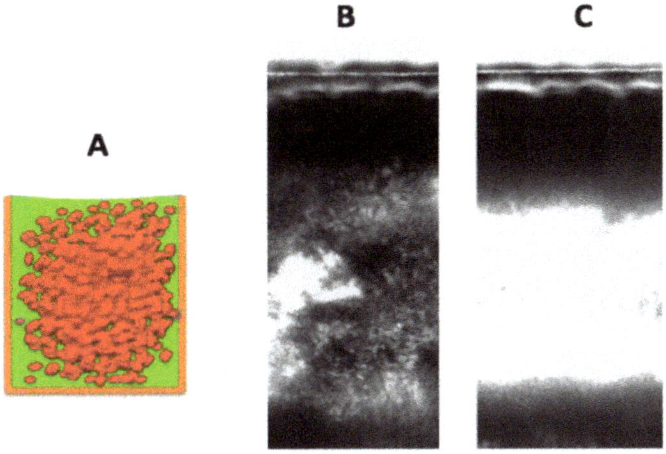

Figure 6.23. (A) A schematic of the whole blood flow; (B) blood plasma starts evaporating during the flow; (C) concentration of RBCs in the corners (often in the form of blood rouleaux). Photographs courtesy of Jean Berthier and Myriam Cubizolles, CEA-Leti.

6.11 Capillary filaments stopped by rounded wedges

Capillary filaments are not always desirable. In this section, we show how rounded corners are used to avoid the formation of capillary filaments.

6.11.1 Triangular open channels

The Concus–Finn condition works only if the inner corner or wedge is 'perfectly sharp.' The notion of 'perfect' is, however, somewhat fuzzy. In the following, we take the example of a triangular V-groove with an imperfect, i.e. rounded, bottom

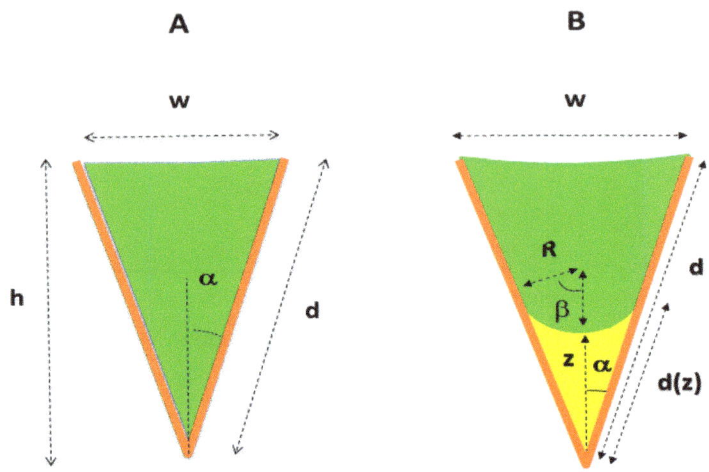

Figure 6.24. (A) A schematic of a perfect V-groove; (B) a schematic of an imperfect V-groove with a rounded inner corner. In this particular case, the rounded bottom of the groove has a shape delimited by a circular arc tangent to the walls.

angle (figure 6.24). In this figure, the rounded part in the corner has been exaggerated for visualization, and its radius R is considered as a parameter.

In this case, using the Cassie formulation developed in chapter 3, the SCF condition is

$$-w + 2[d - d(z)]\cos\theta + 2\beta\, R\cos\theta > 0. \tag{6.18}$$

The circular profile being tangent to the walls, the above relation yields

$$\frac{R}{R+z} = \sin\alpha. \tag{6.19}$$

The radius R can then be expressed as a function of α and z

$$R = \frac{z\sin\alpha}{1 - \sin\alpha}. \tag{6.20}$$

The geometric angles β and α are related by

$$\beta = \frac{\pi}{2} - \alpha. \tag{6.21}$$

Moreover, using (6.20), simple trigonometry yields the relation

$$d(z) = \frac{R}{\tan\alpha} = \frac{z\cos\alpha}{1 - \sin\alpha}. \tag{6.22}$$

Using (6.20), (6.21), and (6.22), the SCF condition (6.18) can be rewritten as

$$\sin\alpha < \cos\theta - 2\frac{z\sin\alpha\cos\theta}{w\,(1 - \sin\alpha)}\left[\cos\alpha - \left(\frac{\pi}{2} - \alpha\right)\sin\alpha\right]. \tag{6.23}$$

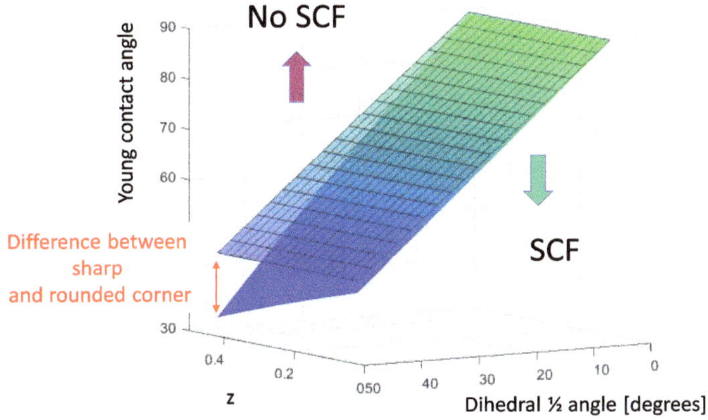

Figure 6.25. A comparison of the SCF conditions for a sharp corner (grid surface on top) and a rounded corner. The difference vanishes when the V-groove angle goes to zero and increases with an increase in the V-groove angle and the thickness of the rounded part.

This relation for SCF in a rounded V-groove can be compared to that for the perfect V-groove (6.4): $\sin\alpha < \cos\theta$. When the distance z goes to zero—to form a perfect corner—relation (6.23) reduces to (6.4). In the case of a small dihedral angle ï ¡, first-order Taylor expansions of (6.23) lead to

$$\cos\theta > \sin\alpha\left(1 + 2\alpha\frac{z}{w}\right). \tag{6.24}$$

Clearly, condition (6.24) is more restrictive than (6.4). The difference between the two conditions is shown in figure 6.25.

Condition (6.23) can be expressed as a function of the curvature radius R as follows:

$$\sin\alpha < \cos\theta - 2\frac{R\cos\theta}{w}\left[\cos\alpha - \left(\frac{\pi}{2} - \alpha\right)\sin\alpha\right]. \tag{6.25}$$

For a small half-dihedral angle α, condition (6.25) becomes

$$\cos\theta > \sin\alpha\left(1 + 2\frac{R}{w}\right). \tag{6.26}$$

This condition is plotted in figure 6.26 for $\alpha = 5°$ and $\alpha = 20°$. As the radius R increases, the contact angle at which an SCF appears decreases. This effect is more pronounced for larger dihedral angles.

In the case of an imperfect V-groove, the contact angle ï ± has to be smaller to observe SCF. This property has been verified using Evolver. Figure 6.27 shows the dependence of the advancing interface of a liquid in a V-groove on the quality of the wedge. The shape of the liquid tip differs considerably as the sharpness of the wedge changes. Even a small curvature is enough to impede capillary filament formation,

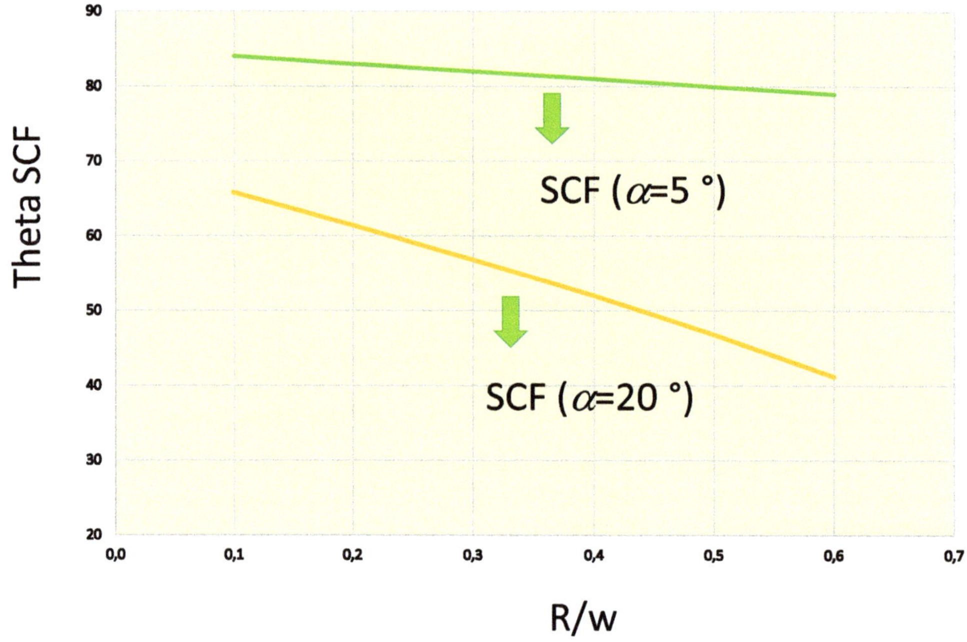

Figure 6.26. The SCF condition as a function of the radius of curvature of the rounded part in the channel dihedral.

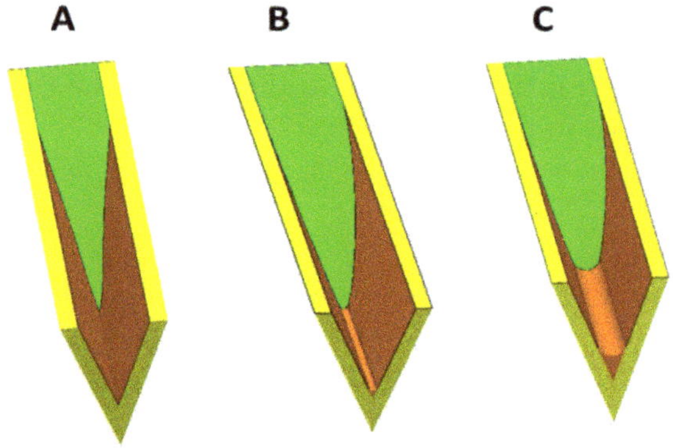

Figure 6.27. A comparison of SCF in (A) a perfect V-groove, (B) a V-groove with a small smooth wedge, (C) and a V-groove with a large rounded bottom. The wedge half-angle α = 19.5° and the contact angle $\theta = 50°$. The curvature radii in the wedge $R = 0$, 6, and 20 µm, the width of the groove $w = 40$ µm, and the slant height of the groove d = 60 µm.

simply because the Concus–Finn relation is no longer satisfied [11]. In the case of a rounded groove bottom, the capillary flow condition is then the general Cassie condition.

Geometric defects can be associated with the fabrication process. The microfabrication of V-grooved channels may be difficult and the result imperfect,

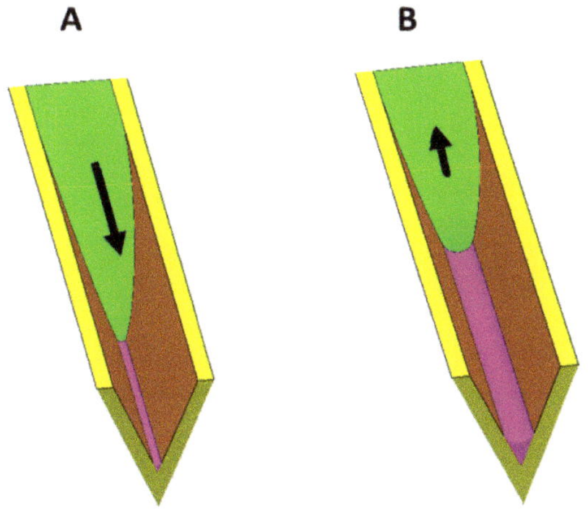

Figure 6.28. A comparison of SCF in: (A) a V-groove with a small deposit in the wedge, (B) a V-groove with a large amount of deposit in the wedge. The wedge half-angle $\alpha = 19.5°$, the contact angle with the wall $\theta = 50°$, the contact angle with the deposit $\theta_d = 70°$, the curvature radii in the wedge $R = 6$ and 20 μm, the width of the groove $w = 40$ μm, and the slant height of the groove $d = 60$ μm.

depending on the material used. Perfectly sharp corners are difficult to achieve. Rounded corners can also be linked to the very functionality of the device. As we have seen in the preceding section, dry deposits accumulate in the corners. We consider here the case in which these obstacles, i.e. the chemical deposits, are fixed and rigid and constitute a surface that is unaffected by the incoming liquid interface. This situation is similar to the preceding case of a geometric defect, except for the difference that the contact angle with the sample fluid is different than that of the walls. Deposits are often hydrophobic, or at best slightly hydrophilic, while the fabricated channel is hydrophilic. Figure 6.28 shows the shape of the interface in a V-groove with two different amounts of deposit. In the case of a very small deposit in the wedge, SCF occurs in the channel. In the case of a larger deposit, the SCF condition is not met, and the flow recedes. This observation is linked to the conditions (6.23) and (6.25).

6.11.2 Rectangular open channels

If the bottom corners of a rectangular open channel are slightly rounded, no capillary filaments can form when $\theta < \pi/4$. Let us assume a slightly rounded channel, as depicted in figures 6.29(A) and (B). The SCF condition is then

$$\cos\theta > \frac{p_F}{p_w} = \frac{w}{w - 2R + 2(h - R) + \pi R}. \tag{6.27}$$

When $R = 0$, one retrieves the SCF condition in perfect rectangular channels.

This condition can be cast in the form

$$\frac{w}{h} < \frac{2\cos\theta}{1 - \cos\theta} - (4 - \pi)\frac{R}{h}\frac{\cos\theta}{1 - \cos\theta}. \tag{6.28}$$

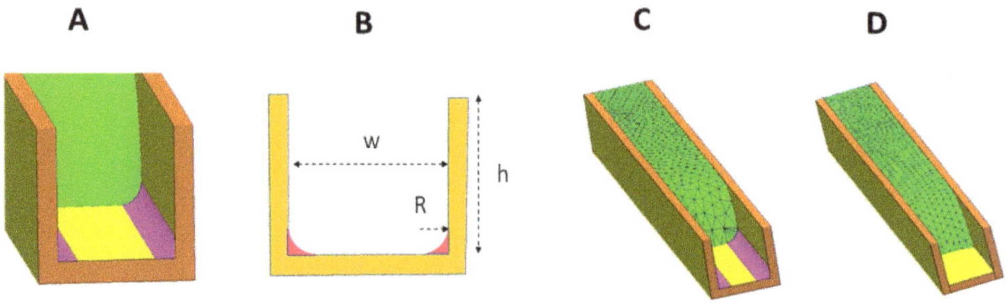

Figure 6.29. (A) An Evolver view of the channel; the rounded corners are shown in violet and the fluid in green. (B) A schematic of the cross-section. (C) An Evolver view of the interface in the channel during the SCF; (D) the same calculation but for a perfect rectangular channel.

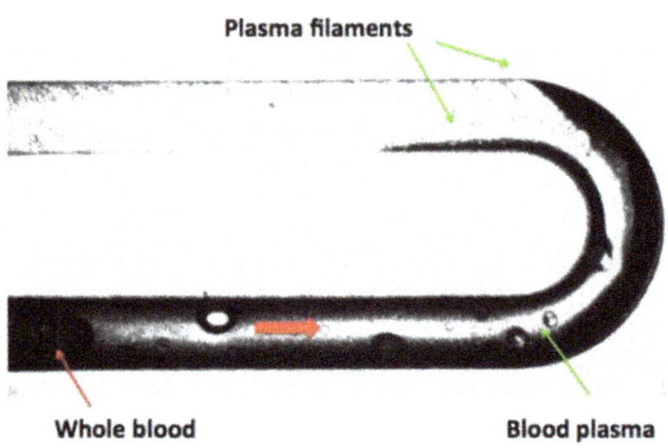

Figure 6.30. A photograph of a whole blood capillary flow in an open rectangular groove, showing the plasma preceding the flow and forming filaments (photograph courtesy of Jean Berthier and Myriam Cubizolles, CEA-Leti).

This last condition is more restrictive due to the second term on the right-hand side, which always creates a negative contribution. In order to observe SCF, the aspect ratio of the rounded rectangular channel w/h should be less than that of a perfect rectangular channel.

6.12 Conclusions

Capillary filaments are an important feature of open capillary flows. They extend endlessly in sharp corners when the Concus–Finn condition is met (and when there is a sufficient volume of liquid). Their velocity is higher than that of the usual bulk SCF.

On the one hand, capillary filaments may be a drawback because they can cause leakage in microdevices, and tight sealing is then required. On the other hand, one can design systems to take advantage of capillary filaments. One of the advantages is shown in figure 6.30. The high velocity of the capillary filament can be used to separate inhomogeneous liquids, such as whole blood. In such a case, filaments concentrate blood plasma, which has a smaller viscosity than that of whole blood.

Capillary filaments may also be used to deposit concentrated reagents in sharp corners. These dried reagents are later dissolved by an incoming capillary flow, but dried filaments in corners impede the formation of new filaments (only SCF is possible). Not all the uses of capillary filaments have been discovered yet, but obviously they constitute an important domain for future research.

References

[1] Concus P and Finn R 1969 On the behavior of a capillary surface in a wedge *Proc. Natl Acad. Sci.* **63** 292–9

[2] Concus P and Finn R 1994 Capillary surfaces in a wedge—differing contact angles *Microgravity Sci. Technol.* **7** 152–5

[3] Berthier J, Brakke K A and Berthier E 2016 *Open Microfluidics* (New York: Wiley)

[4] Berthier J, Brakke K A, Furlani E P, Karampelas I H, Poher V, Gosselin D, Cubizolles M and Pouteau P 2015 Whole blood spontaneous capillary flow in narrow V-groove microchannels *Sensors Actuators* B **206** 258–67

[5] Rye R R, Yost F G and Mann J A 1996 Wetting kinetics in surface capillary grooves *Langmuir* **12** 4625–7

[6] Rye R R, Yost F G and O'Toole E J 1996 Capillary flow in irregular surface grooves *Langmuir* **14** 3937–43

[7] Yost F G, Rye R R and Mann J A 1997 Solder wetting kinetics in narrow V-grooves *Acta Mater.* **45** 5337–45

[8] Lucas R 1918 Ueber das Zeitgesetz des Kapillaren Aufstiegs von Flussigkeiten *Kolloid Z.* **23** 15

[9] Washburn E W 1921 The dynamics of capillary flow *Phys. Rev.* **17** 273–83

[10] Rideal E K 1922 On the flow of liquids under capillary pressure *Philos. Mag. Ser.* **6** 1152–9

[11] Kitron-Belinkov M, Marmur A, Trabold T and Dadheech G V 2007 Groovy-drops: effect of groove curvature on spontaneous capillary flow *Langmuir* **23** 8406–10

[12] Berthier J, Brakke K A, Gosselin D, Navarro F, Belgacem N and Chaussy D 2016 Spontaneous capillary flow in curved, open microchannels *Microfluid. Nanofluid.* **20** 1–9

[13] Jokinen V and Franssila S 2008 Capillarity in microfluidic channels with hydrophilic and hydrophobic walls *Microfluid. Nanofluid.* **5** 443–8

[14] Girardo S, Cingolani R, Chibbaro S, Diotallevi F, Succi S and Pisignano D 2009 Corner liquid imbibition during capillary penetration in lithographically made microchannels *Appl. Phys. Lett.* **94** 171901

[15] Berthier J, A. Brakke K, Gosselin D, Huet M and Berthier E 2014 Metastable capillary filaments in rectangular cross-section open microchannels *AIMS Biophys.* **1** 31–48

[16] Brakke K A 1992 The Surface Evolver *Exp. Math.* **1** 141–65

[17] Berthier J, Gosselin D, Huet M, Sabatte G, Cosnier M-L, Pouteau P, Navarro F and Cubizolles M 2016 Capillary flow of whole blood in microsystems : non-Newtonian blood behavior and substrate reagent-coating effect *Proc. of the 2016 Nanotech Conf. (May 22–25) (Washington DC, USA)*

[18] Huet M, Cubizolles M and Buhot A 2018 Red blood cell agglutination for blood typing within passive microfluidic biochips *High-Throughput* **7** 10

[19] Matsuo T, Tsuchida Y and Morimoto N 2002 Trehalose eye drops in the treatment of dry eye syndrome *Ophthalmology* **109** 2024–9

IOP Publishing

Open-Channel Microfluidics (Second Edition)
Fundamentals and applications
Jean Berthier, Ashleigh B Theberge and Erwin Berthier

Chapter 7

Flow in open channels of nonuniform cross-section

In the preceding chapters, it was assumed that the open channel had a uniform cross-section. In this section, we consider open channels with changes of cross-sectional area. We show that the dynamics of such flows is deeply affected by such geometrical changes. This chapter presents the effect of changing the cross-section of the channel, first from a static or quasi-static point of view, then from a dynamic point of view.

7.1 Static aspects

In this section, we consider open channels with a convergent and/or divergent section. We investigate the spontaneous capillary flow (SCF) condition and analyze the dynamics of the open microflow.

7.1.1 Spontaneous capillary flow in linearly widening and narrowing open channels

Let us first analyze the case of a progressive change of cross-section, as shown in figure 7.1. In such a case, the vertical walls form an angle of 2α, where $\alpha > 0$ for a converging channel and $\alpha < 0$ for a diverging channel.

7.1.1.1 Rectangular open channels

Let us take the example of rectangular open channels with straight vertical walls such that the height h of the walls is constant and the width w progressively increases (figures 7.2(A) and (B)).

Consider the case of figure 7.2(A), in which capillary filaments are absent (the case of figure 7.2(B), in which the flow is preceded by capillary filaments, is more

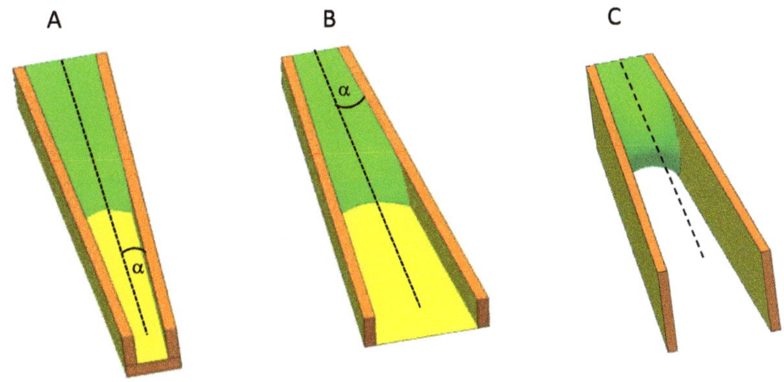

Figure 7.1. (A) Fluid progressing in a converging rectangular open channel; (B) fluid progressing in a diverging rectangular open channel; (C) fluid progressing in a diverging suspended channel. Views obtained using Surface Evolver numerical software [1, 2].

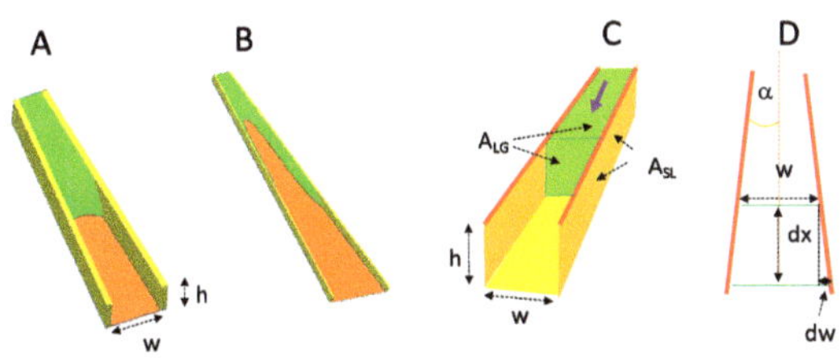

Figure 7.2. (A) Capillary flow in a diverging rectangular open channel; (B) the formation of capillary filaments in a diverging rectangular open channel; (C) schematization of the advancing liquid in the channel; (D) top view of the schematization. Figures (A) and (B) were obtained using Surface Evolver software.

complicated). Let us recall the condition for SCF derived from the Gibbs free energy [3–5]:

$$\frac{dA_{LG}}{dA_{SL}} < \cos\theta, \tag{7.1}$$

where θ is the contact angle, $A_{LG}$the liquid–air surface area, and A_{SL} the liquid–wall surface area (figure 7.2). If the wall's semi-angle is denoted by α, when the liquid front progresses over the distance dx, the changes in surface area are

$$dA_{SL} = 2h\frac{dx}{\cos\alpha} + dx\,(w + dw), \tag{7.2}$$

where the first term on the right-hand side corresponds to the increase in the surface along the vertical walls and the second term to the increase in the surface along the floor. On the other hand, the increase in the liquid–air surface area is the sum of the top surface increase $\sim dx\,(w + dw)$ and the meniscal surface increase $\sim 2h\,dw$, thus

$$dA_{LG} = dx\,(w + dw) + 2h\,dw. \tag{7.3}$$

Noting that $dw = dx \tan \alpha$ and substituting (7.2) and (7.3) into (7.1), we obtain the SCF condition

$$\frac{dA_{LG}}{dA_{SL}} = \frac{w \cos \alpha + 2h \sin \alpha}{w \cos \alpha + 2h} < \cos \theta, \quad (7.4)$$

which leads to the following condition for the aspect ratio

$$q = \frac{w}{h} < \frac{2(\cos \theta - \sin \alpha)}{\cos \alpha \, (1 - \cos \theta)}. \quad (7.5)$$

Note that if $\alpha = 0$, relation (7.5) collapses to the parallel wall condition presented in chapter 2. When α is small, relation (7.5) can be simplified by using the first-order terms of Taylor expansions for $\cos\alpha$ and $\sin\alpha$:

$$q = \frac{w}{h} < \frac{2(\cos \theta - \alpha)}{(1 - \cos \theta)} = \frac{2 \cos \theta}{(1 - \cos \theta)} - \frac{2\alpha}{(1 - \cos \theta)}. \quad (7.6)$$

The second term on the right-hand side of (7.6) constitutes a restriction on the SCF condition for parallel walls $q = \frac{w}{h} < \frac{2 \cos \theta}{(1 - \cos \theta)}$ in the case of a diverging channel ($\alpha > 0$) and conversely a relaxation of the same condition for a converging channel ($\alpha < 0$). In a diverging channel, the capillary flow automatically stops at some distance, because the width w monotonically increases to a value at which condition (7.6) fails. Conversely, in a converging channel, the flow continues up to the channel exit.

Figure 7.3 shows the flow of human whole blood in two diverging rectangular open channels: the first one has sharp inner corners and capillary filaments form in the bottom corners until the flow stops. In the second case, the channel is rounded, and even if the divergence angle is small, the flow stops earlier.

7.1.1.2 Suspended channels

Suspended channels are open channels devoid of a ceiling and a floor [4–7]. Let us take the example of suspended channels with straight vertical walls such that the height h of the walls is constant and the width w progressively increases. The wall-divergence semi-angle is denoted by α. Let us recall the condition for SCF derived from the Gibbs free energy [3–5]:

$$\frac{dA_{LG}}{dA_{SL}} < \cos \theta, \quad (7.7)$$

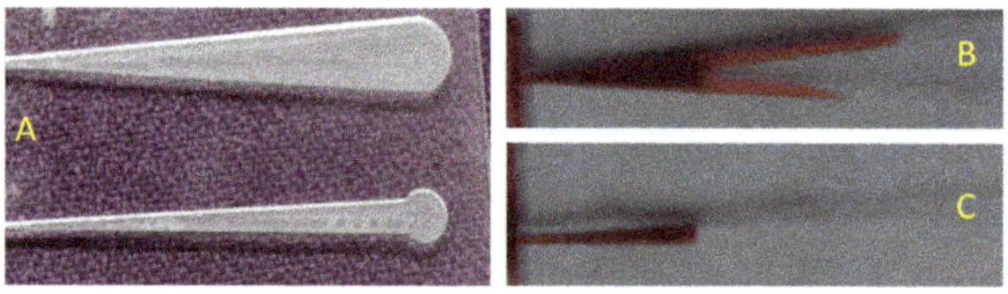

Figure 7.3. (A) Two diverging rectangular open channels milled in PMMA; the top channel has sharp corners, whereas the bottom channel has rounded corners. (B) Whole blood penetrates the channel and halts, preceded by capillary filaments. (C) Whole blood penetration stops in the rounded-corner channel. Photograph (A) courtesy of François Boizot, CEA-Leti; photographs (C) and (D) courtesy of Myriam Cubizolles, CEA-Leti.

where θ is the contact angle, A_{LG} the liquid–air surface area, and A_{SL} the liquid–wall surface area (figure 7.4).

If we assume an infinitesimal progression dx of the front of the fluid, the increase in the solid–liquid surface area is obtained by observing that the increase in the wetting length along a vertical wall is $dx/\cos\alpha$, hence

$$dA_{SL} = 2h\frac{dx}{\cos\alpha}, \tag{7.8}$$

and the increase in the liquid–air surface area is the sum of that of the top surface $\sim 2dx\ (w + dw)$ and the meniscal surface $\sim 2h\ dw$:

$$dA_{LG} \approx 2dx\ (w + dw) + 2h\ dw. \tag{7.9}$$

If we first substitute the relation $dw = dx \tan\alpha$ into (7.8) and (7.9) and combine them with (7.7), we obtain

$$\frac{dA_{LG}}{dA_{SL}} = \frac{w + (h + dx)\tan\alpha}{\frac{h}{\cos\alpha}} < \cos\theta. \tag{7.10}$$

Using the notation $q = w/h$ and neglecting the second-order term, (7.10) can be rewritten as

$$q = \frac{w}{h} < \frac{\cos\theta - \sin\alpha}{\cos\alpha}. \tag{7.11}$$

For $\alpha = 0$, we retrieve the usual condition for parallel walls: $q = w/h < \cos\theta$. Relation (7.11) shows that a diverging channel ($\alpha > 0$) imposes a restrictive limit on the occurrence of SCF for two reasons: first, the function $\frac{\cos\theta - \sin\alpha}{\cos\alpha}$ decreases with α (this is related to the flattening of the front meniscus); second, as w increases ($w\nearrow$) in a divergent channel, condition (7.11) fails after some travel distance. The opposite is true for a converging suspended channel: $\alpha < 0$, $w\searrow$. The function $\frac{\cos\theta - \sin\alpha}{\cos\alpha}$ is plotted in figure 7.3(C).

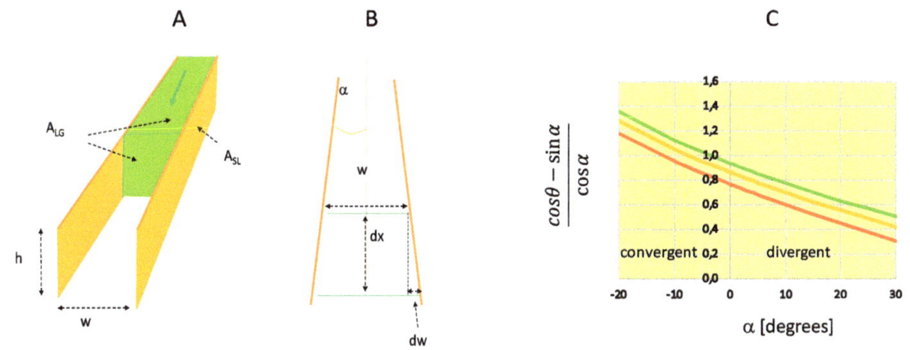

Figure 7.4. (A) A perspective view of a suspended channel; (B) the top view of the suspended channel; (C) function $\frac{\cos\theta - \sin\alpha}{\cos\alpha}$ plotted for three values of the contact angle $\theta = 10°, 20°, 30°$.

In the case of a diverging suspended channel, the width at the front of the flow increases and the flow stops when

$$\frac{w_{max}}{h} = \frac{\cos\theta - \sin\alpha}{\cos\alpha}. \tag{7.12}$$

If the origin of the x-axis is chosen so that it lies at the intersection of the two walls, the width w is linked to the distance x by $w = 2x\tan\alpha$. Substituting this into (7.12) yields the stopping distance

$$x_{max} = \frac{h}{2}\frac{\cos\theta - \sin\alpha}{\sin\alpha}. \tag{7.13}$$

This verifies that when the wall angle decreases to zero ($\alpha \searrow 0$), then the penetration length goes to infinity ($x_{max} \nearrow \infty$).

7.1.1.3 Rails

Rails are open channels resembling suspended channels. The preceding reasoning can be adapted to rails. The SCF condition for rails is

$$\frac{h}{w} < \frac{\cos\theta\cos\alpha}{1 + \sin\alpha}. \tag{7.14}$$

In the case of rails, a convergent flow stops when the width w is such that $w = h\frac{1+\sin\alpha}{\cos\alpha\cos\theta}$. In the case of diverging rails, the capillary flow does not stop. A picture of the capillary flow in converging rails—obtained with Surface Evolver—is shown in figure 7.5.

7.1.2 Sudden enlargement

The picture is somewhat different when the enlargement is not progressive, but abrupt. Two different cases of enlargement can be distinguished: smooth sudden enlargements and sharp-edged enlargements. The difference between the two morphologies pertains to the potentiality of pinning on edges in the second case.

7.1.2.1 Smooth sudden enlargements

Let us first analyze the first case of smooth, sudden enlargements. Due to the geometry of the walls, the curvature of the advancing meniscus decreases and the

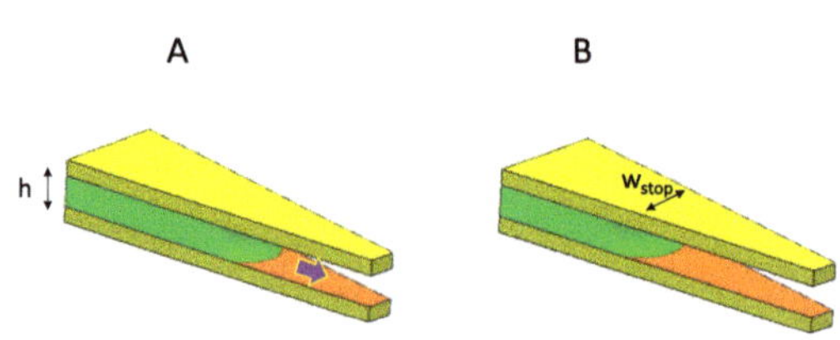

Figure 7.5. The flow stops in converging rails when the SCF condition fails due to the decrease of the width w (Evolver calculation).

Laplace pressure may go to zero, which stops the SCF, assuming quasi-static conditions. Indeed, in a static configuration, the meniscus adopts a minimal-surface shape and the diverging angle of the walls automatically decreases its curvature. Figure 7.6, obtained using the COMSOL software [8], shows a 2D simulation of the leading meniscus passing through a circular enlargement at very low velocity. Due to the 'level set numerical method' used in the simulation, the interface is not infinitely thin, but shows a small smearing (colors light blue, yellow, and orange in the figure). Clearly, in the figure, the meniscal curvature decreases in the diverging section and increases in the converging section. Hence, the driving Laplace pressure is close to zero in the enlargement and its magnitude increases in the converging region.

We shall present the dynamic aspects later in the chapter, but it is interesting to remark that the flow dynamics can change the preceding picture corresponding to a quasi-static state. Figure 7.7 shows a similar COMSOL approach for a flow that has a larger initial velocity.

A 3D picture of the quasi-static open-capillary flow through a cylindrical well (obtained for a quasi-steady state with the software Surface Evolver [1, 2]) is shown in figure 7.8. Four different contact angles have been tested: $\theta = 25$, 43, 50, and 60°. In the two first cases, capillary filaments precede the flow (θ <45°). For $\theta = 50°$, there are no filaments ahead of the flow, but the flow continues. For θ = 60°, the flow stops. The interface shape is such that the Laplace pressure is positive.

Figure 7.9, obtained using the Evolver software, shows the behavior of a capillary flow between rails. In the case of sudden enlargement, the flow continues. In the case of constriction, the outcome depends on the SCF condition in the narrow section.

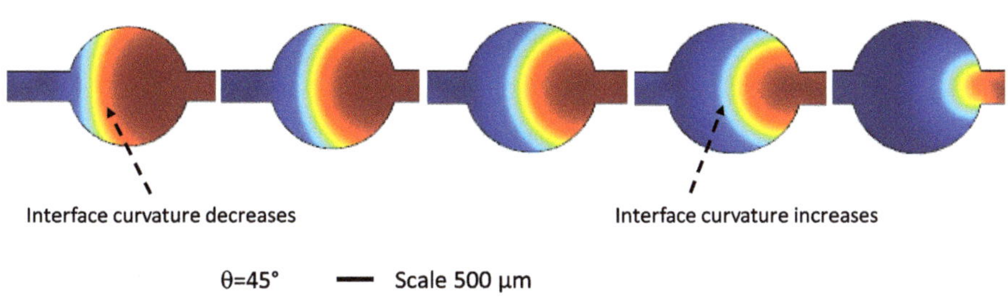

Figure 7.6. (A) Fluid flow passing through a cylindrical well (2D simulation with COMSOL numerical software). Note the change of meniscal shape between the first picture on the left and the last ones.

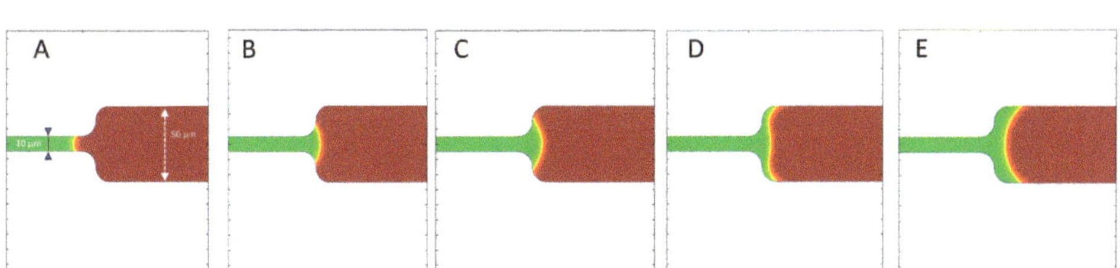

Figure 7.7. Fluid flow in an abruptly diverging channel (2D simulation performed using COMSOL numerical software) under dynamic conditions. Note the deformation of the meniscus due to dynamic forces in panel C. In the enlargement, the flow velocity decreases and the meniscus regains its nearly static shape.

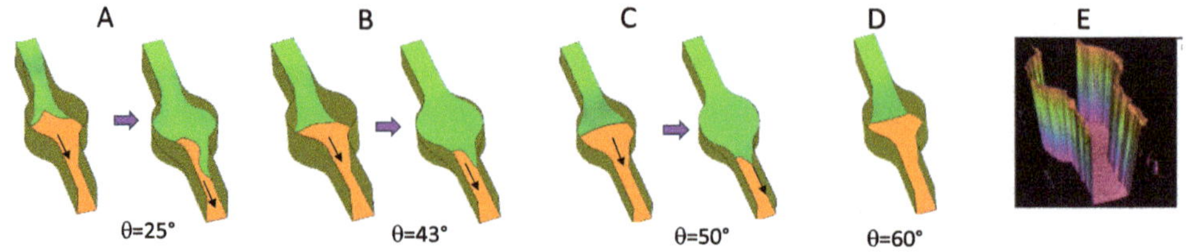

Figure 7.8. SCF through an enlargement in a rectangular open channel ($w = 80$ µm, $h = 60$ µm, $D = 160$ µm). For a contact angle θ less than 45°, filaments precede the bulk flow (A and B); if the contact angle θ is larger than 45°, there are no filaments but the flow continues; (C) if the value of θ is sufficiently large ($\theta > 55°$), the fluid stops (D) because the magnitude of the Laplace pressure of the front meniscus decreases to zero (and equals the upstream pressure). (E) Profilometry image of a similar channel milled in PMMA (photo courtesy of David Gosselin, CEA-Leti).

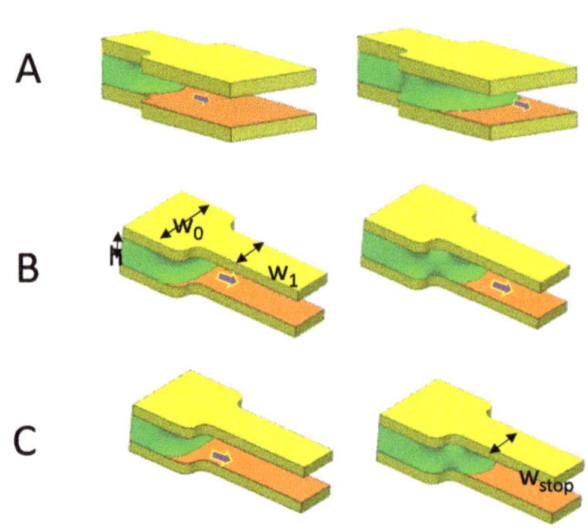

Figure 7.9. (A) SCF between parallel rails with sudden enlargement; (B) SCF between parallel rails with sudden constriction; (C) flow stops at the constriction due to the breach of the SCF condition after the constriction.

7.1.2.2 Sharp enlargements—capillary valves

The case of sudden enlargements of an open channel with sharp edges is very particular. Pinning may occur in such cases, depending on the geometry, materials, and dynamics [9–13].

For a quasi-static flow, pinning takes place if the liquid contact angle α with the non-wetted wall, when the triple line is along the edge, is less than

$$\alpha < \pi - \beta + \theta, \tag{7.15}$$

where θ is the Young contact angle and β the wedge angle. The pinning condition of a liquid on an edge is illustrated in figure 7.10(A), and the pinning of a fluid by a sharp orifice is illustrated in figure 7.10(B).

Hence, SCF in a sharp divergence can be stopped by pinning. This configuration is called a 'capillary valve.' An example is shown in figure 7.11, where a uniform open

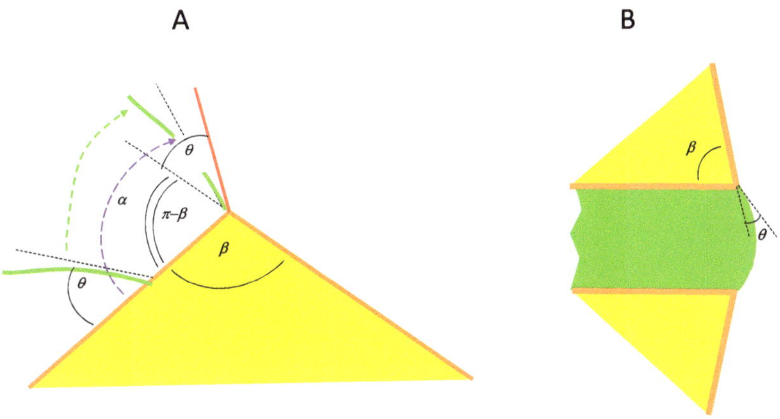

Figure 7.10. (A) The pinning limit: $\alpha < \pi - \beta + \theta$. (B) The pinning limit at a channel orifice.

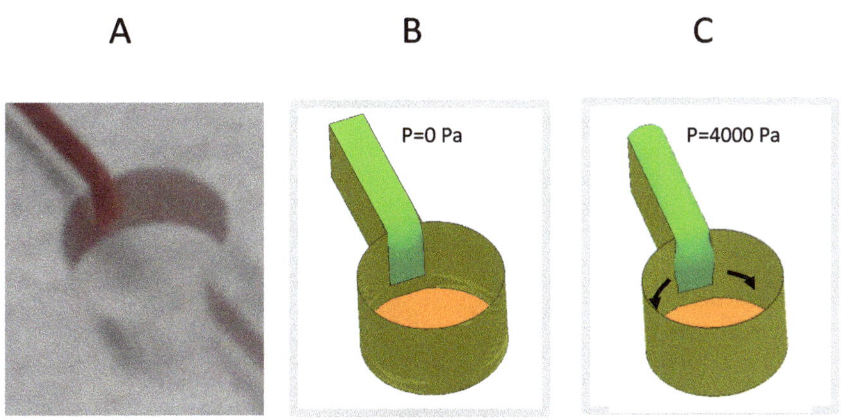

Figure 7.11. (A) A capillary flow of human whole blood stops at a sharp enlargement (valve) opening into a cylindrical well (photo courtesy of Jean Berthier and David Gosselin). (B) SCF simulated (pressure 0 Pa) by Surface Evolver, showing stable pinning. (C) SCF simulated by Surface Evolver, $P = 4000$ Pa, showing broken pinning.

channel opens into a cylindrical well (with a vertical step). Depending on the value of the static contact angle, the dimensions of the channel, and the dynamics of the flow arriving at the gate (opening) or the pressure of the liquid, the pinning may be broken, as shown in figure 7.11(C), where a pressure of 4000 Pa is applied to the liquid.

Note that the valve efficiency is significantly reduced if the pinning is only possible on the two vertical edges and not on the floor, i.e. there is no step between the channel and the exit well, as shown in figure 7.12.

The pinning is also only temporary when capillary filaments form, as shown in figure 7.13 for the capillary flow in a 'flat' channel. For a contact angle greater than 45°, the fluid remains pinned by the two vertical edges. When the contact angle is less than 45°, capillary filaments start flowing alongside the bottom corners, and pinning is broken.

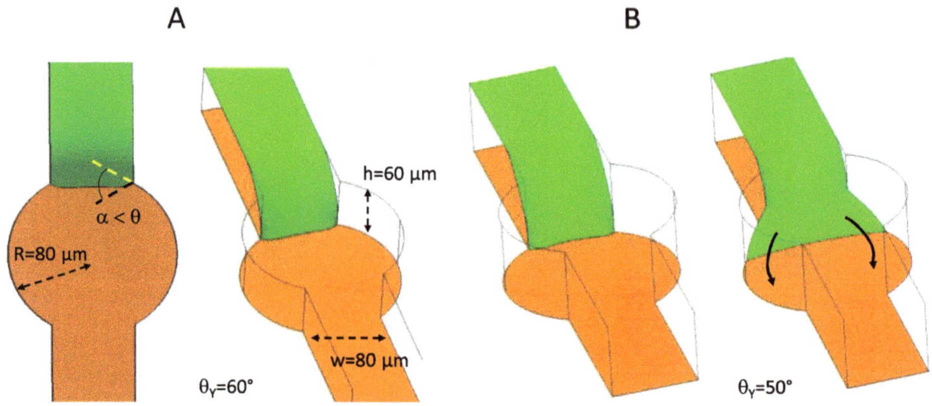

Figure 7.12. (A) The liquid—at zero pressure (SCF)—stops at the 'gate' for a contact angle θ_Y of 60°; (C) for a contact angle $\theta_Y = 50°$, the pinning is temporary and soon broken by the propagation of the liquid along the vertical walls. Figure obtained using Surface Evolver (w = 80 μm, h = 60 μm, R = 80 μm).

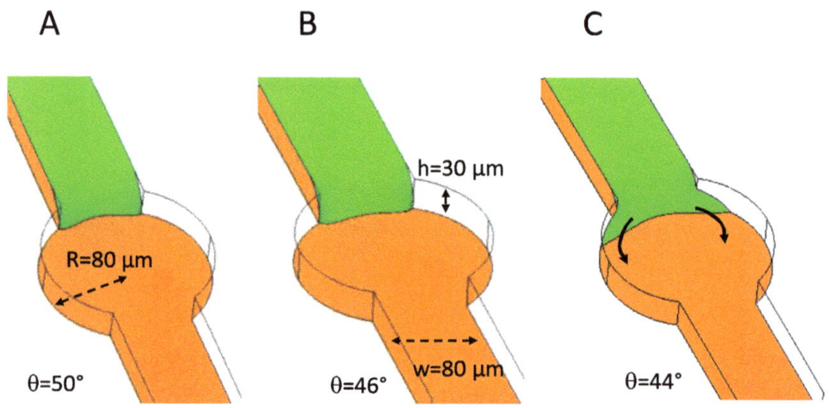

Figure 7.13. (A) and (B) The liquid—at zero pressure (SCF)—stops at the 'gate'; (C) for a contact angle of less than 45°, capillary filaments lead to the depinning of the liquid. Figure obtained using Surface Evolver (w = 80 μm, h = 60 μm, R = 80 μm).

Note that, in the same configuration, the pinning limit is around 55° in the high-wall channel, while it is 45° for the low-wall channel. This is due to the shape of the interface, as shown in figure 7.14. The vertical curvature is very different in both cases, and so is the Laplace pressure. The pinning angle difference corresponds to the difference in the Laplace pressure. In the case of the low-wall channel, the Laplace pressure is high (negative) and only the formation of capillary filaments can break the pinning.

7.1.3 Trigger valves

Trigger valves were first developed for closed channels to release a secondary flow upon the passage of the main flow [14–18]. A capillary flow can be stopped by pinning—or anchoring—on sharp ridges in the flow channel. Trigger valves stop a

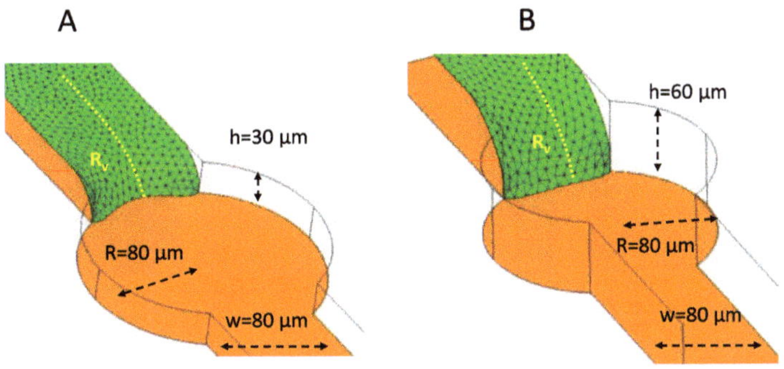

Figure 7.14. The vertical curvature radius R_v is very different depending on the height of the channel. The vertical curvature is larger in the low-wall channel (A) than in the high-wall channel (B).

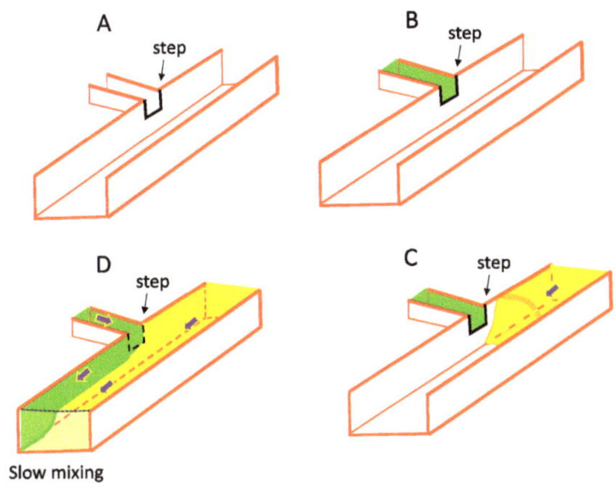

Figure 7.15. The principle of a trigger valve. (A) A sketch of the device, showing the step of the secondary channel that plays the role of a stop valve (B). (C) The arrival of the main flow; (D) release of the trigger valve and coflow (with slow mixing of the two coflowing liquids).

capillary flow in an elevated channel; the flow is released when a capillary sheath flow liberates the pinned meniscus (figure 7.15). Pinning of the flow takes place on the three edges of the 'gate.' This pinning is broken when the secondary sheath flow reaches the intersection and the two liquids form a concurrent flow (figure 7.16) [4].

A similar concept was extended to a large number of secondary channels by Zhang *et al* who developed a multi-microporous membrane-based trigger valve [19]. The principle is illustrated in figure 7.17.

7.1.4 One-way wicking

Microfluidic diodes, e.g. one-way wicking, are of interest for building microfluidic circuits. In an open channel, directional wicking may be obtained by internally structuring the channel. Feng and Rothstein showed that tilted fin-like structures have the effect of orienting the fluid in one direction, which they called the 'easy wicking direction' in opposition to the other direction, the 'hard wicking direction' [20].

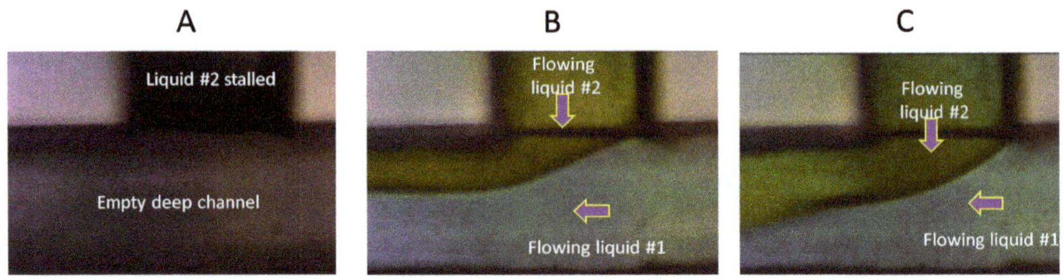

Figure 7.16. The behavior of an open-channel trigger valve milled in PMMA. (A) The secondary liquid is stopped at the channel's abrupt enlargement; (B) and (C) release of the liquid (yellow color) by the arrival of a main liquid. Photographs courtesy of David Gosselin and Jean Berthier (CEA-Leti).

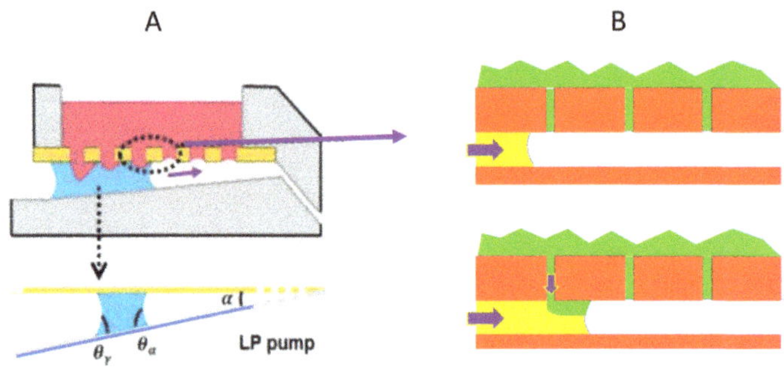

Figure 7.17. (A) The membrane-based trigger valve conceived by Zhang and coworkers. Reproduced from [19]. CC BY 4.0. (B) The principle of the multi-trigger valve.

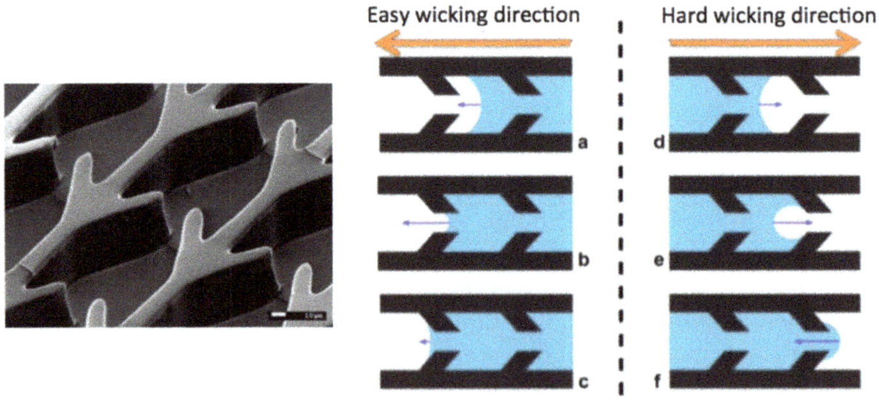

Figure 7.18. Left, a view of the channel and ridges; right, a comparison of the wicking process between the two ridge orientations. The capillary (Laplace) pressure acts against the flow in the 'hard wicking direction,' as shown at the bottom right image. Reprinted from [20], Copyright (2013), with permission from Elsevier.

They showed that the insertion of ridges or fins into an open channel changes the wicking direction. The wicking direction depends on the orientation of the fins relative to the flow direction.

This phenomenon is based on the change in capillary pressure at the narrow section between the fins. The explanation is sketched in figure 7.18. In the 'easy wicking direction,' the Laplace pressure at the advancing front is always negative, as the meniscus is always concave. On the other hand, in the 'hard wicking direction,' the meniscus becomes convex when the liquid reaches the gap between the fins. The positive Laplace pressure blocks the flow.

7.2 Dynamic aspects

In this section, we focus on the dynamics of open-channel microflows in channels of varying cross-section. First, we present the flow dynamics in progressively widening or narrowing open channels, then for a sudden change in cross-section.

7.2.1 Open microflows dynamics in progressively widening and narrowing channels

In this section, the dynamics of open-capillary flows in converging and diverging channels is presented. Let us first consider a rectangular open channel of uniform cross-section, then one with a diverging or converging region, and finally one with a uniform cross-section exit channel (figure 7.19(C)). The cross-section has the shape of a rectangle of width w and height h and has rounded bottom corners to avoid the formation of capillary filaments.

7.2.1.1 The theoretical approach

Here, we assume the flow is non-inertial, i.e. the regime is the viscous regime [21–23]. We shall see later that this assumption is not totally justified. Under this assumption, the equation governing the fluid motion in the convergent/divergent channel results from the balance between the capillary force in the convergent/divergent channel (denoted by the index cd) and the wall friction in the root channel (the channel

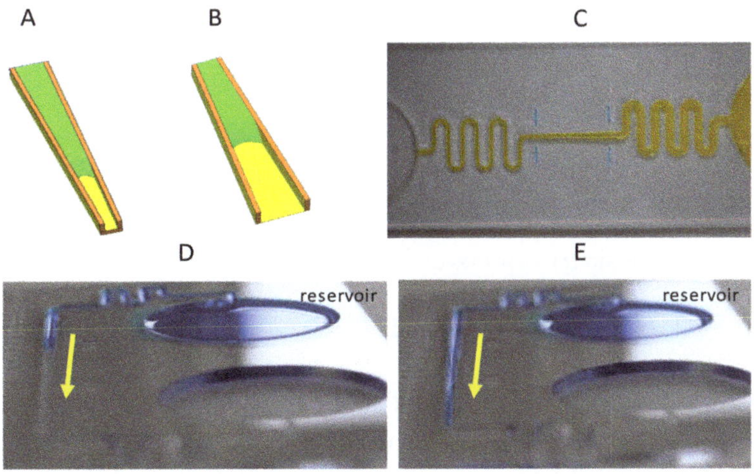

Figure 7.19. (A) A converging rectangular open channel (produced using Evolver); (B) a diverging rectangular open channel (produced using Evolver); (C) the experimental setup, viewed from above, with two reservoirs and a symmetrical design which can be used either as a converging or a diverging channel (the section between the two blue markers); (D) the flow enters the converging section; (E) the flow leaves the converging section. Photographs courtesy of Anika MacManamen and Jodie Tokihiro, University of Washington.

upstream from the convergent/divergent section) and the wall friction in the convergent/divergent channel itself. Hence

$$F_{\text{cap, cd}} = F_{\text{drag, cd}} + F_{\text{drag, root}}. \tag{7.16}$$

In the case of a rectangular open channel, the quasi-static capillary force in a convergent/divergent channel has been derived in section 7.1

$$F_{\text{cap, cd}} = \gamma\, p \cos\theta^* = \gamma\, [2h\,(\cos\theta \pm \sin\alpha) - w\cos\alpha\,(1 - \cos\theta)]. \tag{7.17}$$

The term $\pm\sin\alpha$ depends on whether the channel is converging or diverging. The plus sign corresponds to a converging channel and conversely the minus sign corresponds to a diverging channel. Taking into account a dynamic contact angle (DCA) defined by the molecular kinetic theory (MKT) approach ($\cos\theta = s\theta_0 - \beta V/\gamma$), (7.17) becomes

$$F_{\text{cap, cd}} = \gamma\, [2h\,(\cos\theta_0 \pm \sin\alpha) - w\cos\alpha\,(1 - \cos\theta_0)] - \beta V\,(2h + w\cos\alpha). \tag{7.18}$$

On the other hand, the wall friction is

$$F_{\text{drag, cd}} + F_{\text{drag, root}} = L_0 p_0 \mu \frac{V_0}{\bar{\lambda}_0} + \int_0^z p\mu \frac{V}{\bar{\lambda}} \mathrm{d}z, \tag{7.19}$$

where z is the distance in the convergent/divergent channel, the index 0 refers to the root channel, $\bar{\lambda}_0$ and $\bar{\lambda}$ are the friction lengths in the root channel and the convergent/divergent channel, respectively, and p_0 and p are the total perimeters of the root and convergent/divergent channels, respectively. For rectangular channels, $\bar{\lambda} \cong \frac{1}{3}\frac{w+h}{\frac{w}{h}+\frac{2h}{w}}$ and $p = 2(h + w)$.

Substituting (7.18) and (7.19) into (7.17) yields

$$\begin{aligned}&\gamma\left[2h\,(\cos\theta_0 \pm \sin\alpha) - w\cos\alpha\,(1 - \cos\theta_0)\right] - \beta V_{\text{cd}}\,(2h + w\cos\alpha)\\ &= L_0 p_0 \mu \frac{V_0}{\bar{\lambda}_0} + \int_0^z p\mu \frac{V}{\bar{\lambda}} \mathrm{d}z.\end{aligned} \tag{7.20}$$

Using the conservation of mass, $S_{\text{cd}}\, V_{\text{cd}} = S_0\, V_0 = S\, V$, where the label 'cd' refers to the tip of the flow in the convergent/divergent channel, equation (7.20) can be rewritten as

$$\begin{aligned}&\gamma\left[2h\,(\cos\theta_0 \pm \sin\alpha) - w\cos\alpha\,(1 - \cos\theta_0)\right] - \beta V_{\text{cd}}\,(2h + w\cos\alpha)\\ &= \mu\, S_{\text{cd}}\, V_{\text{cd}}\left[\frac{L_0 p_0}{S_0\,\bar{\lambda}_0} + \int_0^z \frac{p}{S\,\bar{\lambda}} \mathrm{d}z\right].\end{aligned} \tag{7.21}$$

For a given penetration distance z in the convergent/divergent channel, the term $G(z) = \int_0^z \frac{p}{S\,\bar{\lambda}} \mathrm{d}z$ is a known function of z and the geometry of the convergent/divergent channel, and the velocity at the tip of the liquid can be written as

$$V_{\text{cd}} = \frac{\gamma\, [2h\,(\cos\theta_0 \pm \sin\alpha) - w\cos\alpha\,(1 - \cos\theta_0)]}{\mu\, S_{\text{cd}}\left[\frac{L_0 p_0}{S_0\,\bar{\lambda}_0} + G(z)\right] - \beta\,(2h + w\cos\alpha)}. \tag{7.22}$$

7.2.1.2 Experiments

A comparison between the velocities in converging/diverging channels given by the theoretical model and those obtained in experiments is shown in figure 7.20. The experiments presented here were performed using a solution of IPA [24] at a concentration of 50% and a converging/diverging length of 20 mm. The root channel had a rectangular cross-section of $w = 2$ mm, $h = 2$ mm in the case of the experiments with the converging channel, and $w = 1$ mm, $h = 2$ mm in the case of the experiments with the diverging channel.

The velocity of the front meniscus increases in the convergent channel, and decreases in the divergent channel. The theoretical model slightly overshoots the experimental results in the case of the convergent channel. The reason for this is unclear. Two hypotheses are proposed: (1) an inertial effect due to the fact that the velocity and the travel distance increase at the same time; (2) depinning of the liquid from the upper edges of the channel.

7.2.1.3 Conclusions

The velocity of the advancing front meniscus increases in a converging channel and decreases in a diverging channel. This effect becomes more pronounced as the length of the convergent/divergent channel decreases.

7.2.2 Sudden constrictions and enlargements

In the case of closed channels, a sudden change of cross-sectional area completely modifies the velocity of the capillary flow [25–28]. The same is true for open channels

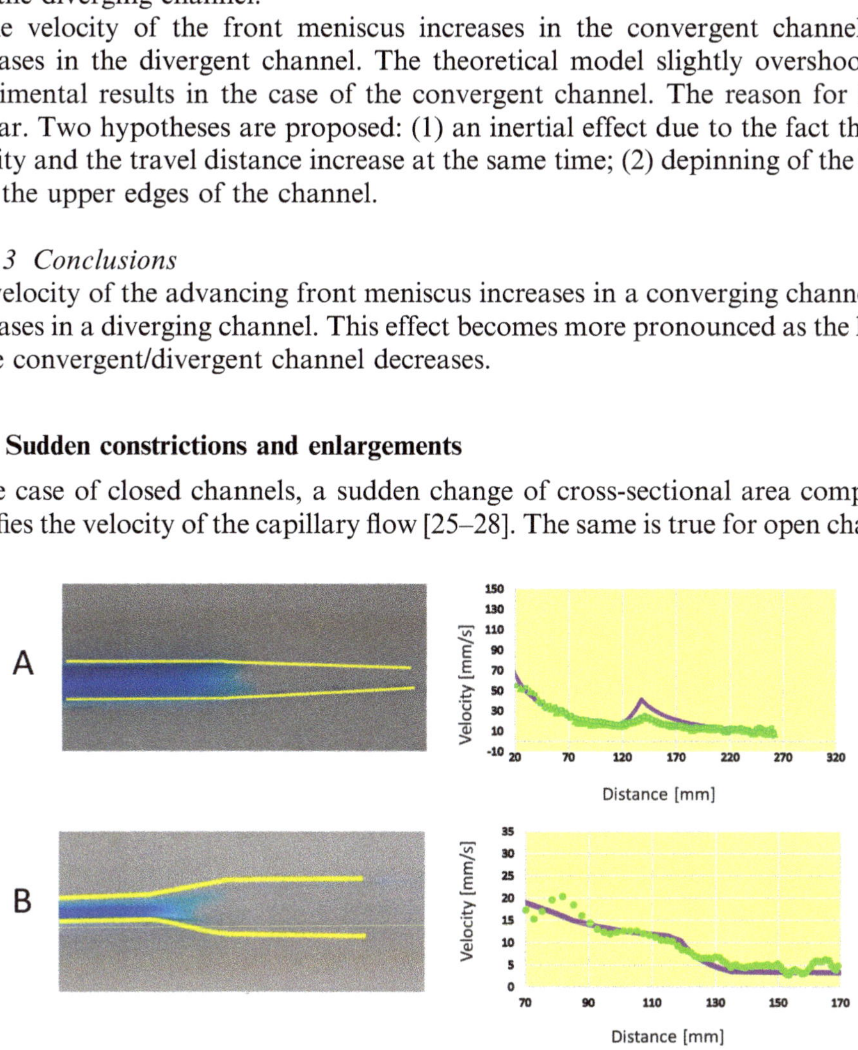

Figure 7.20. (A) A converging open channel; left: the view from the top of the meniscus; right: a comparison between the theoretical model and experimental results. (B) A diverging open channel; left: the view from top of the meniscus; right: a comparison between the theoretical model and experimental results. The purple curve corresponds to the theoretical model results and the green dots to experiments performed using IPA50 in a channel milled in PMMA. Photographs courtesy of Anika MacManamen and Jodie Tokihiro, University of Washington.

[29, 30]. In this section, we consider the case of a constriction and that of an enlargement for a monolithic open channel of rectangular cross-section (figure 7.21). It is shown that a constriction acts as a global flow resistor and an enlargement as a local flow resistor.

In this problem, the capillary force—or the Laplace capillary pressure—is local, i.e. related to the section of the channel where the advancing meniscus is located. For the meniscus in the section numbered k, the capillary force is

$$F_{\text{cap},\,k} = (2h + w_k)\,\gamma\cos\theta - w_k\,\gamma. \tag{7.23}$$

If we denotes $\varepsilon_k = w_k/h$ the aspect ratios of each section, relation (7.23) becomes

$$F_{\text{cap},\,k} = \gamma\,h\,f_k = \gamma\,h\,[(\varepsilon_k + 2)\cos\theta - \varepsilon_k]. \tag{7.24}$$

On the other hand, the friction is 'global,' i.e. it integrates all the sections already wetted by the flow. If we locally approximate the friction by

$$F_{\text{drag, sides}} \approx 6\mu\,\frac{V}{w}\,(2h)z$$

$$F_{\text{drag, bottom}} \approx 6\mu\,\frac{V}{h}\,w\,z, \tag{7.25}$$

then the total friction can be written as

$$F_{\text{drag},\,k} \approx \sum_{i=1,k-1} F_{\text{drag},\,k} + 6\mu\,V_k\,(z - z_{k-1})\left(\frac{w_k}{h} + \frac{2h}{w_k}\right), \tag{7.26}$$

where z_k are the axial locations of the section beginnings/ends. For simplicity, we note

$$C_k = \frac{w_k}{h} + \frac{2h}{w_k} = \varepsilon_k + \frac{2}{\varepsilon_k}, \tag{7.27}$$

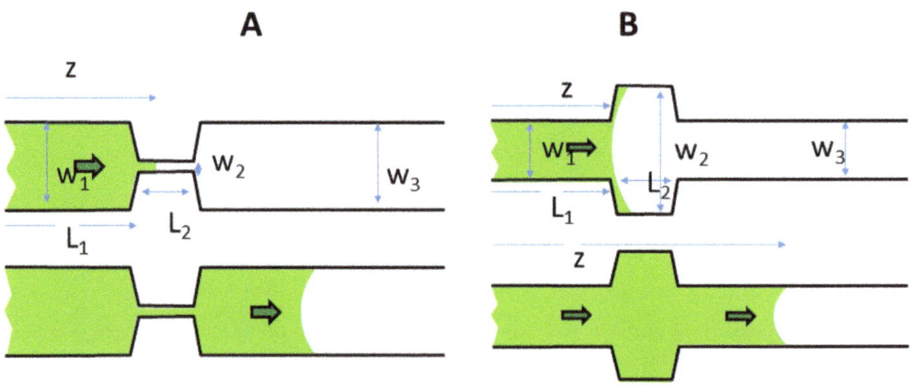

Figure 7.21. A schematic of abrupt changes of cross-section: (A) constriction; (B) enlargement. Note that the depth of the channel is constant and equal to h; also note that the widths after the changes of cross-section have been chosen to equal the initial widths: $w_1 = w_3$.

and we define the resistor's geometrical ratio q as $q = w_2/w_1$. Then, equating the capillary force (7.24) with the wall friction (7.26), the following expressions for the velocities are obtained

$$V_1 = \left(\frac{\gamma}{6\mu}\right)\frac{h}{z}\frac{f_1}{C_1},$$

$$V_2 = \left(\frac{\gamma}{6\mu}\right)h\,\frac{f_2}{\left[q\,L_1\,C_1 + (z - L_1)\,C_2\right]},$$

$$V_3 = \left(\frac{\gamma}{6\mu}\right)h\,\frac{f_1}{\left[\frac{L_2\,C_2}{q} + (z - L_2)\,C_1\right]}. \tag{7.28}$$

These velocities are plotted in figure 7.22 as a function of the travel distance for the two cases of a constriction and an enlargement and compared to the uniform channel velocity plot. A constriction temporarily increases the velocity in the constriction due to a (negative) increase in the capillary pressure but later results in a large decrease in the velocity due to the high friction in the constriction. Conversely, velocity decreases in an enlargement but recovers later because of reduced friction.

Hence, drastic changes of cross-section induce velocity jumps. Let us take the example of a succession of local enlargements in an open channel etched in silicon by deep reactive-ion etching (DRIE) (figure 7.23). The groove width is 50 μm and its depth 120 μm. Cylindrical wells (diameter 200 μm) are regularly etched every two

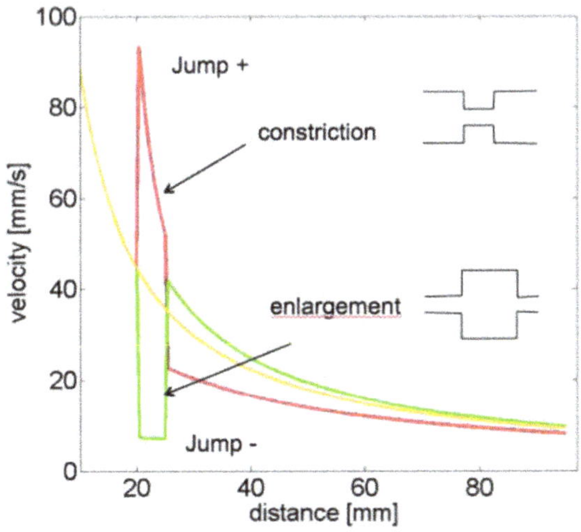

Figure 7.22. Capillary velocity vs. distance in three cases: a uniform channel (yellow), a constriction (red) and an enlargement (green). The constriction acts as a resistor, as it decreases the flow velocity downstream, but the enlargement acts as a local resistor only, as it does not decrease the flow velocity downstream. Reprinted with permission from [29]. Copyright (2015) American Chemical Society.

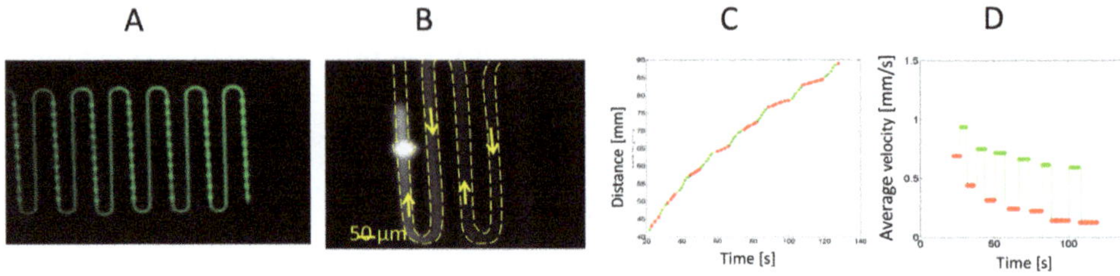

Figure 7.23. (A) A top view of the open channel; (B) a detailed view of the flow passing through an enlarged cylinder; (C) travel distance vs. time, showing stepwise progression (green in the channel, red in the cylinders); (D) velocity vs. time, showing the velocity 'discontinuities' when the flow crosses the cylinders. Panels (A) and (B) courtesy of Jean Berthier and Noémie Villard, CEA-Leti.

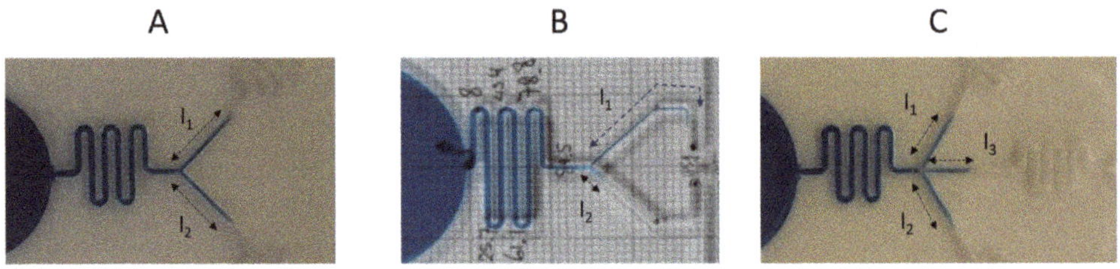

Figure 7.24. Some examples of bifurcation in rectangular open channels: (A) symmetrical bifurcation ($l_1 = l_2$); (B) asymmetrical bifurcation ($l_1 > l_2$). Here, l_1 corresponds to the channel that has the smallest cross-section. (C) Symmetrical trifurcation ($l_1 = l_2 > l_3$). Here, l_3 is smaller because the liquid flows first in the two side channels before entering the middle channel. Photographs courtesy of Jing J Lee and Jean Berthier, University of Washington.

winding lengths. Using the same approach as above, the travel distances can be determined as a function of time and in addition, the velocities can be determined by time derivation. The motion has the typical Lucas–Washburn–Rideal (LWR) shape but shows velocity discontinuities at the crossing of the cylinders.

7.3 Bifurcations and networks

Bifurcations are very common in microfluidic circuitry, at least for closed channels. They can be used for separating particulate flows [31], for droplet splitting [32, 33], or as pumping devices, as we shall see here [34, 35]. The behavior of liquid flows in open bifurcations and networks has, as yet, seen few investigations [36, 37]. Let us first present the flow behavior at a bifurcation.

7.3.1 Bifurcations

Channel bifurcations—or Y-shaped splits—may be symmetrical or asymmetrical. The symmetrical case is a simplification of the asymmetrical case. Figure 7.24 shows two examples of bifurcations and a trifurcation. The physics of the flow in a trifurcation is complicated and has not been studied so far, and we will only consider bifurcations in the following.

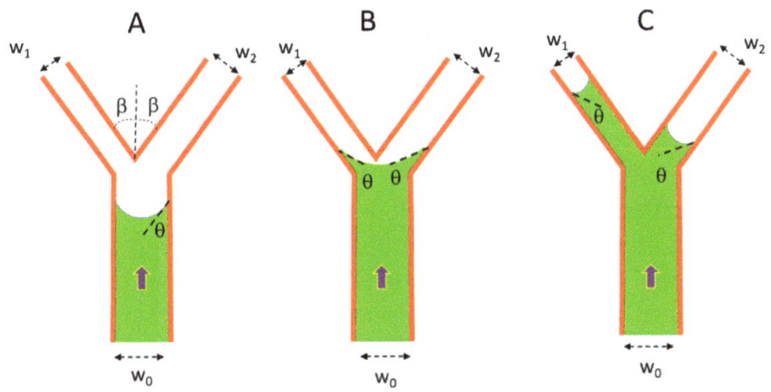

Figure 7.25. Left: a schematic of the flow in a bifurcation. (A) Liquid arrives at the bifurcation; (B) the meniscal curvature decreases in the widening at the fork, resulting in a very temporary decrease in velocity; (C) the liquid flow is divided into two subflows; the liquid in the narrower channel advances first faster due the higher (negative) capillary pressure.

Figure 7.25 sketches the capillary flow in an asymmetrical bifurcation [36]. During the first stage, the liquid advances in the root channel with a contact angle of θ—assumed to be the static Young contact angle. Second, the liquid enters the widening associated with the bifurcation and the meniscal curvature decreases, resulting in a temporary slowing of the liquid. Third, the liquid penetrates the two daughter channels, at first with a larger penetration in the narrower channel, which has the smallest negative capillary pressure (but wall friction quickly compensates for the capillary effect, and the velocity in the larger channel is the highest, as we shall see below).

Let us recall that the capillary pressure in a rectangular closed channel of dimensions w and h and with contact angles of $\{\theta_i,\ i = 1{,}4\}$, is

$$P_{\text{cap}} = \gamma \frac{(w \cos\theta_1 + w \cos\theta_2 + h \cos\theta_3 + h \cos\theta_4)}{w\,h} = \gamma\left(\frac{\cos\theta_1}{h} + \frac{\cos\theta_2}{h} + \frac{\cos\theta_3}{w} + \frac{\cos\theta_4}{w}\right), \tag{7.29}$$

which reduces to

$$P_{\text{cap}} = \gamma \frac{[\cos\theta(2h + w) - w]}{w\,h} \tag{7.30}$$

for a rectangular open channel, and

$$P_{\text{cap}} = 2\gamma \frac{[h\cos\theta - w]}{w\,h} \tag{7.31}$$

for a suspended channel—or rails. Hence, if the depth of the channel stays constant but the width decreases, relations (7.31) and (7.31) indicate that the magnitude of the capillary pressure increases. Indeed, relation (7.30) can be written as

$$P_{\text{cap}} = \gamma\left(\frac{2\cos\theta}{w} + \frac{\cos\theta}{h} - \frac{1}{h}\right), \tag{7.32}$$

which shows that the magnitude of the capillary pressure P_{cap} increases when the channel width decreases. The same reasoning applies in the case of a suspended channel ($P_{cap} = 2\gamma\left(\frac{\cos\theta}{w} - \frac{1}{h}\right)$)—or in the case of rails.

Hence, at first, the travel distance of the liquid in the daughter channel with the smaller width is the largest. However, this phenomenon is transient, because the wall friction in the smaller channel is much larger than that in the larger daughter channel. The travel distance in the larger channel progressively catches up with that of the smaller channel and even overtakes it. Such a behavior is seen in Figure 7.26 for a capillary flow of nonanol [38] in a rectangular open channel.

Let us analyze the flow behavior after the bifurcation in more detail. A sketch of the flow passing the bifurcation is shown in figure 7.27. The index j denotes the node corresponding to the bifurcation.

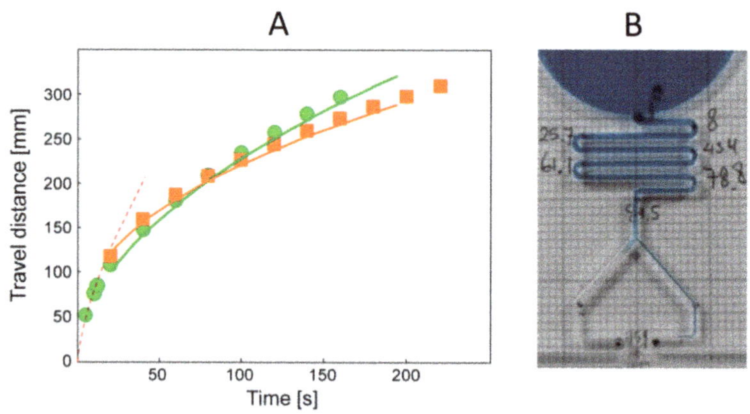

Figure 7.26. A comparison between the travel distances in two daughter channels. (A) The green circles and orange squares correspond respectively to the larger (w = 0.8 mm) and smaller (w = 0.4 mm) channels. The green and orange lines are the results of the numerical approach, and the red dotted line represents the LWR law. Reprinted with permission from [37]. Copyright (2019) American Chemical Society. (B) Image depicting the flow of pentanol (highlighted in blue) following separation through a bifurcation, courtesy Jing Lee and Jodie Tokihiro, University of Washington.

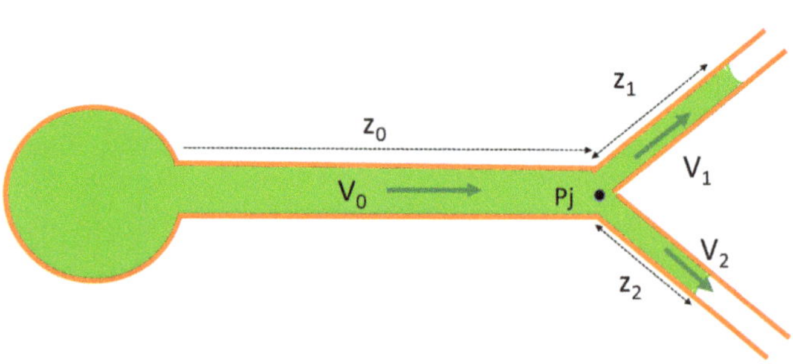

Figure 7.27. A schematic of a bifurcation. The index 0 characterizes the root channel, while indices 1 and 2 refer to the two daughter branches.

The pressure at the bifurcation is given by the difference between the pressure of the reservoir ($P = 0$) and that of the bifurcation (P_j). This difference is due to friction. The pressure at the bifurcation is given by

$$-P_j = z_0\, p_0\, \frac{\bar{\tau}_0}{S_0} \triangleq z_0\, p_0\, \mu \frac{V_0}{\bar{\lambda}_0} \frac{1}{S_0}, \tag{7.33}$$

where S_0 is the cross-sectional area, $\bar{\tau}_0$ the average wall friction, p_0 the total channel perimeter $p_0 = p_{w0} + p_{F0}$, z_0 the length of the root channel, μ the viscosity of the liquid, and $\bar{\lambda}_0$ the average friction length in the root channel. Relation (7.33) stems from the definition of the average friction length $\bar{\lambda}$. Note the negative sign before P_j, which is present because the capillary pressure at node j is negative.

When the liquid has passed the bifurcation and been divided between the two daughter branches, one can write an expression for the balance between the wall friction (from the inlet) and the capillary force (at the front meniscus) for each branch (denoted by 1 and 2):

$$z_1\, p_1\, \mu \frac{V_1}{\bar{\lambda}_1} \frac{1}{S_1} - P_j = \frac{\gamma \left(p_{\mathrm{W1}} \cos\theta - p_{\mathrm{F1}}\right)}{S_1} = \frac{\gamma\, p_1 \cos\theta_1^*}{S_1}, \tag{7.34}$$

where p_1 is the total perimeter of branch 1: $p_1 = p_{\mathrm{W1}} + p_{\mathrm{F1}}$, θ^* the generalized Cassie angle, and

$$z_2\, p_2\, \mu \frac{V_2}{\bar{\lambda}_2} \frac{1}{S_2} - P_j = \frac{\gamma\, p_2 \cos\theta_2^*}{S_2}, \tag{7.35}$$

for branch 2. The contact angle θ is the same everywhere, but the generalized Cassie angle θ^* is not the same if the shape of each daughter channel is different after the bifurcation. In addition, the conservation of mass yields

$$S_0\, V_0 = S_1\, V_1 + S_2\, V_2. \tag{7.36}$$

Noting that $V_i = dz_i/dt$ for each branch and that the pressure P_j is function of time, the system of equations (7.34), (7.35), and (7.36) is a system with three unknowns $\{P_j,\ z_1,\ z_2\}$, which can be solved analytically. If we note that $A_0 = p_0\mu/\bar{\lambda}_0$, $A_1 = p_1\mu/\bar{\lambda}_1$, $A_2 = p_2\mu/\bar{\lambda}_2$, and $B_1 = \gamma p_1 \cos\theta_1^*$, and $B_2 = \gamma\, p_2 \cos\theta_2^*$, manipulation of the system gives the travel distances in the two daughter channels

$$z_1^2 + 2z_0 \frac{A_0}{A_1} \frac{S_1}{S_0^2} (S_1 z_1 + S_2 z_2) = \frac{2B_1}{A_1}\tau. \tag{7.37}$$

and

$$z_2^2 + 2z_0 \frac{A_0}{A_2} \frac{S_2}{S_0^2} (S_1 z_1 + S_2 z_2) = \frac{2B_2}{A_2}\tau. \tag{7.38}$$

where $\tau = t - t_0$ (t_0 is the time at which the flow reaches the bifurcation). In the general case, because of their coupling, solving the system of equations given in (7.37) and (7.38) requires the help of numerical methods. Symmetrical bifurcation—daughter channels of identical cross-sections—is an exception for which a closed-form solution exists.

7.3.1.1 Symmetrical bifurcation

In this case, we have $A_1 = A_2 = A$, $B_1 = B_2 = B$, $S_1 = S_2 = S$, and $z_1 = z_2 = z$, (7.37) and (7.38) reduce to the same polynomial, and we are left with a unique equation:

$$z^2 + \left(4z_0 \frac{A_0}{A} \frac{S^2}{S_0^2}\right) z = \frac{2B}{A}\tau. \tag{7.39}$$

If the length of the root channel z_0 is small (and at the limit $z_0 = 0$), we retrieve the solution given by the generalized LWR model ($z^2 = \frac{2B}{A}t = \frac{\gamma}{\mu}2\bar{\lambda}\cos\theta^* t$). However, if z_0 is not small, the solution is no longer $z^2 \sim t$, because of the first-order term on the left-hand side of (7.39). For simplicity, we note that $K_1 = 4z_0 \frac{A_0}{A}\frac{S^2}{S_0^2}$ and $K_2 = \frac{2B}{A}$, and equation (7.39) becomes

$$z^2 + K_1 z - K_2 \tau = 0. \tag{7.40}$$

Relation (7.40) is a second-order polynomial in z. The solution can be written as

$$z(t) = \frac{-K_1 + \sqrt{K_1^2 + 4K_2\tau}}{2}. \tag{7.41}$$

Let us recall that z is the travel distance beyond the bifurcation, thus

$$Z(t) = z_0 + \frac{-K_1 + \sqrt{K_1^2 + 4K_2(t - t_0)}}{2}. \tag{7.42}$$

The capillary flow velocity after the symmetrical bifurcation is then

$$V(t) = \frac{K_{2,a}}{\sqrt{K_1^2 + 4K_2(t - t_0)}}. \tag{7.43}$$

Figure 7.28 shows the travel distances of nonanol in an open channel milled in PMMA before and after the bifurcation, obtained using relation (7.42). A very small

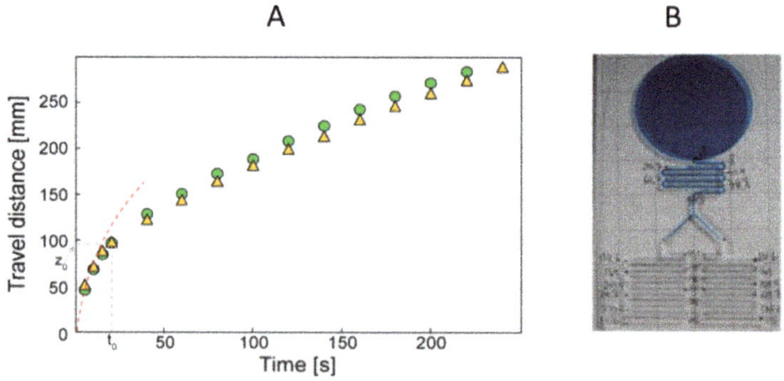

Figure 7.28. (A) Travel distances in a symmetrical bifurcation. The green circles and yellow triangles corresponding to the travel distances in each daughter channel are (nearly) the same; (B) a view of the colored nonanol passing the bifurcation. Reprinted with permission from [37]. Copyright (2019) American Chemical Society.

asymmetry in the travel distances is observed due to the imperfect milling of the bifurcation.

7.3.1.2 Asymmetrical bifurcation

The solution of systems (7.37) and (7.38) for asymmetrical bifurcations is more complicated to obtain. A numerical approach is required. Using the notations $K_1 = z_0 \frac{A_0}{A} \frac{S_1^2}{S_0^2}$, $K_2 = z_0 \frac{A_0}{A} \frac{S_2^2}{S_0^2}$, and $K_{12} = 2 z_0 \frac{A_0}{A} \frac{S_1 S_2}{S_0^2}$, a coupled numerical scheme can be set up, as follows:

$$z_{1,\, i+1} = -K_1 + \sqrt{K_1^2 + K_{12} z_{2,\, i} - \frac{2B_1}{A_1}\tau}$$

$$z_{2,\, i+1} = -K_2 + \sqrt{K_2^2 + K_{12} z_{1,\, i+1} - \frac{2B_2}{A_2}\tau}\,, \tag{7.44}$$

where i is the iteration index. Using an initial value of $z_{1,0} = 0$, for $\tau = 0$, $z_{2,1}$ is obtained by solving the second quadratic equation. Upon inserting $z_{2,1}$ into the first quadratic equation, one obtains $z_{1,1}$. The process is iterated until convergence occurs for each time. It has been observed that convergence is fast (approximately six iterations are required to achieve stabilization at less than 5 μm) if the seed ($z_{1,0}$ or $z_{2,0}$) corresponds to the 'slowest' channel. Divergence may occur if the seed is that of the 'fastest' channel. Once convergence is reached, the travel distances at times τ_i are obtained.

Figure 7.29 shows the asymmetric flow penetration in an asymmetric bifurcation for nonanol in U-grooves. The velocity in the smaller channel is initially the highest, then that of the larger channel quickly becomes the highest.

7.3.2 Networks and capillary pumps

The velocity of capillary flow in closed or open channels decreases as the flow proceeds down the length of the channel, varying in accordance with the inverse of the square root of time or the inverse of travel distance, as stated by the LWR law.

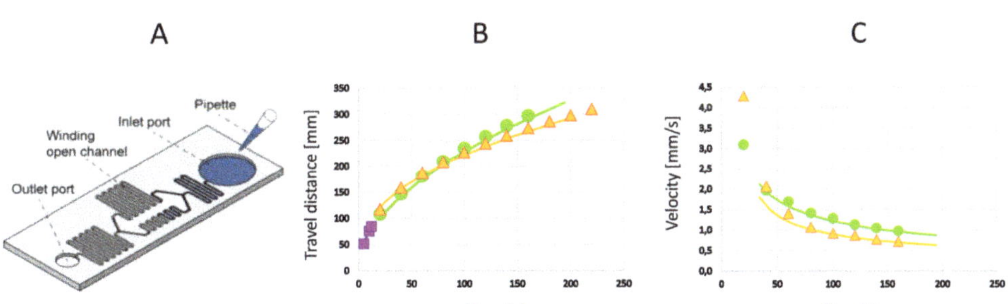

Figure 7.29. (A) A sketch of the experimental device. (B) Travel distances in the asymmetrical bifurcation: the green circles and yellow triangles correspond to the travel distances in each daughter channel, green corresponds to the largest channel. (C) The velocities in the two daughter channels.

In order to increase the flow rate and extend the duration of the flow, capillary pumps have been designed. Their principle is to provide a high internal capillary force—or the largest possible negative capillary pressure—higher than that of the main channel and a sufficient volume to store the pumped liquid. Many different designs have been proposed for closed channels, as shown in figure 7.30 [39–41]. The most common capillary pumps are composed of a maze of small channels or a fiber matrix.

The same types of design can be used in open microfluidic flows. Moreover, in open-microchannel systems, capillary pumping can also be performed by evaporation. For example, Nie and colleagues reported a study at the boundary between closed and open microfluidics in which they used a grid pierced with holes on top of the outlet port heated by a thin conductive layer to evaporate the liquid once it had reached the outlet port [41]. They found a linear relationship between the temperature rise and the flow rate. To date, no evaporation pump dedicated to open channels has been reported, to our knowledge.

In this section, an open-capillary pump—based on the geometry of a capillary tree—is presented [34, 35]. The design of the pump is shown in figure 7.31.

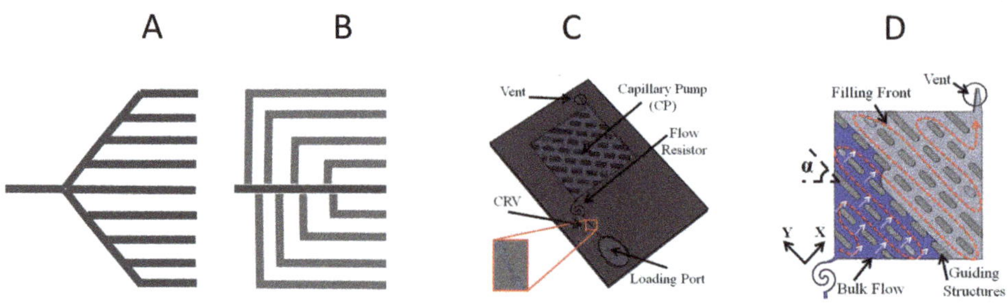

Figure 7.30. (A) and (B) Typical capillary pumps. Reproduced from [39] with permission of The Royal Society of Chemistry. (C) and (D) Capillary pumps consisting of arrays of pillars. Reprinted from [40], Copyright (2015), with permission from Springer.

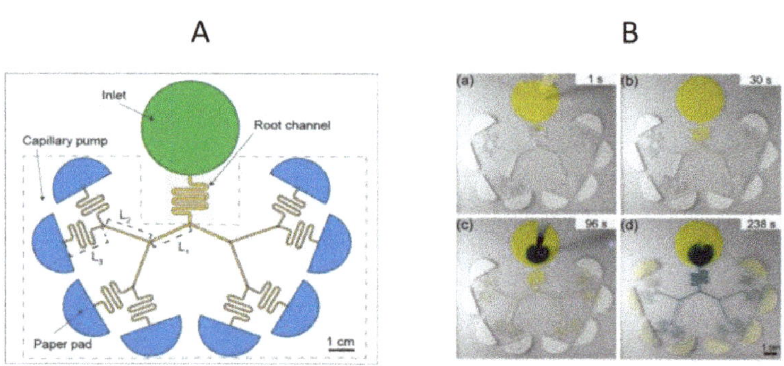

Figure 7.31. (A) A sketch of the pump, showing the inlet reservoir, the 'root' channel—channel of interest, the tree-like channels, and additional paper pad pumps. (B) The progression of nonanol in the network. The inlet reservoir is first filled with nonanol tinted yellow; once the yellow nonanol reaches the paper pads, blue-tinted nonanol is added. Reprinted from [35], with the permission of AIP Publishing.

The device consists of serpentine channels—including the root channel or channel of interest, where the flow velocity has to be maintained as high as possible—and a large inlet where the liquid is introduced using a pipette. Downstream from the root channel, the device has three branching levels and semicircular paper pads at the extremities of the last set of channel branches (figure 7.31(A)). All the channels have a rounded rectangular cross-section (U-grooves). The widths and depths of the channels are homothetically reduced by a factor of 0.85 after each bifurcation. The turns in the winding channels do not affect the capillary flow in the absence of capillary filaments, as the rounded bottom avoids the formation of filaments observed in channels of rectangular cross-section. The winding channels are sufficiently far from the entrance that the liquid velocity is sufficiently low that no Dean-type flows—i.e. centrifugal forces—develop [42].

The experimental approach consists in first filling the inlet reservoir with yellow-tinted nonanol (figure 7.31(B)). The travel distance of the yellow nonanol is measured and the flow velocity at each branch level is determined by time derivation. When the yellow nonanol reaches the pads, blue-tinted nonanol is added to the inlet channel. The two fluids, yellow and blue nonanol, undergo minimal mixing (figure 7.32), due to the laminar flow in small channels. This phenomenon is called Taylor–Aris dispersion [43, 44]. The length L necessary for complete mixing is

$$L \cong w\, Pe = w\, \frac{w\, V}{D}, \tag{7.45}$$

where w is the width of the channel, V the average velocity, Pe the Peclet nondimensional number, and D the diffusion coefficient. Numerically, in the case of nonanol, $D \sim 2 \times 10^{-9}$ m^2 s^{-1}, so that for a channel of width $w = 1$ mm where

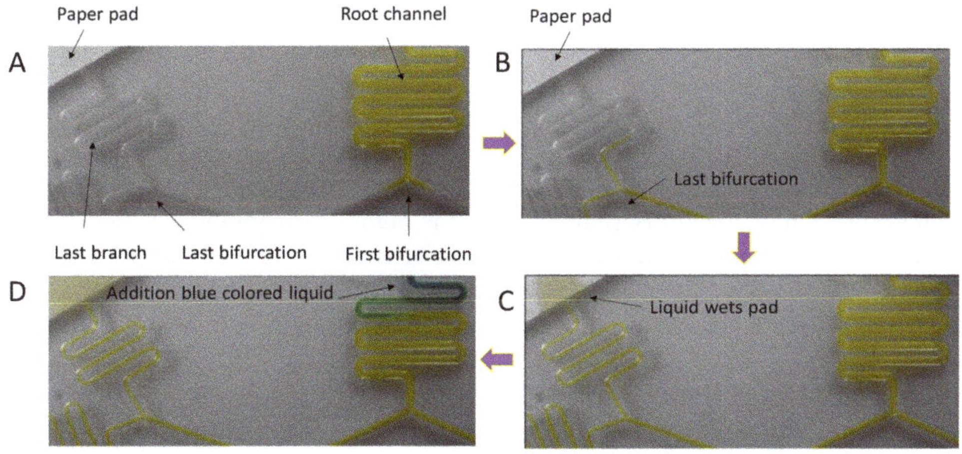

Figure 7.32. Details of the upper-left corner of the device during the progression of the liquid. (A) The flow fills the root channel and reaches the first bifurcation; (B) the flow reaches the last bifurcation, and the symmetry of the branches is approximately respected; (C) yellow-tinted liquid arrives at the paper pads; (D) the addition of blue-tinted liquid. Photographs courtesy of Jodi Tokihiro, Jing J Lee, and Wan-Chen Tu, University of Washington.

liquids flow at 3 mm s^{-1}, L^{-1}~1.5 m or 1500 mm, which is much larger than the total length of the tree. As there is little mixing of the two liquids in the channels, the travel distance of the blue-colored nonanol gives the velocity in the root channel due to the pumping performed by the paper pads.

The method used to determine the flow dynamics in the tree is based on the symmetric bifurcation model presented in the preceding section. It is found that the travel distance in branch level #n is given by

$$z_n = A_n\left[-1 + \sqrt{1 + \frac{\alpha^n C}{A_n^2}\left(t - t_{n-1}\right)}\right], \tag{7.46}$$

where $C = \frac{2\lambda\gamma\cos\theta_0^*}{\mu}$, α is the homothetic ratio, t_{n-1} is the time at which the liquid enters the nth channel, and A_n is a geometrical factor which depends on the channel lengths L_0 to L_{n-1} and α. The index 0 corresponds to the root channel. Once the flow enters the pads, Darcy's law is used to calculate the travel distance z_p in the paper pads (it is assumed that the flows in the tree branches are synchronized). Note that Darcy's law is detailed later in chapter 8.

$$\frac{dz_p}{dt} = V_p = -\frac{K}{\mu\phi}\nabla P = \frac{K}{\mu\phi}\frac{P_{cap} - P_j}{z_p}, \tag{7.47}$$

where P_{cap} is the capillary pressure of the paper, P_j is the pressure at the channel–paper pad junction, K is the permeability of the pad, and ϕ is its porosity. The index p refers to the paper, and the triplet (P_{cap}, K, ϕ) characterizes the paper strip. The complete analytical model for the flow in the tree and pads can be found at reference [35]. Here, we just show the efficiency of such a pump milled in PMMA for three liquids: IPA 50%, pentanol, and nonanol (figure 7.33). Velocities in the region of 3 mm s^{-1} are obtained over 150 s for the very viscous nonanol (μ~11 Pa s).

7.4 Filters

Filters can easily be inserted into open-channel capillary flows due to their accessibility. Figure 7.34 shows an enlargement of an open channel into which cellulose fibers have been inserted. Cellulose is highly hydrophilic, so that aqueous liquid is easily wicked by the whole matrix. The capillary flow continues under the action of the pressure difference between the tip of the flow—negative capillary pressure—and the inlet pressure, which is kept constant and equal to zero.

7.5 Open deterministic lateral devices

Deterministic lateral devices (DLDs) are one of the most efficient tools with which to separate particulate flows [45–49]. The principle is shown in figure 7.35(A) [50]. DLDs have rows of pillars—embedded microposts—at an angle with the carrier liquid direction. A critical diameter (D_c) is defined that depends on the geometry—the diameter of the posts, the distance between posts, and the distance between the rows of posts. This critical diameter determines the particle behavior. A particle

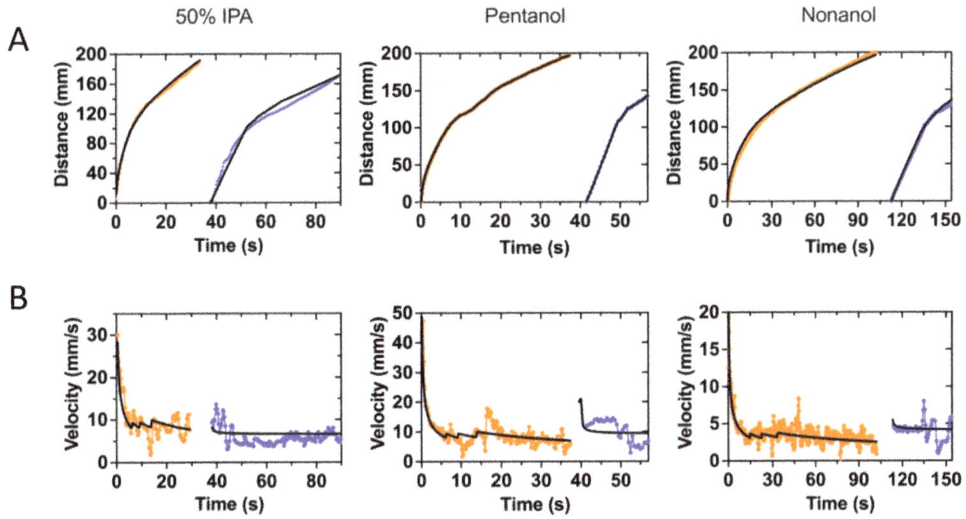

Figure 7.33. The travel distance and velocity over time for a 50% (v/v) IPA solution, pentanol, and nonanol in an open-channel capillary tree with paper pads placed at the branch extremities. (A) Experimentally determined travel distances in the bifurcating capillary tree vs. time for the first phase of the flow (orange dots) and the second phase of the flow (blue dots) after the flow has reached the pads and blue-tinted solution has been added to the inlet. The black lines are the theoretical results. (B) The velocities in the root channel vs. time when the tip of the flow is in the capillary tree (orange dots) and in the paper pads (blue dots). Reprinted from [35], with the permission of AIP Publishing.

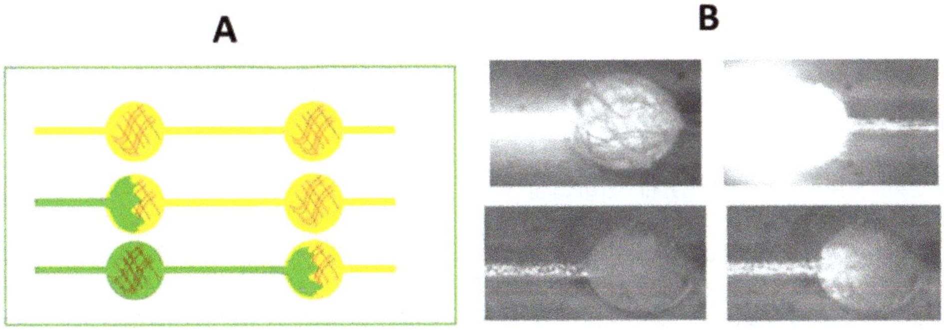

Figure 7.34. An open-capillary flow traversing a cellulose filter: the aqueous liquid is marked with fluorescent beads 2 μm in diameter. The capillary flow first wets the first chamber, which traps a large number of the beads, and then continues towards the second chamber. The second chamber traps the remaining beads. The capillary flow continues further in the channel. Photographs courtesy of Noémie Villard and Jean Berthier, CEA-Leti.

whose diameter is larger than the critical diameter follows the direction of the rows of pillars—i.e. is deviated laterally by the posts—and a particle whose diameter is smaller than the critical diameter follows a zig-zag mode roughly similar to that of the carrier flow.

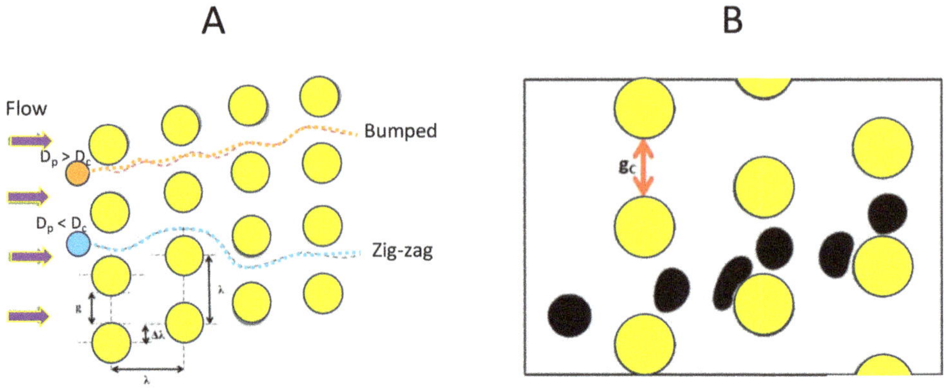

Figure 7.35. (A) The principle of a DLD with a cut-off diameter of D_c; (B) deformable cells in a DLD. Panels (A) and (B) reprinted from [50] with the permission of AIP Publishing.

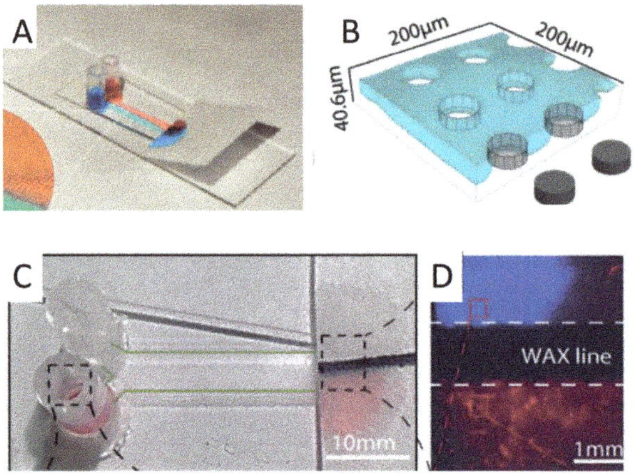

Figure 7.36. (A) An open DLD actuated by a paper pump; (B) details of an open DLD; (C) a top view of the device with RBCs (red color); (D) the separation of RBCs and MCF7 cells. Reprinted from [51]. CC BY 3.0.

It has been observed that the use of closed DLDs—DLDs with a bottom plate and a cover plate—is often hampered by the propensity for the device to be clogged due to the small channel sizes and especially when sorting 'soft,' deformable objects, such as cells. In particular, blood cells clog DLDs when the blood is not diluted enough. The advantages of open DLDs [51] are that, in cases of clogging, cleaning might be possible, depending on the types of particles; moreover, paper pads can be used as capillary pumps: the different fractions of separated particles or cells are deposited in the paper at different locations and can be handled independently by simply cutting the paper for analysis.

Figure 7.36 shows the open DLD designed by Tran *et al*, demonstrating the separation of red blood cells (RBCs) from MCF7 cells (a subtype of breast cancer cells) for a critical diameter D_c of 14 μm.

The authors concluded that sorting based on deterministic lateral displacement performs equivalently well in comparison with standard covered devices. Successful

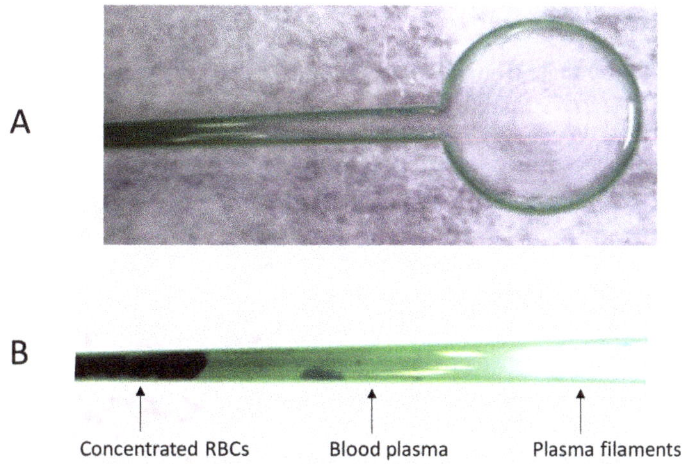

Figure 7.37. Capillary filaments of blood plasma (colored in green) in a diverging open channel: (A) the capillary filaments merge in the cylindrical outlet reservoir corners; (B) plasma flows ahead of whole blood due to the decrease in velocity in the divergent channel, leaving concentrated RBCs behind. Photographs courtesy of Myriam Cubizolles and Jean Berthier.

separation of cancer cells and parasites from blood with good viability was achieved in the open DLD.

7.6 Example of blood plasma separation in a diverging channel

The separation of blood plasma from red blood cells, known as blood plasma fractionation, is a fundamental operation in biology. The separation is complicated, partly due to the non-Newtonian properties of whole blood [52–56]. Blood plasma analysis is of great importance in finding viruses, metabolites, etc. In order to better visualize its flow and obtain sharper contrast with the red blood cells, human blood plasma can be colored green.

Figure 7.37 shows the behavior of whole blood in a diverging rectangular open channel ending in a circular reservoir. As the flow velocity decreases in the divergent channel, the less viscous component—blood plasma in this case—flows ahead of the bulk, which increases in RBC concentration, accelerating the effect. The capillary filaments then constitute plasma only.

7.7 Conclusions

The dynamics of microflows in open channels of varying cross-section is still a subject of investigation. In this chapter, the dynamics of capillary flow in characteristic geometries—such as converging and diverging open channels, bifurcations, and networks—has been presented. The understanding of this dynamics is of the utmost importance for the elaboration and design of capillary-driven microfluidic devices. Much is still to be done.

References

[1] Brakke K A 1992 Minimal surfaces, comers, and wires *J. Geom. Anal.* **2** 12–35
[2] Brakke K A 1992 The surface evolver *Exp. Math.* **1** 141–65
[3] Gibbs J W 1873 A method of geometrical representation of the thermodynamic properties of substances by means of surfaces *Trans. Conn. Acad. Arts Sci.* **2** 382–404
[4] Berthier J, Brakke K A and Berthier E 2016 *Open Microfluidics* (New York: Wiley)
[5] Berthier J, Brakke K A and Berthier E 2014 A general condition for spontaneous capillary flow in uniform cross-section microchannels *Microfluid. Nanofluid.* **16** 779–85
[6] Casavant B P *et al* 2013 Suspended microfluidics *Proc. Natl Acad. Sci.* **110** 10111–6
[7] Berthier J, Brakke K A, Gosselin D, Bourdat A-G, Nonglaton G, Villard N, Laffite G, Boizot F, Costa G and Delapierre G 2015 Suspended microflows between vertical parallel walls *Microfluid. Nanofluid.* **18** 919–29
[8] COMSOL Multiphysics (https://comsol.com)
[9] Berthier J and Brakke K A 2012 *The Physics of Microdroplets* (Beverly, MA: : Scrivener Publishing LLC)
[10] Ondarçuhu T and Raphaël E 1992 Etalement d'un ruban liquide à cheval entre deux substrats solides différents *C.R. Acad. Sci. Paris* **314** 453–6
[11] Ondarçuhu T 1995 Total or partial pinning of a droplet on a surface with a chemical discontinuity *J. Phys. II France* **5** 227–41
[12] Langbein D 2002 Canthotaxis/wetting barriers/pinning lines *Capillary Surfaces. Springer Tracts in Modern Physics* ed D Langbein (Berlin: Springer) vol 178
[13] Berthier J, Loe-Mie F, Tran V-M, Schoumacker S, Mittler F, Marchand G and Sarrut N 2009 On the pinning of interfaces on micropillar edges *J. Colloid Interface Sci.* **338** 296–303
[14] Olanrewaju A O, Robillard A, Dagher M and Juncker D 2016 Autonomous microfluidic capillaric circuits replicated from 3D-printed molds *Lab Chip* **16** 3804–14
[15] Menges J, Meffan C, Dolamore F, Fee C, Dobson R and Nock V 2021 New flow control systems in capillarics: off valves *Lab Chip* **21** 205–14
[16] Zimmermann M, Hunziker P and Delamarche E 2008 Valves for autonomous capillary systems *Microfluid. Nanofluid.* **5** 395–402
[17] Wang S, Zhang X, Ma C, Yan S, Inglis D and Feng S 2021 A review of capillary pressure control valves in microfluidics *Biosensors* **11** 405
[18] Melin J, Roxhed N, Gimenez G, Griss P, van der Wijngaart W and Stemme G 2004 A liquid-triggered liquid microvalve for on-chip flow control *Sensors Actuators* B **100** 463–8
[19] Zhang Y, Li Y, Luan X, Li X, Jiang J, Fan Y, Li M, Huang C, Zhang L and Zhao Y 2022 A 3D capillary-driven multi-micropore membrane-based trigger valve for multi-step biochemical reaction *Biosensors* **13** 26–39
[20] Feng J and Rothstein J P 2013 One-way wicking in open micro-channels controlled by channel topography *J. Colloid Interface Sci.* **404** 169–78
[21] Quéré D 1997 Inertial capillarity *Europhys. Lett.* **39** 533–8
[22] Young W-B 2004 Analysis of capillary flows in non-uniform cross-sectional capillaries *Colloids Surf.* A **234** 123–8
[23] Berthier J, Gosselin D and Berthier E 2015 A generalization of the Lucas–Washburn–Rideal law to composite microchannels of arbitrary cross-section *Microfluid. Nanofluid.* **19** 497–507
[24] Lebo R B 1921 Properties of mixtures of isopropyl alcohol and water (Contribution from the Laboratory of Chemical Products Division, Standard Oil Co.) pp 1005–16

[25] Berthier J, Brakke K A, Gosselin D, Navarro F, Belgacem N, Chaussy D and Berthier E 2017 On the halt of spontaneous capillary flows in diverging open channels *Med. Eng. Phys.* **48** 75–80

[26] Chen G-L, He G-Y, Sheng Y-J and Tsao H-K 2023 Imbibition dynamics in a U-groove microchannel with sudden enlargement *Langmuir* **39** 10993–1002

[27] Erickson D, Li D and Park C B 2002 Numerical simulations of capillary-driven flows in non-uniform cross-sectional capillaries *J. Colloid Interface Sci.* **250** 422–30

[28] Safavieh R and Juncker D 2013 Capillarics: pre-programmed, self-powered microfluidic circuits built from capillary elements *Lab Chip* **13** 4180

[29] Berthier J, Gosselin D, Pham A, Delapierre G, Belgacem N and Chaussy D 2016 Capillary flow resistors: local and global resistors *Langmuir* **32** 915–21

[30] Berthier J, Gosselin D, Pham A, Boizot F, Delapierre G, Belgacem N and Chaussy D 2016 Spontaneous capillary flows in piecewise varying cross section microchannels *Sensors Actuators* B **223** 868–77

[31] Roman S, Merlo A, Duru P, Risso F and Lorthois S 2016 Going beyond 20 μm-sized channels for studying red blood cell phase separation in microfluidic bifurcations *Biomicrofluidics* **10** 034103

[32] Shen Z, Coupier G, Kaoui B, Polack B, Harting J, Misbah C and Podgorski T 2016 Inversion of hematocrit partition at microfluidic bifurcations *Microvasc. Res.* **105** 40–6

[33] Chowdurhy I U, Mahapatra P S, Sen A K, Pattamatta A and Tiwari M K 2021 Autonomous transport and splitting of a droplet on an open surface *Phys. Rev. Fluids* **6** 094003

[34] Lee J J, Berthier J, Kearney K E, Berthier E and Theberge A B 2020 Open-channel capillary trees and capillary pumping *Langmuir* **36** 12795–803

[35] Tokihiro J C, Tu W-C, Berthier J, Lee J J, Dostie A M, Khor J W, Eakman M, Theberge A B and Berthier E 2023 Enhanced capillary pumping using open-channel capillary trees with integrated paper pads *Phys. Fluids* **35** 082120

[36] Mehrabian H, Gao P and Feng J J 2011 Wicking flow through microchannels *Phys. Fluids* **23** 122108

[37] Lee J J, Berthier J, Theberge A B and Berthier E 2019 Capillary flow in open microgrooves: Bifurcations and networks *Langmuir* **35** 10667

[38] PubChem: 5-Nonanol https://pubchem.ncbi.nlm.nih.gov/compound/5-nonanol

[39] Zimmermann M, Schmid H, Hunziker P and Delamarche E 2007 Capillary pumps for autonomous capillary systems *Lab Chip* **7** 119–25

[40] Safavieh R, Tamayol A and Juncker D 2015 Serpentine and leading-edge capillary pumps for microfluidic capillary systems *Microfluid. Nanofluid.* **18** 357–66

[41] Nie C, Frijns A J H, Mandamparambil R and de Toonder J M J 2015 A microfluidic device based on an evaporation-driven micropump *Biomed. Microdevices* **17** 47

[42] Nivedita N, Ligrani P and Papautsky I 2017 Dean flow dynamics in low-aspect ratio spiral microchannels *Sci. Rep.* **7** 44072

[43] Taylor G and Geoffrey 1953 Dispersion of soluble matter in solvent flowing slowly through a tube *Proc. Royal Soc. London, Ser.* A **219** 186–203

[44] Aris R 1956 On the dispersion of a solute in a fluid flowing through a tube *Proc. R. Soc.* A **235** 67–77

[45] Huang L R, Cox E C, Austin R H and Sturm J C 2004 Continuous particle separation through deterministic lateral displacement *Science* **304** 987–90

[46] Davis J A, Inglis D W, Morton K J, Lawrence D A, Huang L R, Chou S Y, Sturm J C and Austin R H 2006 Deterministic hydrodynamics: taking blood apart *Proc. Natl Acad. Sci.* **103** 14779–84

[47] Inglis D W, Davis J A, Austin R H and Sturm J C 2006 Critical particle size for fractionation by deterministic lateral displacement *Lab Chip* **6** 655–8

[48] Long B R, Heller M, Beech J P, Linke H, Bruus H and Tegenfeldt J O 2008 Multi-directional sorting modes in deterministic lateral displacement devices *Phys. Rev.* E **76** 046304

[49] Pariset E, Pudda C, Boizot F, Verplanck N, Berthier J, Thuaire A and Agache V 2017 Anticipating cutoff diameters in deterministic lateral displacement (DLD) microfluidic devices for an optimized particle separation *Small* **13** 1701901

[50] Aghilinejab A, Aghaamoo M and Chen X 2019 On the transport of particles/cells in high-throughput deterministic lateral displacement devices: implications for circulating tumor cell separation *Biomicrofluidics* **13** 034112

[51] Tran T S H, Ho B D, Beech J P and Tegenfeldt J O 2017 Open channel deterministic lateral displacement for particle and cell sorting *Lab Chip* **17** 3592

[52] Gosselin D, Huet M, Cubizolles M, Rabaud D, Belgacem N, Chaussy D and Berthier J 2016 Viscoelastic capillary flow: the case of whole blood *AIMS Biophys.* **3** 340–57

[53] Fedosov D A, Pan W, Caswell B, Gompper G and Karniadakis G E 2011 Predicting human blood viscosity in silico *Proc. Natl Acad. Sci.* **108** 11772–7

[54] Fu T, Carrier O, Funfschilling D, Ma Y, Li H and Chen G 2016 Newtonian and non-Newtonian flows in microchannels: in line rheological characterization *Chem. Eng. Technol.* **39** 987–92

[55] Steffe J F 1982 *Rheological Methods in Food Process Engineering* 2nd edn (San Francisco, CA: Freeman)

[56] Herschel W H and Bulkley R 1926 Konsistenz-messungen von Gummi-Benzollösungen *Kolloid-Z* **39** 291–300

IOP Publishing

Open-Channel Microfluidics (Second Edition)
Fundamentals and applications
Jean Berthier, Ashleigh B Theberge and Erwin Berthier

Chapter 8

Capillary flow in fibrous media

Fibrous media—such as paper pads or threads—are widely used in biotechnology. In particular, paper pads (paper strips) are commonly used as sensors for different applications, such as pregnancy tests, glucose monitoring, and virus detection. These applications use only a drop of sample liquid of approximately 4 μl [1, 2]. In the case of whole blood monitoring, the sample is obtained by a finger prick. Such devices are highly compatible with electrochemical detection [3]. Paper pads and threads are also used for blood plasma extraction [4, 5] and *in vivo* monitoring [6]. The development of smart textiles—textiles with incorporated sensors—is progressing quickly and requires determination of the capillary flow in structured woven fibers [7, 8].

Biosensors are the most common applications of fibrous media. Figure 8.1 sketches typical biosensing devices: paper pads and cotton thread. The principle is the same in both cases. A drop of sample is pipetted into the inlet hole and the liquid wicks a sample pad containing reagents. The sample continues its capillary motion towards a test zone where detection occurs, and the flow continues into the absorbent pad.

Flow in fibrous media falls in the category of open microfluidics because the interface with the surrounding air constantly increases with the propagation of the fluid between the fibers. In this case, the flow channel is approximately defined by the envelope of the wetted fibers. Figure 8.2 shows some examples of fibrous media and illustrates the complexity of the approach required for flow field determination.

8.1 Parameters characterizing the capillary flow in fibrous media

Fibrous media have a complicated topology that depends on the fiber compaction, orientation, and homogeneity. They are characterized by three general main properties—porosity, permeability, and capillary pressure—and some more specific properties, such as tortuosity, pore size, and fiber alignment or randomness [11, 12].

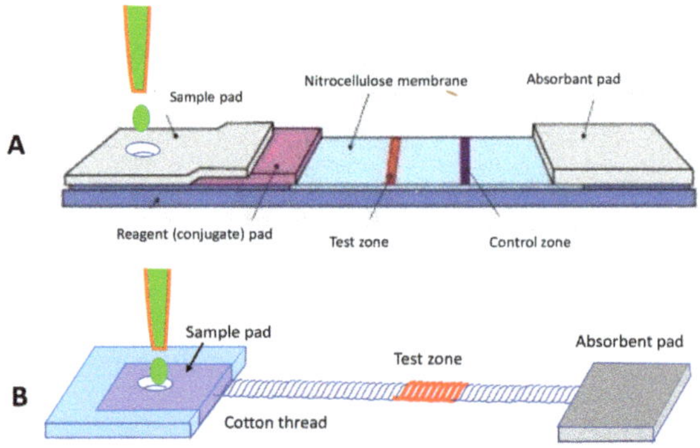

Figure 8.1. (A) A sketch of the use of a paper pad for biosensing, showing the different paper components from the inlet (left) to the sensing strip (red) and the absorbent pad (right); (B) a sketch of a thread used for biosensing, showing similarities with the paper pad.

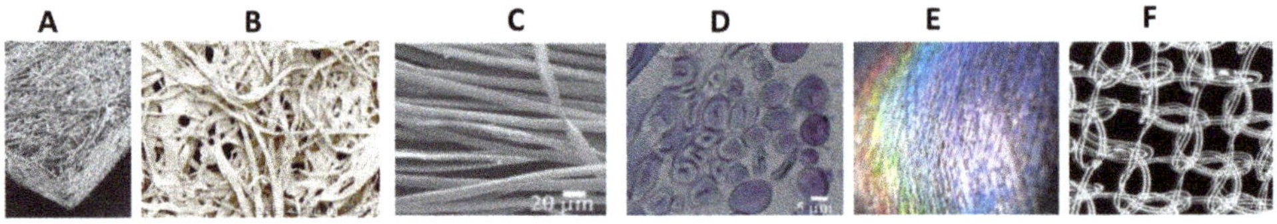

Figure 8.2. Some examples of fibrous media: (A) and (B) fibers in paper pads (photographs from https://www.handprint.com and https://www.blog.boisechoices.com); (C) and (D) threads used for blood plasma extraction (reprinted from [8], with the permission of AIP Publishing); (E) thermofibers for smart bandages; (F) woven tissue for tissue engineering. Panels (E) and (F) reprinted from [10], Copyright (1998), with permission from Elsevier.

Porosity—usually denoted by ϕ—is the ratio between the free volume and the total volume of a fibrous medium. It is a nondimensional number that lies between zero and one.

$$\phi = 1 - \frac{V_{\text{fiber}}}{V_{\text{total}}} = 1 - \frac{\rho}{\rho_{\text{f}}}, \tag{8.1}$$

where V_{fiber} and V_{total} are the volume of the fibers and the total volume in a control volume, respectively, and ρ and ρ_{f} the density of the porous media and the density of the solid fibers, respectively.

Permeability—usually denoted by K—depends on the average free area between fibers. It has the dimension of a surface area and is expressed in square meters (m^2). In microfluidics, the pore size (the average distance between fibers) is generally of the order of 1–10 μm, so that the value of K in square meters is in the range 10^{-12}–10^{-10} m^2. Note that in the case of an anisotropic medium, for example, if the fibers in the paper matrix are not randomly disposed—which is often the case, the permeability depends on the orientation of the considered cross-section. In the case of highly inhomogeneous media, permeability is a symmetrical matrix $[K]$.

The capillary pressure—usually denoted by P_{cap} or P_c—is the pressure differential across the different advancing meniscuses in the medium. Because of the different pore sizes that the advancing interface encounters during its progression, the capillary pressure varies locally and the flow front shows geometric fluctuations (figure 8.3) [13, 14].

However, an equivalent capillary pressure can be determined that gives the average imbibition length. First, we recall that the capillary pressure in a cylindrical tube is

$$P_{cap} = \frac{F_{cap}}{S} = \frac{2\pi R\,\gamma\cos\theta}{\pi R^2} = \frac{2\gamma\cos\theta}{R}. \tag{8.2}$$

This equation can also be derived from the Laplace pressure across the liquid–vapor interface in a tube of radius R (figure 8.4):

$$P_{cap} = \Delta P = \gamma\frac{2}{\frac{R}{\cos\theta}}. \tag{8.3}$$

Considering the flow in the fibrous medium as a collection of approximately parallel flow channels and making an analogy with the Washburn law, it is usual to approximate the capillary pressure by the relation [15, 16]

$$P_{cap} = \frac{2\gamma\cos\theta}{r_{eq}}, \tag{8.4}$$

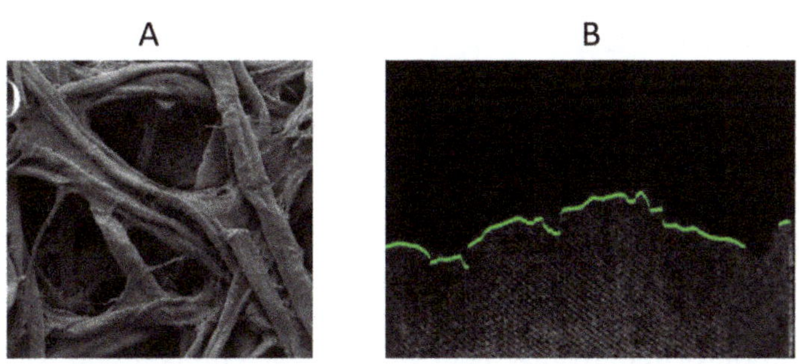

Figure 8.3. (A) A microscopic view of fibers in a paper pad; (B) the geometrical variability of the pore size leads to an uneven liquid front. Panel (A) reprinted from [13], Copyright (2016), with permission from Elsevier; panel (B) reprinted from [14], Copyright (2017), with permission from Springer.

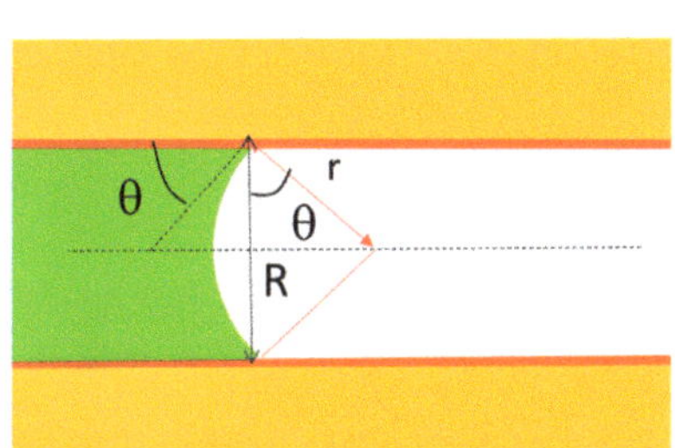

Figure 8.4. The relation between the curvature radius r and the tube radius R for a cylinder.

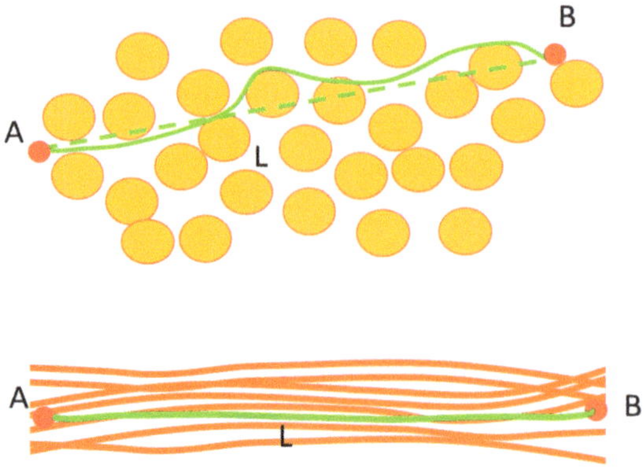

Figure 8.5. A sketch of the tortuosity in a granular medium (top) and in a thread (bottom): $\tau = L/\| AB \|$.

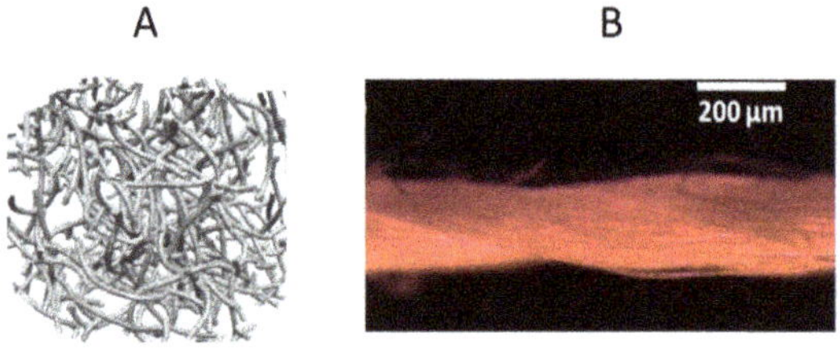

Figure 8.6. The structural difference between randomly oriented fibers (A) and fibers aligned in a thread (B). Panel (B) reprinted from [17], Copyright (2012), with permission from Springer.

where r_{eq} is the equivalent pore radius. The pore radius was at first somewhat nebulous and has been the subject of many investigations. The notion of equivalent pore radius is developed in section 8.5.2.

In addition, there are also other parameters associated with a fibrous medium. For example, tortuosity—usually denoted by τ—is the ratio of the shortest distance between two points in the medium and the length of the straight line. Tortuosity gives an indication of the arrangement of fibers or grains inside the medium. For example, in a compact granular medium, tortuosity can be high, whereas it is rather small in threads (if measured along the orientation of the fibers), as shown in figure 8.5. Note that the tortuosity depends on the degree of fiber alignment.

The alignment of the fibers is also a crucial factor: the dynamics of the flow are not the same in random fibrous media and in nearly aligned threads (figure 8.6). The relations between porosity, permeability, tortuosity, and fiber arrangement are presented in section 8.5.3.

8.2 Flow dynamics in fibrous media

Fluid penetration into papery or porous media is a highly complex process due to the complexity of the porous media. As underlined by Castro [18], the detailed physics of the flow take place within the micro and macro structures of porous media. However, from the perspective of the design and conception of fibrous systems and sensors, the area of interest is the macroscopic regime. We next present the two different approaches to the dynamics of flows in fibrous media.

8.2.1 The Lucas–Washburn analogy

The first approach considers the 1D flow in fibrous media to be a collection of parallel flow channels with an adapted equivalent radius—r_{eq}—[19–21]. This approach seems logical when the fibers are more or less aligned but less so when the fibers are entangled (figure 8.7). In this last case, the difficulty lies in the determination of the equivalent radius—r_{eq}.

According to the developments of chapter 2, the momentum balance for the capillary flow inside a vertical tube of inner radius R is given by [21]

$$\rho\frac{\mathrm{d}}{\mathrm{d}t}\left(h\frac{\mathrm{d}h}{\mathrm{d}t}\right) = \frac{2\gamma\cos\theta}{R} - \frac{8\mu}{R^2}\left(h\frac{\mathrm{d}h}{\mathrm{d}t}\right) - \rho g h, \tag{8.5}$$

where h is the imbibition height, ρ the liquid density, μ the liquid viscosity, and θ the contact angle. The left-hand side of (8.5) corresponds to inertia, the first term on the right-hand side to the capillary force, the second term to wall friction, and the third term to gravity. As we have seen in chapter 2, inertia is only important in the very early stages of imbibition. Hence, for a horizontal fibrous medium, replacing the height h by the travel distance z gives

$$\frac{2\gamma\cos\theta}{R} = \frac{8\mu}{R^2}\left(z\frac{\mathrm{d}z}{\mathrm{d}t}\right). \tag{8.6}$$

The term $2\gamma\cos\theta/R$ is the capillary pressure. Equation (8.6) is the Lucas–Washburn expression, i.e.

$$z = \sqrt{\frac{R\gamma\cos\theta}{2\mu}t}. \tag{8.7}$$

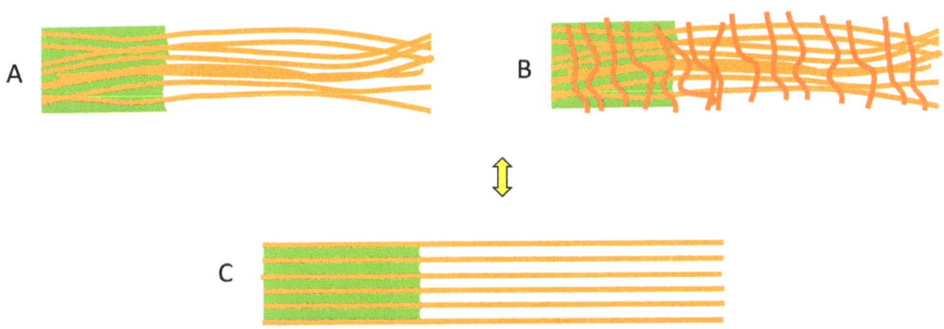

Figure 8.7. The equivalence between fibrous media—a thread (A), a pad (B), and a bundle of parallel tubes of equivalent radius r_{eq} (C).

The equivalence shown in figure 8.7 replaces the tube radius R with an equivalent capillary radius r_{eq}. Then

$$z = \sqrt{\frac{r_{\mathrm{eq}}\gamma \cos\theta}{2\mu} t}. \tag{8.8}$$

In this case, the term $2\gamma \cos\theta / r_{\mathrm{eq}}$ is the capillary pressure. The fluid velocity is then

$$V = \frac{1}{2}\sqrt{\frac{r_{\mathrm{eq}}\gamma \cos\theta}{2\mu}}\sqrt{\frac{1}{t}}. \tag{8.9}$$

Eliminating the time from (8.8) and (8.9) yields

$$V = \frac{r_{\mathrm{eq}}\gamma \cos\theta}{4\mu}\frac{1}{z}. \tag{8.10}$$

The difficulty here is the determination of the equivalent radius as a function of the morphology of the fibrous medium.

8.2.2 Darcy's law

The second approach makes use of Darcy's law [22, 23]. It is assumed here that the saturation—i.e. the percentage of liquid in the pores—is total (100%) behind the advancing fluid front (the so-called 'sharp front' approach [24]). In this case, Darcy's law links the average flow velocity V (Darcy's velocity) to the pressure gradient between the inlet reservoir and the advancing front meniscus

$$\overline{V} = -\frac{K}{\mu}\nabla P, \tag{8.11}$$

where K is the medium's permeability (in units of m^2) and μ the viscosity of the liquid (in units of Pa s or kg m^{-1} s^{-1}). Darcy's law has the same form as Fick's law for mass transport and Fourier's law for heat flux [25]. Note that the real velocity in the pores is

$$V = \frac{V}{\phi} = -\frac{K}{\mu\phi}\nabla P, \tag{8.12}$$

where ϕ is the porosity of the medium. Since the porosity ϕ is smaller than one, the Darcy velocity is larger than the average velocity (figure 8.8).

For a paper strip of length L, width w, and thickness h, the volumetric flow rate is then

$$Q = -\frac{K}{\mu} wh \frac{\Delta P}{L}, \tag{8.13}$$

where ΔP is the pressure difference between the inlet and the liquid front in the paper. Usually, the inlet pressure is null and the pressure at the front is the capillary

Figure 8.8. (A) A sketch of the advancing interface in a fibrous medium; (B) the Darcy velocity (yellow arrows) is not always the same due to local inhomogeneities. Panel (B) reprinted from [14], Copyright (2017), with permission from Springer.

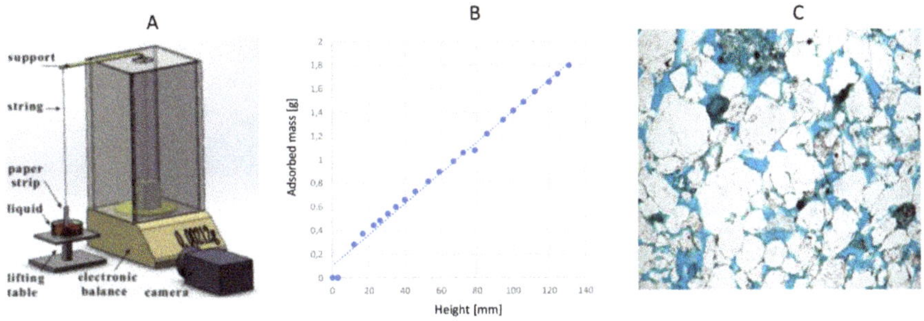

Figure 8.9. (A) A device used to measure porosity by the weighing method. Reprinted from [26], Copyright (2015), with permission from Elsevier. (B) The correlation between the adsorbed liquid mass and height. (C) The optical method; the pores are shown tinted in blue (courtesy PERM Inc. https://perminc.com/).

pressure. Hence, a porous medium is characterized by its porosity, permeability, and capillary pressure:

$$V = \frac{K}{\mu\phi}\frac{P_{\text{cap}}}{z}, \tag{8.14}$$

where z is the imbibition length. The solution of (8.9) and (8.13) requires the determination of r_{eq}, ϕ, K, and P_{cap} to be determined. Upon comparing relations (8.4) and (8.14), a similarity between the Lucas–Washburn and Darcy approaches appears. We investigate this similarity in the following section.

8.3 Determining porosity, permeability, and capillary pressure

In this section, we present an investigation of the different means of determining the three main parameters required to predict the capillary flow in fibrous media.

8.3.1 Porosity

The most usual method for determining the porosity of a fibrous medium is the experimental weighing method. The principle is shown in figure 8.9(A). The fibrous pad is placed vertically and contacts the surface of a liquid reservoir. The adsorbed weight is measured using an electronic balance [26]. The relation between the adsorbed mass and the height of the liquid is

$$m = \rho \, \phi \, ewh, \tag{8.15}$$

where m is the adsorbed mass, ρ the liquid density, e and w the thickness and width of the pad, respectively, and h the imbibed height. Note that paper swelling due to the imbibed liquid, partial saturation, and evaporation can perturb the measurement. However, if the different measurements corresponding to different heights are aligned (as shown in figure 8.9(B) for an absorbent pad of a commercial paper strip), the porosity can be simply deduced from the slope of the linear relation $\phi = \text{slope}/\rho ew$.

Another means for porosity measurement is the optical method (figure 8.9(C)). This method consists in filling the pores with a colored liquid or gel and analyzing the fraction of colored surface in a section. A comparison between the two methods can be found in [27].

8.3.2 Tortuosity

As shown in figure 8.5, tortuosity depends on the porosity of the porous medium and the arrangement of the fibers. Different methods have been proposed to measure tortuosity. The most usual method is the diffusion method, in which the diffusion of a chemical component in the porous medium is monitored over time. For example, Delgado [27] and Diao and coworkers [28] used the diffusion of salt in water contained in the porous media. In agreement with intuition (figure 8.10(B)), it can be shown that tortuosity is a monotonically decreasing function of porosity [29] (figure 8.10(C)).

8.3.3 Permeability and capillary pressure

Permeability and capillary pressure can be approached via theoretical of experimental means. Let us first present the theoretical approach.

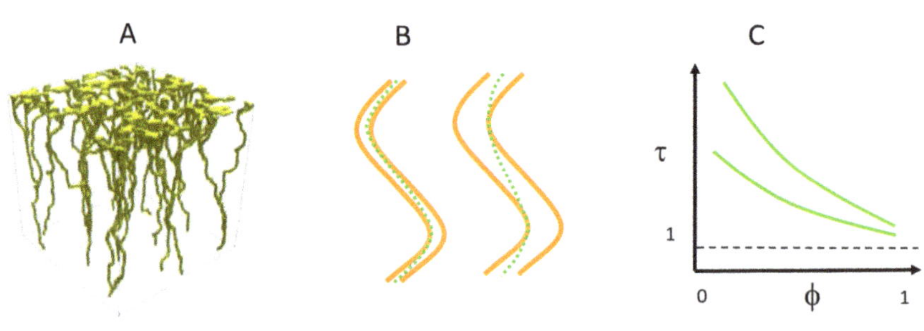

Figure 8.10. (A) A visualization of liquid channels in a compact porous medium (the solid structure has been removed), reprinted from [30] Copyright (2021), with permission from Elsevier; (B) tortuosity in channels of different sizes: the length of the green dotted line is longer for a narrow channel (small porosity) than for a wider channel (high porosity); (C) typical plots of tortuosity vs. porosity for two different porous media.

8.3.3.1 The theoretical approach

Let us consider relation (8.14), obtained using the Darcy's law, and substitute in the expression for capillary pressure (8.4):

$$V = \frac{K}{\mu\phi}\frac{P_{\text{cap}}}{z} = \frac{K}{\mu\phi}\frac{2\gamma\cos\theta}{r_{\text{eq}}\, z}. \tag{8.16}$$

Comparing this expression to the Lucas–Washburn equivalence expression (8.10) yields

$$V = \frac{r_{\text{eq}}\gamma\cos\theta}{4\mu}\frac{1}{z} = \frac{K}{\mu\phi}\frac{2\gamma\cos\theta}{r_{\text{eq}}\, z}. \tag{8.17}$$

The permeability K is deduced from (8.17)

$$K = \phi\,\frac{r_{\text{eq}}^2}{8}. \tag{8.18}$$

On the other hand, the capillary pressure is also a function of the capillary radius r_{eq}, as stated in relation (8.4):$P_{\text{cap}} = 2\gamma\cos\theta/r_{\text{eq}}$.

Many investigations have been performed to determine the value of the capillary radius as a function of the interior topology of a porous medium. For threads (aligned fibers), Melciu and coworkers [31] worked out the relation

$$r_{\text{eq}} \cong \frac{r_{\text{f}}}{2}\left(\sqrt{\frac{2\pi}{3\sqrt{3}(1-\phi)}} - 1\right), \tag{8.19}$$

where r_{f} is the radius of the fibers. Given that the porosity of fibrous media is usually of the order of 0.8–0.95, we obtain the approximation $4r_{\text{f}} > r_{\text{eq}} > 0.7r_{\text{f}}$.

However, in the general case, the capillary radius is a nebulous concept for granular as well as fibrous media. This is why correlations for the permeability have been investigated using the fiber radius r_{f} as a reference, instead of the equivalent capillary radius r_{eq}. The most known and used correlation was that proposed by Kozeny and Carman (CK) [32, 33]. The CK correlation was initially developed for granular porous media and has been extended to fibrous media [29].

$$K = \frac{1}{C_{\text{CK}}}\frac{\phi^3}{S^2\,\tau^2}, \tag{8.20}$$

where S is the specific solid surface area (dimension 1 m^{-1})—the ratio between the wetted surface S_{wetted} and the volume of solid V_{solid}, and C_{CK} is the CK constant. Although approximate because of the C_{CK} coefficient that depends on the topology of the porous media, this relation is widely used in the domain of granular porous media. We just indicate here the main characteristics of the CK approach for fiber bundles. If r_{f} denotes the fiber radius, S_{f} the wetted surface of the fibers, V_{ol} the total volume, V_{f} the volume of the fibers, and S the specific surface as the ratio between the wetted surface S_{f}—the surface of the fibers—and the volume of the solid V_{f}, one finds

$$S = \frac{S_{\text{f}}}{V_{\text{f}}} = \frac{S_{\text{f}}}{(1-\phi)V_{\text{ol}}} = \frac{2}{r_{\text{f}}}, \tag{8.21}$$

under the assumption of cylindrical fibers. Using Poiseuille's law, the liquid velocity is

$$V = -\frac{D_h^2}{32\mu}\nabla P, \tag{8.22}$$

where D_h is the hydraulic diameter given by

$$D_h = \frac{4\phi\ V_{ol}}{S_f} = \frac{4\phi\ V_{ol}}{(1-\phi)S_f} = \frac{4\phi}{(1-\phi)S} = \frac{2\phi r_f}{(1-\phi)}. \tag{8.23}$$

Then, substituting this into (8.22) and comparing the result to Darcy's law, we obtain

$$V = -\frac{4\phi^2 r_f^2}{32\mu\,(1-\phi)^2}\nabla P = -\frac{K}{\phi\mu}\nabla P, \tag{8.24}$$

and

$$\frac{K}{d_f^2} = \frac{\phi^3}{32(1-\phi)^2}, \tag{8.25}$$

where d_f is the diameter of the fibers ($d_f = 2r_f$). In order to extrapolate this to disordered fibers, the coefficient 32 must be replaced by a coefficient ψ_{CK}:

$$\frac{K}{d_f^2} = \frac{\phi^3}{\psi_{CK}\,(1-\phi)^2}. \tag{8.26}$$

Carman [33] linked the coefficient ψ_{CK} to the tortuosity by

$$\psi_{CK} = C\,\tau^2, \tag{8.27}$$

where C is a coefficient called the 'pore shape factor.' The Carman–Kozeny expression for fibrous media is then

$$\frac{K}{d_f^2} = \frac{\phi^3}{C\,\tau^2\,(1-\phi)^2}. \tag{8.28}$$

This expression is similar to that for granular media (8.20). The difficulty now lies in determining ψ_{CK} or C. The value of ψ_{CK} is 32 for straight cylinders. Generally, depending on the topology of the porous media, ψ_{CK} varies between 80 and 320 [29]. Hence, an experimental approach is still needed for the determination of K.

Many correlations similar to the Carman–Kozeny correlation have been proposed, depending on the particular topology of the fibrous medium. For example, Crawford *et al*, Nogai *et al*, and Zhu *et al* [34–36] proposed a correlation of the form

$$\frac{K}{d_f^2} = Cste\frac{\phi^3}{(1-\phi)^\beta}, \tag{8.29}$$

where d_f is the fiber diameter and β is a coefficient that depends on the medium; it usually lies between 1.5 and 2. In particular, Zhu and coworkers found the approximation [36]

$$\frac{K}{d_f^2} \approx 0.022 \frac{\phi^3}{(1-\phi)^2}, \tag{8.30}$$

where the coefficient 0.022 is an average for different types of fibers.

8.3.3.2 The experimental approach

Let us now focus on the experimental measurement of the permeability and capillary pressure of fibrous media.

Consider a paper pad—or a fibrous medium—placed vertically in contact with a wetting liquid. In such a case, the liquid rises in the vertical fibrous media until an equilibrium is reached between the hydrostatic pressure and the capillary pressure. Using equation (8.5), for a height h close to equilibrium, the velocity V is small, i.e. $V \to 0$, the inertia (the term on the left-hand side of the equation) is negligible and so is the friction (the second term on the right-hand side of the equation). Then, we are left with

$$P_{cap} = \rho\, g\, h_{eq}, \tag{8.31}$$

where ρ is the density of the fluid, g the gravitational constant, and h_{eq} the equilibrium height (maximum capillary rise). Note that it takes a long time for the liquid to reach its equilibrium height—sometimes a few hours—and evaporation should be eliminated by controlling the surrounding atmosphere.

Generally, experiments are needed to estimate the permeability and capillary pressure of fibrous media. The usual method consists in placing the fibrous medium—such as the paper pad—vertically in contact with a liquid and monitoring the capillary rise as a function of time. If we first consider a cylinder of interior radius R, using (8.5) and neglecting inertia, the capillary rise is given by

$$\frac{2\gamma \cos\theta}{R} = \frac{8\mu}{R^2}\left(h\frac{dh}{dt}\right) - \rho g h, \tag{8.32}$$

Pezron and Quéré [37] decomposed (8.32) in two regimes: the initial regime, in which gravity is negligible, and the asymptotic regime close to the equilibrium height. The full solution of (8.32) was given by Fries and Dreyer [38]:

$$h = \frac{a}{b}\left[1 + W\left(-e^{-1-\frac{b^2}{a}t}\right)\right], \tag{8.33}$$

where $a = R\gamma \cos\theta/4\mu$ and $b = \rho g R^2/8\mu$ and W is the Lambert W function, i.e. the solution $y = W(x) \Leftrightarrow x = y(x)e^{y(x)}$. The Lambert W function is already programmed in some software. If no such software is available, one can use the formula proposed by Barry *et al* [39]: $W(x) \cong -1 + \sqrt{2+2ex}/\left(1 + \frac{4.135\,01\sqrt{2+2ex}}{12.7036 + \sqrt{2+2ex}}\right)$, where

e is Euler's number $e = 2.718$, or one can follow a numerical method by discretizing equation (8.32).

For fibrous media, Darcy's law states that

$$\frac{\mathrm{d}h}{\mathrm{d}t} = \frac{K}{\phi\,\mu}\left(\frac{P_{\mathrm{cap}}}{h} - \rho g\right), \tag{8.34}$$

whose solution is [23]:

$$h = \frac{P_{\mathrm{cap}}}{\rho g}\left(1 + W\left(-e^{-1-\frac{\rho^2 g^2 K}{\phi\,\mu P_{\mathrm{cap}}}t}\right)\right). \tag{8.35}$$

Relation (8.35) is relation (8.32), as transposed to fibrous media. P_{cap} and K are the two parameters in relation (8.35). If the height $h(t)$ is measured, P_{cap} and K can be determined (approximately) by a best fit, as we shall see next.

8.3.3.3 Example

Take the example of figure 8.1(A) and consider the two pads, i.e. nitrocellulose and conjugate pads. These pads are currently used in commercial devices. Figure 8.11 shows the fitted plots. The liquid used in the experiment is water. Depending on the pad material, the equilibrium height is of the order of 3.5–15 cm, and the equilibrium time is of the order of 600–9000 s (10 min to 2 h and 30 min). The fitted values of the permeability and capillary pressure are listed in table 8.1. The permeability K_{Zhu} is obtained using (8.30) with a coefficient of 0.022, and the permeability K_{CK} is obtained using (8.28) with CK coefficients of 120 and 80 for conjugate and nitrocellulose pads, respectively.

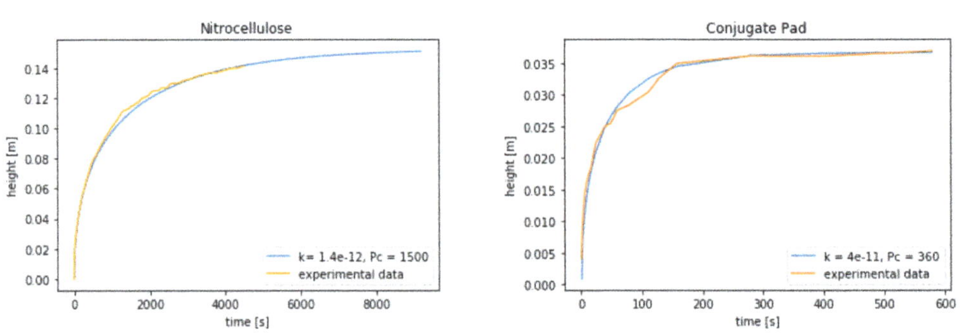

Figure 8.11. Left: the capillary rise in the nitrocellulose pad; right: the capillary rise in the conjugate pad (orange line). The blue line corresponds to the fit with equation (8.33) for the capillary rise.

Table 8.1. The capillary pressure and permeability of conjugate and nitrocellulose pads

	h_{eq} (mm)	$P_{\mathrm{cap,\,exp}}$ (Pa)	K_{exp} (μm^2)	ϕ	d_{f} (μm)	K_{Zhu} (μm^2)	K_{CK} (μm^2)
Conjugate	36	360	42	0.89	9	47	49
Nitrocellulose	150	1500	1.4	0.89	1.5	1.5	1.3

8.3.3.4 Radial imbibition

The dynamics of radial imbibition can be used to determine the equivalent radius and consequently the capillary pressure using (8.4). Danino and Marmur [40] and Hyväluoma *et al* [41] have shown that the time dependence of the imbibition radius R is given by

$$\left(\frac{R}{R_0}\right)^2\left(\ln\frac{R}{R_0}-\frac{1}{2}\right)+\frac{1}{2}=r_{eq}\,\frac{\gamma\cos\theta}{6\mu\,R_0^2}t. \qquad (8.36)$$

Monitoring the wicking radius R as a function of time t, as shown in figure 8.12, gives the value of the equivalent radius r_{eq}. Substituting this into (8.4) yields the capillary pressure.

8.3.4 Compression (compaction)

Fibrous media are not rigid and incompressible. They can be compressed—or compacted—through the application of an external force, or they may swell by the imbibition of liquid, depending on their rigidity. Let us consider a vertical plate compressing a horizontal paper pad, as shown in figure 8.13.

Locally in the compressed zone, the physical properties of the paper pad are modified. In particular, the porosity and permeability decrease, while the capillary pressure increases. Let us note that $e = h/h_0$ is the relative compression. Following Woudberg and coworkers [42], the volume available for the liquid after compression is

$$V_{liq} = Ah - Ah_0\left(1-\phi_0\right), \qquad (8.37)$$

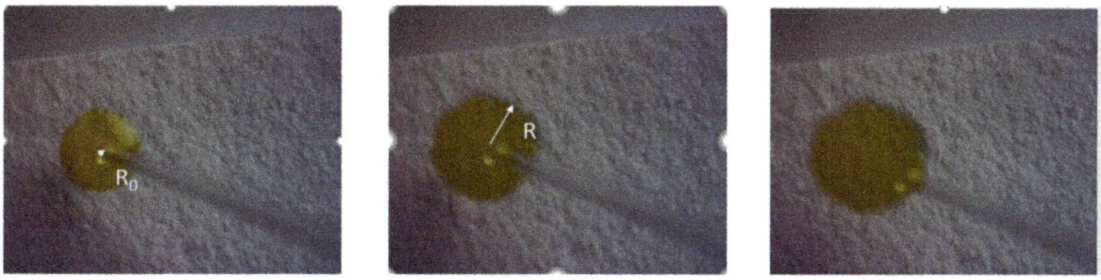

Figure 8.12. Radial imbibition in a paper pad (photograph courtesy of Jing J Lee, University of Washington).

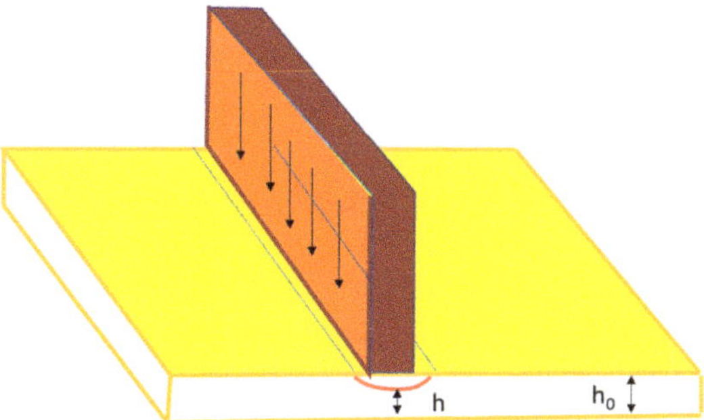

Figure 8.13. A sketch of the compression of the pad.

where A denotes the constant base area of the porous medium. The first term, Ah, corresponds to the total volume of compressed material, while the second term, $Ah_0\,(1-\phi_0)$, corresponds to the volume of fibers, which remains the same during compression. As $V_{\text{liq}} = Ah\,(1-\phi)$, rearranging (8.37) yields

$$e = \frac{h}{h_0} = \frac{1-\phi_0}{1-\phi}. \tag{8.38}$$

Hence the porosity is a function of the compression factor e

$$\phi = 1 - \frac{1-\phi_0}{e}. \tag{8.39}$$

Porosity is plotted as a function of the compression factor in figure 8.14(A). The compression factor $e = h/h_0$ cannot go to zero because there is an (unknown) minimum porosity corresponding to full contact of the fibers.

Using the correlation (8.30), the ratio between the compressed-state permeability and the initial-state permeability is

$$\frac{K}{K_0} \cong \frac{\phi^3}{\phi_0^3}\frac{(1-\phi_0)^\beta}{(1-\phi)^\beta} = G(\phi) = G(\phi(e)). \tag{8.40}$$

The relative permeability $G(\phi(e))$ is plotted in figure 8.14(B) as a function of the compression factor $e = h/h_0$ for a coefficient value of $\beta = 2$.

Finally, let us examine the variation of the capillary pressure with compression. If we make use of relations (8.4) and (8.18), an approximate relation for the capillary pressure is [43]

$$P_{\text{cap}} \approx \frac{1}{2\sqrt{2}}\gamma\cos\theta\left(\frac{\phi}{K}\right)^{\frac{1}{2}} = 0.35\gamma\cos\theta\left(\frac{\phi}{K}\right)^{\frac{1}{2}}. \tag{8.41}$$

Using (8.39), (8.40), and (8.41), the term $\frac{K}{\phi}P_{\text{cap}}$ is the following function of the relative compression e:

$$\frac{K}{\phi}P_{\text{cap}} = \frac{1}{2\sqrt{2}}\gamma\cos\theta\left(\frac{K}{\phi}\right)^{\frac{1}{2}} = \frac{1}{2\sqrt{2}}\gamma\cos\theta\left[\frac{K_0}{1-\frac{1-\phi_0}{e}}\frac{\phi^3}{\phi_0^3}\frac{(1-\phi_0)^\beta}{(1-\phi)^\beta}\right]^{\frac{1}{2}}. \tag{8.42}$$

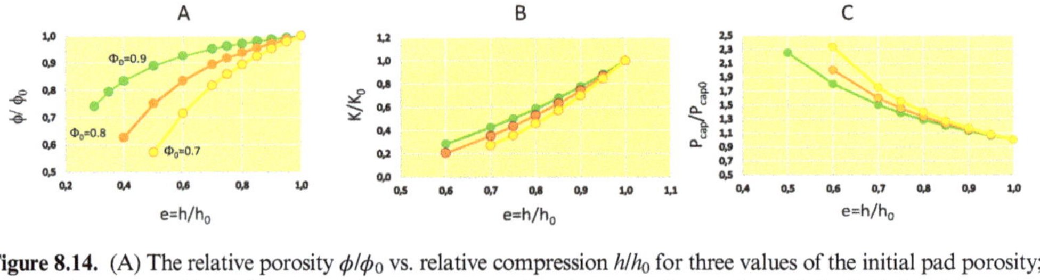

Figure 8.14. (A) The relative porosity ϕ/ϕ_0 vs. relative compression h/h_0 for three values of the initial pad porosity; (B) relative permeability K/K_0 vs. relative compression for the same three values of the initial pad porosity; (C) relative capillary pressure vs. relative compression for the same three values of the initial pad porosity.

The relative value of $\frac{K}{\phi}P_{\text{cap}}$ is plotted in figure 8.15 as a function of the compression factor h/h_0, for three values of the initial porosity (ϕ_0= 0.9, 0.8, and 0.7). This term is divided by roughly two when the pad thickness is compressed by a factor of two. Hence, referring to Darcy's law, the flow velocity decreases when the pad is compressed.

8.4 Equivalent permeability

Paper strips used as sensors are usually composed of different pads placed in series, as shown in figure 8.16, each one of which has specific properties. It is convenient to use the concept of equivalent permeability to calculate the flow dynamics in the different pads. Each pad is characterized by its porosity ϕ_i, permeability K_i, length L_i, cross-section S_i, and capillary pressure P_{ci}.

The pressure at the extremity of a pad is denoted by $_{Pi}$. P_0 is the pressure at the inlet (here, at atmospheric pressure, $P_0 = 0$). Assume that the tip of the capillary flow (the advancing front) is located in the nth pad. The conservation of total mass—assuming that evaporation is negligible—yields

$$V_1\, S_1\, \varnothing_1 = V_2\, S_2\, \varnothing_2 = \dots = V_n\, S_n\, \varnothing_n, \tag{8.43}$$

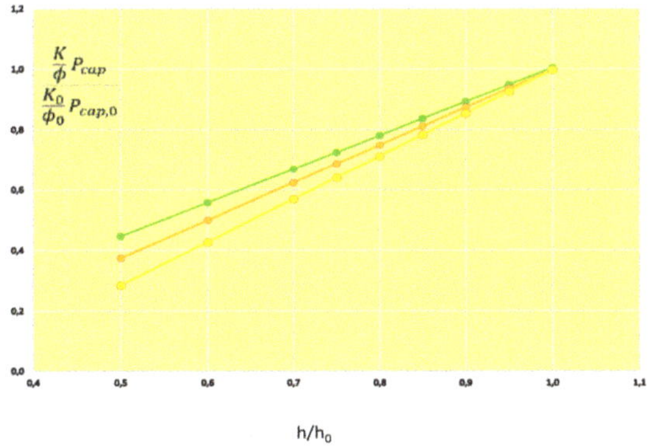

Figure 8.15. The relative decrease of $\frac{K}{\phi}P_{\text{cap}}$ vs. the compression factor h/h_0

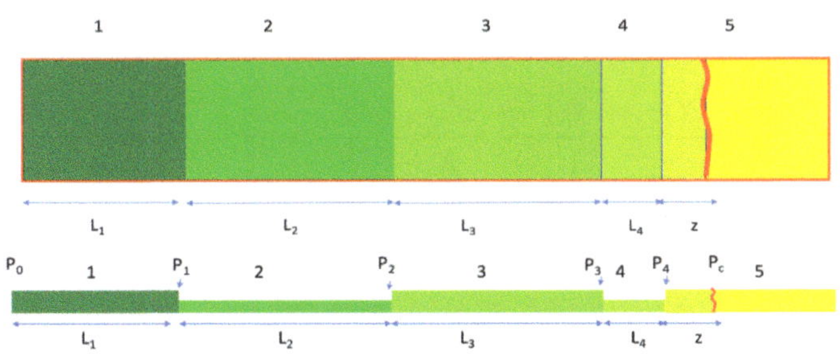

Figure 8.16. A sketch of the composite paper strip: top, view from above; bottom, side view. The red line symbolizes the location of the interface.

where V_i is the average flow velocity in strip i. On the other hand, Darcy's law states that

$$V_i = -\frac{K_i}{\varnothing_i\,\mu}\nabla P_i = \frac{K_i}{\varnothing_i\,\mu}\,\frac{P_i - P_{i-1}}{L_i}, \tag{8.44}$$

for each index i corresponding to a pad completely wetted, and

$$V_n = -\frac{K_n}{\varnothing_n\,\mu}\nabla P_n = \frac{K_n}{\varnothing_n\,\mu}\,\frac{P_{cn} - P_{n-1}}{z}, \tag{8.45}$$

when the advancing front is in the nth pad, z being the travel distance in this pad. Combining (8.43), (8.44), and (8.45) yields

$$\begin{aligned} V_n\,S_n\,\varnothing_n &= \frac{S_n\,K_n}{\mu}\,\frac{(P_{cn} - P_{n-1})}{z} = \ldots = V_i\,S_i\,\varnothing_i \\ &= \frac{S_i\,K_i}{\mu}\,\frac{(P_i - P_{i-1})}{L_i} = \ldots = V_1\,S_1\,\varnothing_1 = \frac{S_1\,K_1}{\mu}\,\frac{(P_1 - P_0)}{L_1}. \end{aligned} \tag{8.46}$$

If we $K_{eq,n}$ denotes the equivalent permeability when the front is in the nth pad, we obtain, by definition

$$V_n \triangleq \frac{K_{eq,\,n}}{\varnothing_n\,\mu}\,\frac{P_{cn} - P_0}{\left(\sum_{i=1,n-1} L_i\right) + z}. \tag{8.47}$$

Then, using (8.46) and (8.47), we obtain

$$\begin{aligned} K_{eq,\,n}\,S_n\frac{(P_{cn} - P_0)}{\left(\sum_{i=1,n-1} L_i\right) + z} &= K_n S_n\,\frac{(P_{cn} - P_{n-1})}{z} \\ &= \ldots = K_i S_i\,\frac{(P_i - P_{i-1})}{L_i} = \ldots = K_1 S_1\,\frac{(P_1 - P_0)}{L_1} \end{aligned}. \tag{8.48}$$

We deduce the relations

$$\begin{aligned} \frac{(P_{cn} - P_0)}{\left(\sum_{i=1,n-1} L_i\right) + z}\,\frac{K_{eq,\,n}}{K_n}S_n\,\frac{z}{S_n} &= (P_{cn} - P_{n-1}), \\ \frac{(P_{cn} - P_0)}{\left(\sum_{i=1,n-1} L_i\right) + z}\,\frac{K_{eq,\,n}}{K_i}S_n\frac{L_i}{S_i} &= (P_i - P_{i-1}), \\ \frac{(P_{cn} - P_0)}{\left(\sum_{i=1,n-1} L_i\right) + z}\,\frac{K_{eq,\,n}}{K_1}S_n\frac{L_1}{S_1} &= (P_1 - P_0). \end{aligned} \tag{8.49}$$

Adding all the equations of system (8.49) yields

$$\frac{K_{\mathrm{eq},\,n}\,S_n}{\left(\sum\limits_{i=1,n-1} L_i\right)+z}\left[\frac{z}{K_n\,S_n}+\ldots+\frac{L_i}{K_i\,S_i}+\ldots+\frac{L_1}{K_1S_1}\right]=1. \tag{8.50}$$

Hence, the equivalent permeability is

$$K_{\mathrm{eq},\,n}=\frac{1}{S_n}\frac{\left(\sum\limits_{i=1,n-1} L_i\right)+z}{\left[\frac{z}{K_n\,S_n}+\ldots+\frac{L_i}{K_i\,S_i}+\ldots+\frac{L_1}{K_1S_1}\right]}=\frac{1}{S_n}\frac{\left(\sum\limits_{i=1,n-1} L_i\right)+z}{\left[\frac{z}{K_n\,S_n}+\sum\limits_{i=1,n-1}\frac{L_i}{K_i\,S_i}\right]}. \tag{8.51}$$

Note that if all the cross-sections are identical, the equivalent permeability defined by (8.51) is identical to the expression given by Ashraf *et al* and Prakash *et al* [44, 45]

$$K_{\mathrm{eq}}=\frac{\sum\limits_{i=1,n} L_i}{\sum\limits_{i=1,n}\frac{L_i}{K_i}}. \tag{8.52}$$

Substituting (8.52) into (8.47), we obtain

$$V_n=\frac{1}{S_n\varnothing_n\,\mu}\frac{P_{cn}-P_0}{\left[\frac{z}{K_n\,S_n}+\sum\limits_{i=1,n-1}\frac{L_i}{K_i\,S_i}\right]}. \tag{8.53}$$

As $V_n = \mathrm{d}z/\mathrm{d}t$, equation (8.53) becomes

$$\frac{\mathrm{d}z}{\mathrm{d}t}\left[\frac{z}{K_n\,S_n}+\sum\limits_{i=1,n-1}\frac{L_i}{K_i\,S_i}\right]=\frac{P_{cn}-P_0}{S_n\varnothing_n\,\mu}. \tag{8.54}$$

Multiplying both sides of (8.54) by $K_n\,S_n$ yields

$$\frac{\mathrm{d}z^2}{\mathrm{d}t}+2K_n\,S_n\left(\sum\limits_{i=1,n-1}\frac{L_i}{K_i\,S_i}\right)\frac{\mathrm{d}z}{\mathrm{d}t}-2K_n\frac{P_{cn}-P_0}{\varnothing_n\,\mu}=0. \tag{8.55}$$

As $P_0 = 0$, (8.55) can be rewritten as

$$\frac{\mathrm{d}z^2}{\mathrm{d}t}+2K_n\,S_n\left(\sum\limits_{i=1,n-1}\frac{L_i}{K_i\,S_i}\right)\frac{\mathrm{d}z}{\mathrm{d}t}-2K_n\frac{P_{cn}}{\varnothing_n\,\mu}=0. \tag{8.56}$$

The integration of (8.56) leads to

$$z^2 + 2K_n S_n\left(\sum_{i=1,n-1}\frac{L_i}{K_i S_i}\right)z - 2K_n\frac{P_{cn}}{\varnothing_n \mu}(t - t_{n-1}) = 0, \tag{8.57}$$

where t_{n-1} is the time at which the tip of the flow enters substrip n. The solution of (8.57) is

$$z = -K_n S_n\left(\sum_{i=1,n-1}\frac{L_i}{K_i S_i}\right) + \sqrt{(K_n S_n)^2\left(\sum_{i=1,n-1}\frac{L_i}{K_i S_i}\right)^2 + 2K_n\frac{P_{cn}}{\varnothing_n \mu}(t - t_{n-1})} \tag{8.58}$$

The velocity in section n is then

$$\begin{aligned} V = \frac{\mathrm{d}z}{\mathrm{d}t} &= \frac{K_n P_{cn}}{\varnothing_n \mu}\frac{1}{\sqrt{(K_n S_n)^2\left(\sum_{i=1,n-1}\frac{L_i}{K_i S_i}\right)^2 + 2K_n\frac{P_{cn}}{\varnothing_n \mu}(t - t_{n-1})}} \\ &= \frac{K_n P_{c,n}}{\varnothing_n \mu}\frac{1}{\left(z + K_n S_n\sum_{i=1,n-1}\frac{L_i}{K_i S_i}\right)} \end{aligned} \tag{8.59}$$

Relation (8.59) shows that the velocity is not proportional to $1/z$, as in the Lucas–Washburn case, but to the outcome of the preceding sections: $z + K_n S_n \sum_{i=1,n-1}\frac{L_i}{K_i S_i}$.

An example of flow velocity in a typical assembly of pads is shown in figure 8.17. Note that the characteristics of the conjugate pad have been chosen such that the velocity considerably decreases in this pad to give enough time for the traveling sample to load the reagents immobilized in the conjugate pad.

8.5 Flow velocity in paper strips of varying width

Paper pads may have different widths, whereas their thicknesses are usually the same due to the fabrication process used. We first analyze the case of piecewise varying width, then that of progressively varying width.

8.5.1 Paper pads of piecewise varying width

Let us now consider a paper strip with N piecewise varying widths, as shown in figure 8.18. Applying Darcy's law to each section gives the relations

$$Q_i = -\frac{K}{\mu}w_i h \nabla P_i = \frac{K}{\mu}w_i h\frac{\Delta P_i}{L_i}, \tag{8.60}$$

where h is the thickness of the paper strip, K the permeability, μ the viscosity of the fluid, and p_i the pressure at the nodes i (located at the changes of cross-section).

A

	Sample pad	Conjugate pad	nitrocellulose	Absorbent pad
Porosity	0.74	0.89	0.88	0.9
Permeability [m²]	$1.\ 10^{-12}$	$42.\ 10^{-12}$	$5.5\ 10^{-12}$	$6.\ 10^{-12}$
Capillary pressure [Pa]	6900	360	1100	3000
Pad thickness [mm]	0.39	0.22	0.3	0.81
Pad lengths [mm]	15.0	8.3	25.0	12.0
Total cross-sectional area [mm²]	3.9	2.2	3.0	8.1

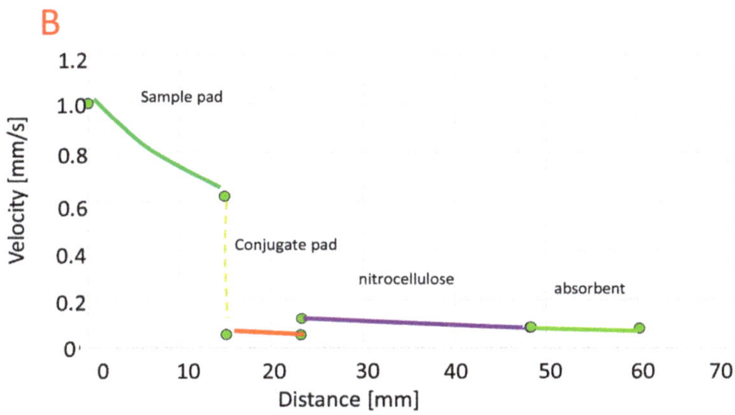

Figure 8.17. (A) The definitions of the different pads; (B) the velocities of the tip of the advancing flow at different distances.

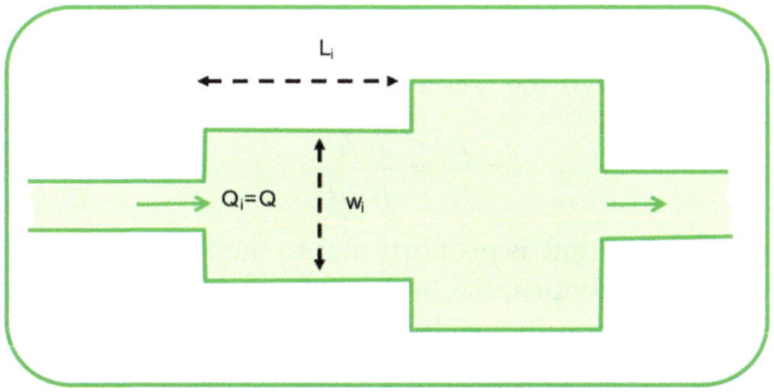

Figure 8.18. A sketch of a paper strip of piecewise varying cross-section.

If the timescale is sufficiently small, so that we can neglect evaporation, then the conservation of mass imposes the following:

$$Q_i = Q \text{ for any index } i = 1,\ N. \tag{8.61}$$

The total pressure drop is the sum of all local pressure drops:

$$\Delta P = \sum_{i=1,N} \Delta P_i. \tag{8.62}$$

Combining (8.60), (8.61), and (8.62) yields

$$\Delta P = \frac{\mu}{Kh} \sum_{i=1,N} \frac{L_i}{w_i} Q_i = \frac{\mu}{Kh} Q \sum_{i=1,N} \frac{L_i}{w_i}. \tag{8.63}$$

Hence

$$Q = \frac{Kh}{\mu \sum_{i=1,N} \frac{L_i}{w_i}} \Delta P. \tag{8.64}$$

The flow velocity in each section is given by the relation

$$V_i = \frac{Q_i}{w_i h} = \frac{Q}{w_i h} = \frac{K}{\mu \, w_i \sum_{i=1,N} \frac{L_i}{w_i}} \Delta P. \tag{8.65}$$

Relation (8.65) states that the velocity V_i in each section is inversely proportional to its width w_i (since the term $\frac{K}{\mu \sum_{i=1,N} \frac{L_i}{w_i}} \Delta P$ is independent of the index i). Note that the thickness h of the paper layer does not appear in the expression for the velocity. This observation has already been reported by Fu and colleagues [46]. In the case of a single-width strip of length L, $i = 1$ in relation (8.63) and we obtain

$$V = \frac{K}{\mu \, L} \Delta P \tag{8.66}$$

The transport time τ through the whole length of the strip is

$$\tau = \frac{L}{V} = \frac{\mu \, L^2}{K \Delta P}. \tag{8.67}$$

As expected, the transport time is proportional to the viscosity of the liquid imbibed by the pad and inversely proportional to the paper pad's permeability. In the general case of a strip of piecewise varying width, we have to refer to relation (8.65), and the total transport time is given by

$$\begin{aligned} \tau &= \sum_{i=1,N} \tau_i = \sum_{i=1,N} \frac{L_i}{V_i} = \frac{\mu}{K \Delta P} \sum_{i=1,N} w_i \, L_i \sum_{j=1,N} \frac{L_j}{w_j} \\ &= \frac{\mu}{K \Delta P} \left(\sum_{i=1,N} w_i \, L_i \right) \left(\sum_{j=1,N} \frac{L_j}{w_j} \right) \end{aligned} \tag{8.68}$$

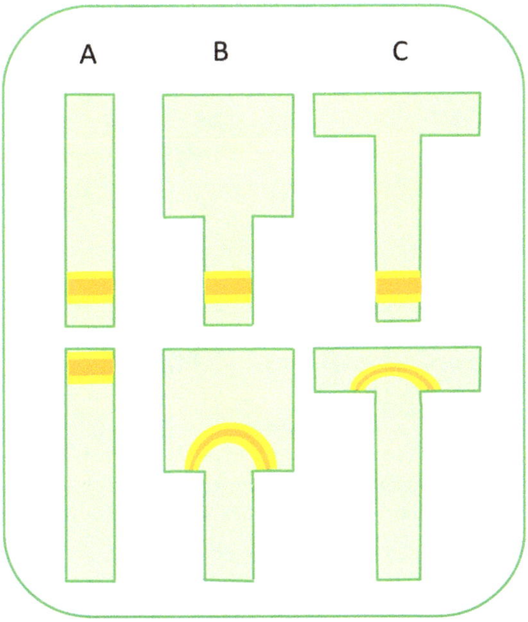

Figure 8.19. The transport time of a chemical species (orange and yellow color) in a strip of piecewise varying width [18]: the fastest transport is geometry A, followed by C, and finally B.

For example, for strips that contain two constant-width segments whose lengths are L_1 and L_2 and whose widths are w_1 and w_2, respectively, the transport time is given by

$$\tau = \frac{\mu}{K\Delta P}(w_1\, L_1 + w_2\, L_2)\left(\frac{L_1}{w_1} + \frac{L_2}{w_2}\right). \tag{8.69}$$

The equation is symmetric in L_1 and L_2, and a local maximum exists at $L_1 = L_2$, indicating in the case of piecewise different areal values, the greatest transit time occurs in a strip with equal-length segments. This property is sketched in figure 8.19.

8.5.2 Triangular circular section pads

Recently, triangular (circular section) paper pads have appeared for different applications. Rectangular and triangular pads are compared in figure 8.20. The velocity in the pad is given by Darcy's law:

$$V_\mathrm{p} = -\frac{K}{\mu\phi}\nabla\mathrm{P} = \frac{K}{\mu\phi}\frac{P_\mathrm{cap}}{z}, \tag{8.70}$$

where z is the travel distance in the paper pads, as shown in figure 8.20. The velocity at the pad entrance (upstream boundary) is related to the velocity in the paper by the conservation of mass:

$$S_0 V_0 = \phi S_\mathrm{p} V_\mathrm{p}, \tag{8.71}$$

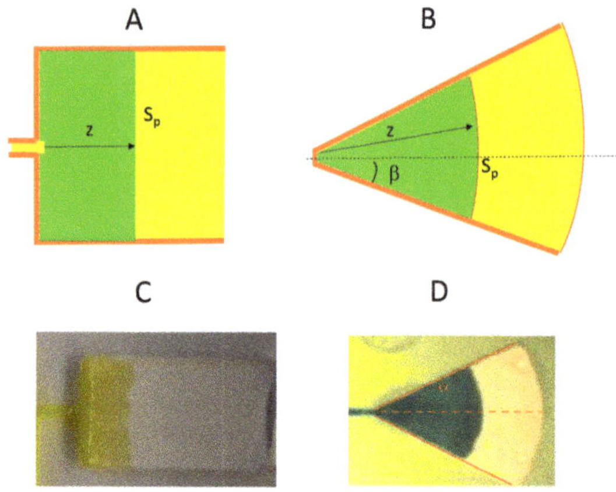

Figure 8.20. (A) and (B) Sketches of rectangular and triangular pads. (C) and (D) Experimental images. Photograph (C), imbibition in Whatman #1 paper, courtesy of J.J. Lee (University of Washington, WA, USA). Photograph (D) reproduced from [47] with permission of The Royal Society of Chemistry.

where S_p is the surface area of the advancing liquid front. Then

$$V_0 = \frac{\phi S_p V_p}{S_0} = \frac{S_p}{S_0}\frac{K}{\mu}\frac{P_{cap}}{z}. \tag{8.72}$$

In the case of a rectangular pad, S_p is constant and relation (8.72) yields

$$V_0 \sim \frac{cste}{z}, \tag{8.73}$$

and the inlet velocity decreases with the imbibition length z. In the case of a triangular pad, the surface S_p is given by

$$S_p = 2\beta\, z\, h_p, \tag{8.74}$$

where h_p is the pad thickness. Substituting (8.74) into (8.72) yields

$$V_0 = \frac{2\beta\, h_p}{S_0}\frac{K}{\mu} P_{cap}. \tag{8.75}$$

Hence, for a triangular pad

$$V_0 = \text{a constant}. \tag{8.76}$$

The advantage of using a triangular pad in a capillary circuit is that it produces a constant pumping velocity.

8.6 Open channels connected to paper pads

Open channels can easily be connected to paper pads. In such cases, paper pads act as pumps. Two examples are shown in figure 8.21. Tran and coworkers have used pads as pumps for an open deterministic lateral device (DLD) [48]. DLDs are

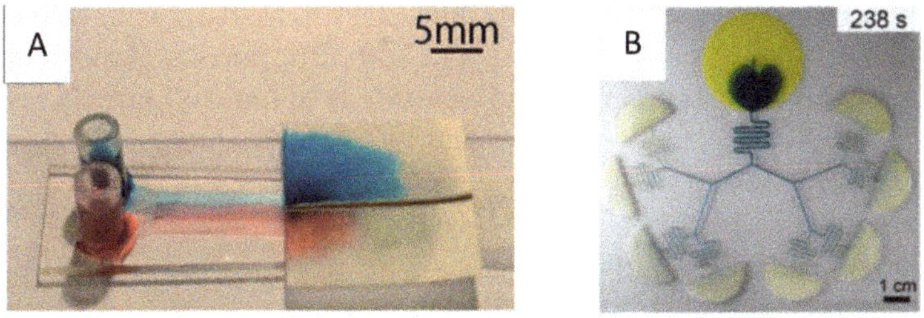

Figure 8.21. (A) An open DLD with pads placed downstream that pump the liquid; (B) a capillary tree with successive bifurcations and paper pads placed at the extremity of the tree. Photograph (A) reprinted from [48], CC BY 3.0. Photograph (B) reprinted from [50], with the permission of AIP Publishing.

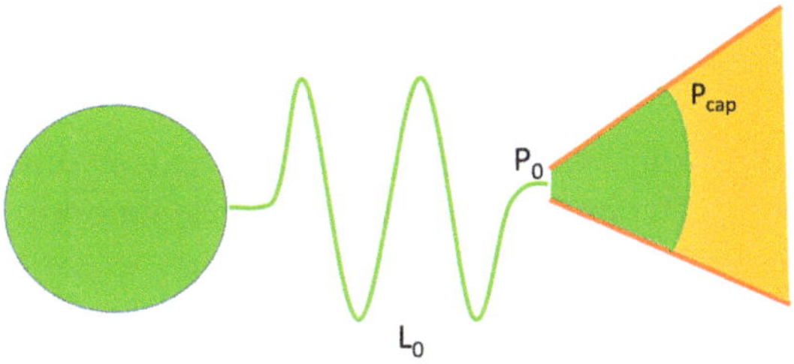

Figure 8.22. A sketch of a device with an inlet reservoir, a connection channel (the root channel), and a pad.

devices used to separate objects—particles, beads, cells—depending on their size. There is an abundant literature that analyzes the physics of DLDs. The second example—Lee *et al* and Tokihiro *et al* [49, 50]—is a capillary tree connected to paper pads that maintains high velocity in the root channel (the first channel connected to the reservoir).

Let us consider an open channel connected to a paper pad, as shown in figure 8.22.

8.6.1 The root channel

First, consider the flow in the open channel or root channel. Neglecting the initial inertial regime, referring to chapter 4, the relation between travel distance and elapsed time is

$$z = \sqrt{\frac{\gamma}{\mu} 2\bar{\lambda} \cos\theta^* t}\ , \tag{8.77}$$

where z is the travel distance, t the time, γ the surface tension of the liquid with air, μ its dynamic viscosity, $\bar{\lambda}$ the average friction length, and $\cos\theta^*$ the generalized Cassie angle. The velocity of the liquid—averaged over its cross-section—decreases with time according to

$$V = \frac{1}{2}\sqrt{\frac{\gamma}{\mu}\frac{2\bar{\lambda}\cos\theta^*}{t}}\ . \tag{8.78}$$

It has been found that connecting the channel to microporous paper pads can partially palliate this velocity decrease [47, 50]. In porous paper pads—or any microporous media—the wicking liquid velocity is given by Darcy's law

$$V_p = -\frac{K}{\mu\phi}\nabla P, \tag{8.79}$$

where K is the paper permeability, ϕ the porosity, and ∇P the pressure gradient. Let us consider the device sketched in figure 8.22 and illustrated in figure 8.23, where a single capillary channel is connected to a paper pad. In the channel, the flow is governed by relations (8.77) and (8.78), and the time at which the liquid reaches the pad is

$$t_0 = \frac{\mu L_0^2}{2\gamma\bar{\lambda}\cos\theta^*}. \tag{8.80}$$

where L_0 is the length of the winding root channel. After time t_0, the liquid starts wicking the paper pad. If P_0 denotes the pressure at the junction between the channel and the paper and P_{cap} the capillary pressure in the paper pad, Darcy's law can be cast in the form

$$V_p = -\frac{K}{\mu\varphi}\nabla P = \frac{K}{\mu\varphi}\frac{P_{cap} - P_0}{z_p}, \tag{8.81}$$

where z_p is the wicking distance in the paper pad. The triplet (P_{cap}, K, φ) characterizes the paper pad. In order to solve equation (8.81), the pressure P_0 must first be determined.

Assuming a spontaneous capillary flow (zero upstream pressure), the pressure at the junction between the channel and the paper is determined using the wall friction in the whole length of the channel:

$$P_0 = \frac{p_0 L_0}{S_0}\mu\frac{V_0}{\bar{\lambda}}, \tag{8.82}$$

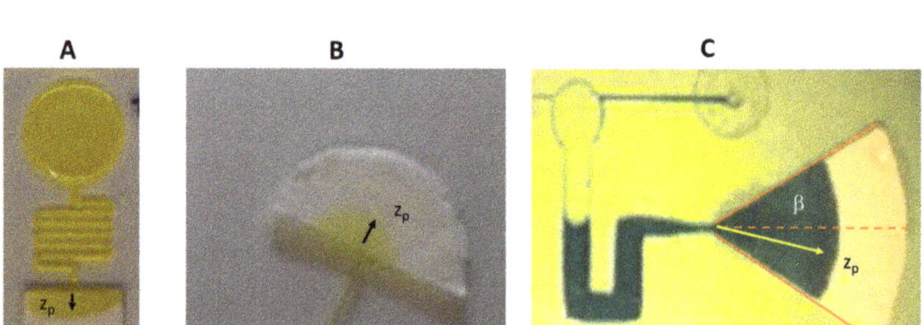

Figure 8.23. (A) A photograph of the imbibition in a rectangular paper pad; (B) imbibition in a semicircular paper pad; (C) wicking in a triangular paper pad (photographs (A) and (B) courtesy of Jing. J. Lee, University of Washington, WA, USA). Photograph (C) reprinted from [47] with permission of The Royal Society of Chemistry.

where S_0 is the cross-sectional area of the channel and p_0 the cross-section's perimeter (including the 'free' perimeter in the case of an open channel). Substituting (8.82) into (8.81) yields

$$V_p = \frac{K}{\mu\phi}\frac{P_{cap} - \mu\frac{p_0 L_0}{\bar{\lambda}\, S_0} V_0}{z_p}. \tag{8.83}$$

The velocities V_p and V_0 are linked by the conservation of mass:

$$V_0 S_0 = V_p S_{p,\,F}(z_p) = V_p\, \phi\, S_p(z_p), \tag{8.84}$$

where $S_{p,\,F}$ and S_p are the free and total cross-sectional areas of the paper strip at the distance z_p. Let us consider two cases: (1) a rectangular pad, (2) a triangular pad.

8.6.2 Rectangular pads

In the case of a rectangular pad (figure 8.23(A)), the paper pad's cross-section is constant, and relation (8.84) becomes

$$V_0 S_0 = V_p\, \phi\, S_p. \tag{8.85}$$

Inserting (8.85) into (8.83) yields

$$V_p = \frac{K}{\mu\phi}\frac{P_{cap} - \mu\frac{p_0 L_0}{\bar{\lambda}\, S_0^2} V_p\, \phi\, S_p}{z_p} = \frac{K}{\mu\phi}\frac{P_{cap}}{z_p} - -K\frac{p_0 L_0}{\bar{\lambda}\, S_0^2}\frac{V_p\, S_p}{z_p}. \tag{8.86}$$

Because $V_p = dz_p/dt$, relation (8.86) becomes the differential equation

$$z_p\frac{dz_p}{dt} = \frac{K}{\mu\phi}\left(P_{cap} - K\frac{p_0 L_0}{\bar{\lambda}\, S_0^2}\frac{dz_p}{dt}\, S_p\right), \tag{8.87}$$

or

$$\frac{dz_p^2}{dt} + 2K\frac{p_0 L_0}{\bar{\lambda}\, S_0^2} S_p \frac{dz_p}{dt} = \frac{2K}{\mu\phi}\, P_{cap}, \tag{8.88}$$

whose solution is the root of the polynomial

$$z_p^2 + 2K\frac{p_0 L_0}{\bar{\lambda}\, S_0^2} S_p z_p - \frac{2K}{\mu\phi}\, P_{cap}\tau = 0. \tag{8.89}$$

In (8.89), τ is the time at which the liquid contacts the paper pad ($\tau = t - t_0$). Finally,

$$z_p = -KS_p\frac{p_0 L_0}{\bar{\lambda}\, S_0^2} + \sqrt{\left(KS_p\frac{p_0 L_0}{\bar{\lambda}\, S_0^2}\right)^2 + \frac{2K}{\mu\phi}\, P_{cap} t}\,. \tag{8.90}$$

8.6.3 Triangular (circular section) pads

In the second case of a triangular paper pad, if we assume a perfectly rounded interface—as shown in figure 8.23(B) and (C), the cross-sectional area of the paper pad is

$$S_p(z_p) = 2\beta\, h_p\, z_p, \tag{8.91}$$

where h_p is the thickness of the paper pad and α the cone semi-angle. Successively substituting (8.88) into (8.84) and then into (8.85) yields

$$V_p = \frac{K}{\mu\phi}\,\frac{P_{cap} - \mu\frac{p_0 L_0}{\bar{\lambda}\, S_0^2}\, V_p\, \phi 2\beta\, h_p\, z_p}{z_p} = \frac{K}{\mu\phi}\,\frac{P_{cap}}{z_p} - -K\frac{p_0 L_0}{\bar{\lambda}\, S_0^2}\, V_p 2\beta\, h_p. \tag{8.92}$$

Again, using the relation $V_p = \mathrm{d}z_p/\mathrm{d}t$, and multiplying (8.92) by z_p, the travel distance z_p is the solution of the differential equation

$$\frac{1}{2}\frac{\mathrm{d}z_p^2}{\mathrm{d}t}\left(1 + K\frac{p_0 L_0}{\bar{\lambda}\, S_0^2}2\beta\, h_p\right) = \frac{K\, P_{cap}}{\mu\phi}. \tag{8.93}$$

The solution is then

$$z_p = \sqrt{\frac{2K\, P_{cap}}{\mu\phi\left(1 + K\frac{p_0 L_0}{\bar{\lambda}\, S_0^2}2\beta\, h_p\right)}\tau}, \tag{8.94}$$

where τ is the time at which the liquid contacts the paper pad ($\tau = 0$, $z_p = 0$). The velocity of the fluid in the pad is then

$$V_p = \sqrt{\frac{K\, P_{cap}}{2\mu\phi\left(1 + K\frac{p_0 L_0}{\bar{\lambda}\, S_0^2}2\beta\, h_p\right)}\frac{1}{\tau}}. \tag{8.95}$$

Using the equation for the conservation of mass, the velocity in the root channel when the liquid wicks the paper is given by

$$V_0 = V_p\frac{\phi 2\beta\, h_p\, z_p}{S_0} = \frac{2\beta\, h_p\, K\, P_{cap}}{\mu\left(1 + K\frac{p_0 L_0}{\bar{\lambda}\, S_0^2}2\beta\, h_p\right)S_0} = cste. \tag{8.96}$$

We obtain this important result that once the liquid has reached the pad, the root channel velocity is constant in the case of a triangular pad.

Note that there is a jump of velocity when the liquid reaches the paper pad. Indeed, just before contact with the pad, the velocity can be deduced from (8.77) and (8.78):

$$V_0^- = \frac{\gamma}{\mu}\frac{\bar{\lambda}\cos\theta^*}{L_0}. \tag{8.97}$$

When the fluid wicks the pad, the velocity V_0 is constant and given by (8.96). There is a jump of velocity

$$\Delta V_0 = V_0^+ - V_0^- = \frac{1}{\mu}\left[\frac{2\beta\, h_\mathrm{p}\, K\, P_\mathrm{cap}}{\left(1 + K\frac{p_0 L_0}{\bar{\lambda}\, S_0^2} 2\beta\, h_\mathrm{p}\right) S_0} - \frac{\bar{\lambda}\gamma\cos\theta^*}{L_0}\right]. \tag{8.98}$$

8.6.4 Numerical application

Consider the typical case of an open channel etched in PMMA and connected to a paper pad (for example, Whatman paper). Let us consider nonanol to be a liquid [50], due to its small volatility. The paper characteristics have been partly furnished by the manufacturer and partly measured. The characteristics of the problem are listed in table 8.2.

Table 8.2. Numerical values of the parameters

Physical and geometrical properties	Symbol	Value	Unit
Paper pad			
Permeability	K	5×10^{-12}	m^2
Porosity	ϕ	0.8	—
Capillary pressure	P_cap	8000	Pa
Cross-section	$S_{\mathrm{p}0}$	4×10^{-6}	m^2
Pad thickness	h_p	5×10^{-4}	m
Root channel			
Friction length	$\bar{\lambda}$	2×10^{-4}	m
Perimeter	p_0	4×10^{-3}	m
Cross-section	S_0	1×10^{-6}	m^2
Length of root channel	L_0	0.1	m
Liquid			
Viscosity of nonanol	μ	10	mPa s
Surface tension	γ	30	mN m^{-1}
Generalized contact angle	θ^*	55	°
Coefficient C = $\sqrt{\frac{\gamma}{\mu} 2\bar{\lambda}\cos\theta^*}$	C	25	$mm^2\ s^{-1}$
Coefficients			
$a = K\frac{p_0 L_0}{\bar{\lambda}\, S_0}$	a	0.02	mm
$b = \frac{2K}{\mu\phi} P_{cap}$	b	10	mm^2
$\delta = \frac{S_0}{2\beta h_p}$, $\beta=\pi/6$	δ	2	mm
$\delta = \frac{S_0}{2\beta h_p}$, $\beta=\pi/3$	δ	1	mm
$\delta = \frac{S_0}{2\beta h_p}$, $\beta=\pi/2$	δ	0.6	mm

A comparison of the liquid velocities in the root channel as a function of the pad shape (rectangular or triangular) is shown in figure 8.24(A). In figures 8.24(B) and (C), a comparison is made between the present model and the experimental results of Kokjalj *et al* [47].

8.6.5 Capillary trees connected to paper pads

An efficient design for a capillary pump has been proposed by Lee *et al* and Tokihiro *et al* [49, 50] and is shown in figure 8.25. The design comprises a capillary tree (already described in chapter 4) connected to paper pads. In itself, the capillary tree constitutes a pumping system. Coupled to pads, the system is a powerful pumping system. The example of 50% wt isopropyl alcohol (IPA50) is shown in figure 8.26. Velocities of the order of 3 mm s^{-1} are obtained for 75 s (the surface tension of IPA50 is 30 mN m^{-1} and its viscosity is 3 mPa s).

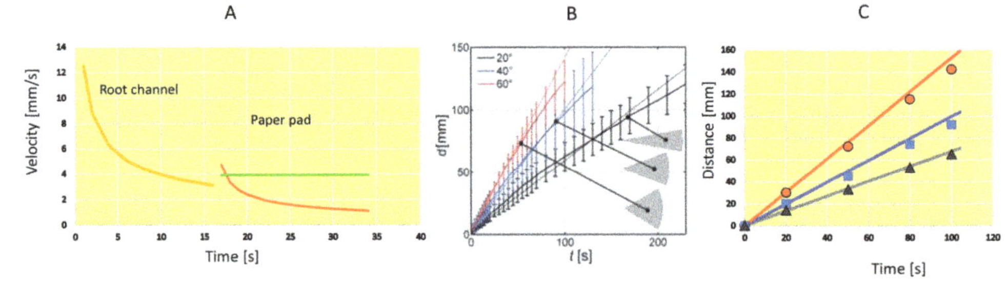

Figure 8.24. (A) Velocity vs. time; the velocity in the root channel decreases as the inverse of time (orange line). At $t = 16$ s, the liquid starts wicking the pad. The case of the rectangular pad is shown in brown, while the green line corresponds to a triangular pad of semi-angle $\beta=\pi/6$. A velocity jump appears at the transition between the channel and the pad. (B) The root channel velocity is the slope of the linear segments (red, blue, and black). Reprinted from [47] with permission of The Royal Society of Chemistry. (C) A fit of the plots of figure B produced using the model under discussion.

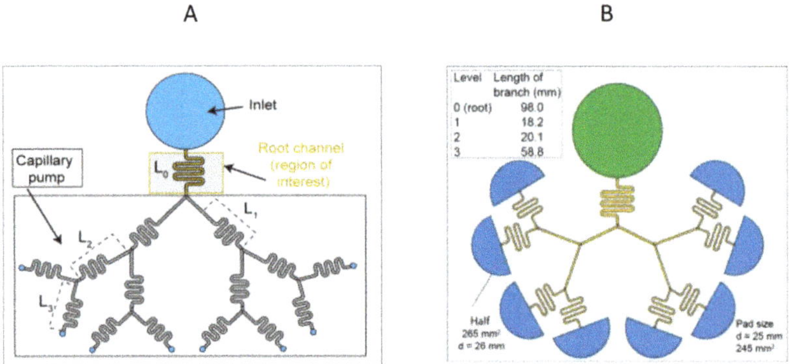

Figure 8.25. (A) A sketch of a capillary tree (see chapter 4) with a root channel and three levels of branches; (B) the same capillary tree connected to semicircular paper pads. Panels (A) and (B) reprinted with permission from [49]. Copyright (2020) American Chemical Society.

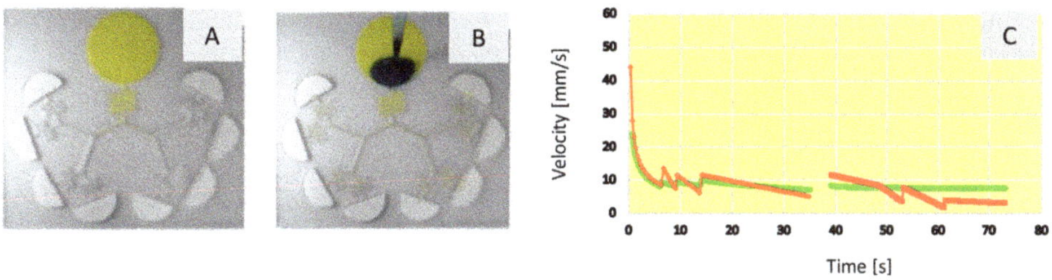

Figure 8.26. (A) and (B) The flow of IPA50 in the capillary tree (see figure 8.25); (C) the root channel velocity of IPA50 (experimental results are shown in red and the results of the model in green). Photos (A) and (B) reprinted from [50], with the permission of AIP Publishing.

8.7 Conclusions

Capillary flow in fibrous media can be considered to be open microfluidic flow. This chapter focused on paper-based imbibition. The main characteristics of capillary flows in fibrous media were reviewed, and a parallel between the Lucas–Washburn approach and Darcy's law was outlined, précising the notion of the equivalent capillary radius.

Microflows in assemblies of paper pads were investigated. In particular, we showed how the notion of equivalent permeability simplifies the prediction of liquid velocities in the different pads. The flow in triangular pads was investigated and it was shown that a constant flow velocity is established at the pad entrance.

It was shown that open channels and paper pads can easily be connected. Such assemblies can be used to obtain powerful pumping devices.

References and further reading

[1] Martinez A W, Phillips S T, Carrilho E, Thomas S W, Sindi H and Whitesides G M 2008 Simple telemedicine for developing regions: camera phones and paper-based microfluidic devices for real-time, off-site diagnosis *Anal. Chem.* **80** 3699

[2] Yager P, Edwards T, Fu E, Helton K, Nelson K, Tam M R and Weigl B H 2006 Microfluidic diagnostic technologies for global public health *Nature* **442** 412–8

[3] Nie Z, Nijhuis C A, Gong J, Chen X, Kumacheva A, Martinez A W, Narovlyansky M and Whitesides G M 2010 Electrochemical sensing in paper-based microfluidic devices *Lab Chip.* **10** 477–83

[4] Weng X, Kang Y, Guo Q, Peng B and Jiang H 2019 Recent advances in thread-based microfluidics for diagnostic applications *Biosensors Bioelectron.* **132** 171–85

[5] Kim J H, Woenker T, Adamec J and Regnier F E 2013 Simple, miniaturized blood plasma extraction method *Anal. Chem.* **85** 11501–8

[6] Mostafalu P, Akbari M, Alberti K A, Xu Q, Khademhosseini A and Sonkusale S R 2016 A toolkit of thread-based microfluidics, sensors, and electronics for 3D tissue embedding for medical diagnostics *Microsyst. Nanoeng.* **2** 16039

[7] Li Y, Fischer R, Zboray R, Boillat P, Camenzind M, Toncelli C and Rossi R M 2020 Laser-engraved textiles for engineering capillary flow and application in microfluidics *ACS Appl. Mater. Interfaces* **12** 29908–16

[8] Nilghaz A, Ballerini D R and Shen W 2013 Exploration of microfluidic devices based on multi-filament threads and textiles: a review *Biomicrofluidics* **7** 051501

[9] Commonwealth Scientific and Industrial Research Organisation (CSIRO) (https://csiro.au)

[10] Young C D, Wu J R and Tsou T L 1998 High-strength, ultra-thin and fiber-reinforced pHEMA artificial skin *Biomaterials* **19** 1745–52

[11] Cai J, Perfect E, Cheng C-L and Hu X 2014 Generalized modeling of spontaneous imbibition based on Hagen–Poiseuille flow in tortuous capillaries with variably shaped apertures *Langmuir* **30** 5142–51

[12] Ashari A and Vahedi Tafreshi H 2009 A two-scale modeling of motion-induced fluid release from thin fibrous porous media *Chem. Eng. Sci.* **64** 2067–75

[13] Malekghasemi S, Kahveci E and Duman M 2016 Rapid and alternative fabrication method for microfluidic paper based analytical devices *Talenta* **159** 401–11

[14] Parada M, Vontobel P, Rossi R M, Derome D and Carmeliet J 2017 Dynamic Wicking Process in Textiles *Transp. Porous Med.* **119** 611–32

[15] Dullien F A L 1991 Characterization of porous media—pore level *Transpo. Porous Media* **6** 581–606

[16] Huinink H 2016 *Fluids in Porous Media—Transport and Phase Changes* (San Rafael, CA: Morgan & Claypool Publishers)

[17] Wei Y-C, Fu L-M and Lin C-H 2013 Electrophoresis separation and electrochemical detection on a novel thread-based microfluidic device *Microfluid. Nanofluid.* **14** 583–90

[18] Castro C 2017 *Imbibition in paper-based devices* UC Riverside https://escholarship.org/uc/item/67f3692n

[19] Hamraoui A and Nylander T 2002 Analytical approach for the Lucas–Washburn equation *J. Colloid Interface Sci.* **250** 415–21

[20] Chen X, Vroman P, Lewandowski M, Perwuelz A and Zhang Y 2009 Study of the influence of fiber diameter and fiber blending on liquid absorption inside nonwoven structures *Textile Res. J.* **79** 1364–70

[21] Fries N and Dreyer M 2008 The transition from inertial to viscous flow in capillary rise *J. Colloid Interface Sci.* **327** 125–8

[22] Darcy H 1856 *Les fontaines publiques de la ville de Dijon* (Paris: Victor Dalmont Editeur)

[23] Whitaker S 1986 Flow in porous media I: a theoretical derivation of Darcy's law *Transp. Porous Med.* **1** 3–25

[24] Zarandi M A F, Pillai K M and Kimmel A S 2018 Spontaneous imbibition of liquids in glass-fiber wicks. Part I: usefulness of a sharp-front approach *AIChE J.* **64** 294–305

[25] Bejan A 2013 *Convection Heat Transfer* (New York, NY: John Wiley & Sons, Inc.) 4th edn first published

[26] Liu Z, Hu J, Zhao Y, Qu Z and Xu F 2015 Experimental and numerical studies on liquid wicking into filter papers for paper-based diagnostics *Appl. Thermal Eng.* **88** 280e287

[27] Delgado J M P Q 2006 A simple experimental technique to measure tortuosity in packed beds *Can. J. Chem. Eng.* **84** 651–5

[28] Diao K, Zhang L and Zhao Y 2017 Measurement of tortuosity of porous Cu using a diffusion diaphragm cell *Measurement* **110** 335–8

[29] Yazdchi K, Srivastava S and Luding S 2011 On the validity of the Carman–Kozeny equation in random fibrous media *II International Conference on Particle-based Methods—Fundamentals and Applications, Particles (Barcelona, 26-28 October 2011)* E Oñate and D R J Owen (Barcelona: CIMNE (International Center for Numerical Methods in Engineering))

[30] Fu J, Thomas H R and Li C 2021 Tortuosity of porous media: image analysis and physical simulation *Earth Sci. Rev.* **212** 1034–9
[31] Melciu I C and Pascovici M D 2016 Imbibition of liquids in fibrous porous media *7th Int. Conf. on Advanced Concepts in Mechanical Engineering IOP Conf. Series: Material Science and Engineering* 147 *(Iasi, Romania, 9–10 June 2016)* 012041
[32] Kozeny J 1927 Ueber kapillare Leitung des Wassers im Boden *Sitzungsber Akad. Wiss., Wien* **136** 271–306
[33] Carman P C 1937 Fluid flow through granular beds *Trans., Inst. Chem. Eng. (London)* **15** 150–66
[34] Crawford R, Jones G F, You L and Wua Q 2011 Compression-dependent permeability measurement for random soft porous media and its implications to lift generation *Chem. Eng. Sci.* **66** 294–302
[35] Nogai T and Ihara M 1978 Study on air permeability of fiber assemblies oriented unidirectionally *J. Text. Mach. Soc. Japan, Trans.* **31** 166–70
[36] Zhu Z, Wang Q and Wu Q 2017 On the examination of the Darcy permeability of soft fibrous porous media; new correlations *Chem. Eng. Sci.* **173** 525–36
[37] Pezron I, Bourgain G and Quéré D 1995 Imbibition of a fabric *J. Colloid Interface Sci.* **173** 319–27
[38] Fries N and Dreyer M 2008 An analytical solution of capillary rise restrained by gravity *J. Colloid Interface Sci.* **320** 259–63
[39] Barry D A, Parlange J-Y, Sander G C and Sivaplan M 1993 A class of exact solutions for Richards' equation *J. Hydrol.* **142** 29–46
[40] Danino D and Marmur A 1994 Radial capillary penetration into paper: limited and unlimited liquid reservoirs *J. Colloid Interface Sci.* **166** 245–50
[41] Hyväluoma J, Raiskinmäki P, Jäsberg A, Koponen A, Kataja M and Timonen J 2006 Simulation of liquid penetration in paper *Phys. Rev. E: Stat., Nonlinear, Soft Matter Phys.* **73** 036705
[42] Woudberg S, van Heyningen M C, Le Coq L, Legrand J and Du Plessis J P 2014 Analytical determination of the effect of compression on the permeability of fibrous porous media *Chem. Eng. Sci.* **112** 108–15
[43] Hao L and Cheng P 2012 Capillary pressures in carbon paper gas diffusion layers having hydrophilic and hydrophobic pores *Int. J. Heat Mass Transfer* **55** 133–9
[44] Prakash K and Sridharan A 2013 Permeability of layered soils: an extended study *Geotech. Geol. Eng.* **31** 1639–44
[45] Ahsraf S and Pirhani J 2019 A generalized model for spontaneous imbibition in a horizontal, multi-layered porous medium *Chem. Eng. Sci.* **209** 115175
[46] Fu E, Ramsey S A, Kauffman P, Lutz B and Yager P 2011 Transport in two-dimensional paper networks *Microfluid. Nanofluidics* **10** 29–35
[47] Kokalj T, Park Y, Vencelj M, Jenko M and Lee L P 2014 Self-powered imbibing microfluidic pump by liquid encapsulation: SIMPLE *Lab Chip* **14** 4329
[48] Tran T S H, Ho B D, Beech J P and Tegenfeldt J O 2017 Open channel deterministic lateral displacement for particle and cell sorting *Lab Chip* **17** 3592
[49] Lee J J, Berthier J, Kearney K E, Berthier E and Theberge A B 2020 Open-channel capillary trees and capillary pumping *Langmuir* **36** 12795–803
[50] Tokihiro J C, Tu W-C, Berthier J, Lee J J, Dostie A M, Khor J W, Eakman M, Theberge A B and Berthier E 2023 Enhanced capillary pumping using open-channel capillary trees with integrated paper pads *Phys. Fluids* **35** 082120

Chapter 9

Biomimetics—open microfluidics in nature

9.1 Introduction

In the last ten years, philosophers and scientists have found that closer links to nature can hold great interest. Bioinspired structures can help in the development of new concepts. This movement was initiated by Otto Herbert Schmidt and theorized by Janine Benyus and has been called biomimetics [1, 2]. While biomimetics is limited to the '*in vitro*' reproduction of living beings, bioinspiration is more general and consists of innovations inspired by the observation and analysis of the living world.

In this chapter, we present a few open-microfluidics topologies found in nature. These topologies are surprisingly very elaborated and sophisticated, and they are the origins of new concepts such as anisotropic microfluidics and directional microfluidics.

9.2 Open channels on *Dryopteris marginata* leaves

In arid environments, some plants have developed external structures to collect water with the least evaporation loss, and in wet environments, these structures favor the drainage of excess water. Sharma and coworkers [3] studied the topology of the leaves of *Dryopteris marginata*, an evergreen fern native to the temperate regions between Nepal and Taiwan. They showed that the surface of D. marginata leaves integrates a system of multiscale channels so that water spreads quickly and is very efficiently transported via open channels (Ofigure 9.1).

The capillary flow in the open channels on the leaf surface is increased by the presence of the striations. Indeed, the associated capillary force is larger due to an increase in the wetted perimeter. The authors made a replica using soft lithography in an epoxy substrate. They compared the wetting contact angles (table 9.1) and verified the increase of drainage triggered by the open-channel longitudinal structures (figure 9.2).

doi:10.1088/978-0-7503-5507-0ch9

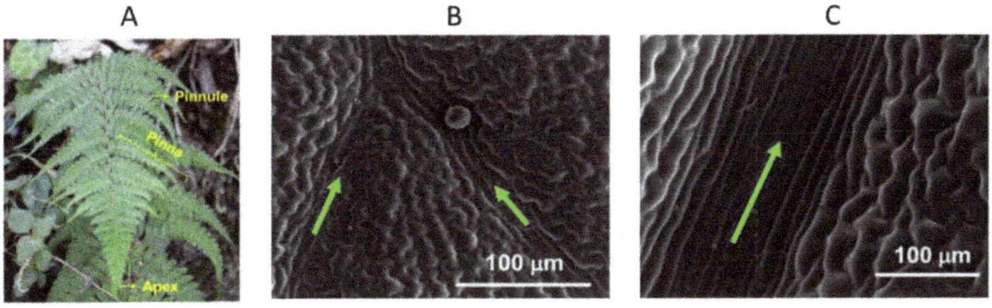

Figure 9.1. (A) A photograph of a *D. marginata* leaf; (B) two open channels merging; (C) detail of an open channel on a *D marginata* leaf; notice the parallel striation of the channel bottom. Reproduced from [3]. CC BY 4.0.

Table 9.1. A comparison of the contact angles.

	Static contact angle (°)	Advancing contact angle (°)	Receding contact angle (°)
Dryopteris leaf	39	44	29
Replica (epoxy)	45	49	28

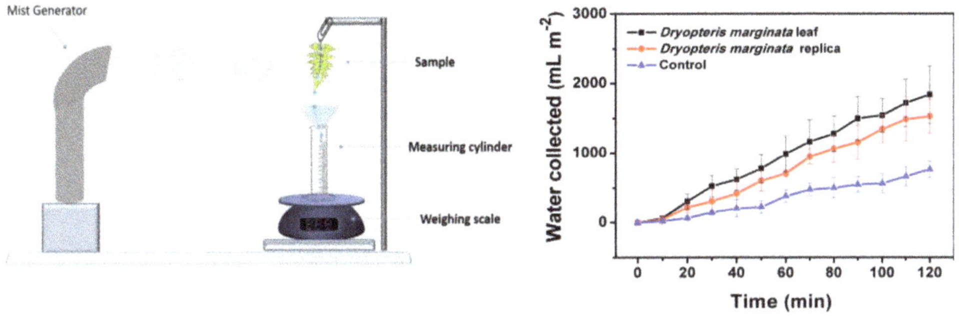

Figure 9.2. Left: a sketch of the device with a fog generator and the leaf placed vertically; right: the amount of water collected as a function of time. Reproduced from [3]. CC BY 4.0.

Figure 9.2 shows the device used to collect fog droplets and the collection efficiency of the *D. marginata* leaves.

In conclusion, it is observed for *D. marginata* leaves that the insertion of rivulets (striations) at the channel bottom promotes capillary flow. This corresponds to the analysis already presented in chapter 5.

Coincidentally, the principle used to collect water droplets described above has some similarity to the design of a portable air sampling device (figure 9.3) developed by Lee *et al* [4]. A fog of microdroplets is emitted by a piezo crystal and captures aerosols transported by the ambient air. The droplets are then collected on inclined

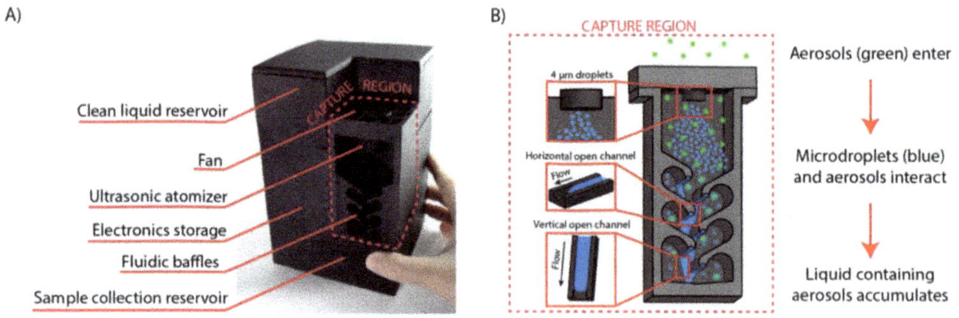

Figure 9.3. (A) A sketch of the device with the fog generator and the fluidic baffles collecting the liquid; (B) the fog is captured by the inclined fluidic baffles and drained via microgrooves. Reprinted with permission from [4]. Copyright (2021) American Chemical Society.

fluidic baffles—analogous to leaves—whose surfaces are etched with microgrooves. The liquid is collected at the bottom of the device.

9.3 Flow alongside *Sarracenia* trichomes

Trichomes are appendages on plants and algae. The name comes from the ancient Greek 'trikhoma,' meaning hair. Remember that capillarity comes from the Latin 'capillaris,' also meaning hair. An example of trichomes is shown in figure 9.4.

Let us analyze the capillary flow along trichomes. Trichomes have the shape of slender solid cones with a very small apical angle.

First, let us recall that the Laplace pressure of a droplet on the outside of a thin uniform fiber (figure 9.5) was calculated by Carroll [5] and McHale *et al* [6] for the case of droplets of dimension smaller than the capillary length ($R_0 < \sqrt{\gamma/\rho g}$)—or, equivalently, characterized by a Bond number smaller than one ($Bo = \rho g\ R_0^2/\gamma < 1$). For a zero contact angle ($\theta = 0°$), a complicated algebraic development led them to the Laplace pressure

$$\Delta P = \frac{2\gamma}{r + R_0}, \tag{9.1}$$

where r is the radius of the fiber and R_0 the apparent radius of the droplet. Note that when $r \to 0$, then relation (9.1) is the Laplace pressure of a sphere of radius R_0: $\Delta P = 2\gamma/R_0$, and when $R_0 \to r$, relation (9.1) produces the Laplace pressure of a cylinder of radius r: $\Delta P = \gamma/r$.

First consider a slightly conical solid fiber with a hydrophilic surface, as sketched in figure 9.6. In the case of a conical wire of half-apex angle α, Lorenceau and Quéré showed [9] that the change of Laplace pressure with the distance z from the cone apex is

$$\frac{\mathrm{d}P}{\mathrm{d}z} = -\frac{2\gamma}{(r + R_0)^2}\frac{\mathrm{d}r}{\mathrm{d}z}. \tag{9.2}$$

Figure 9.4. Trichomes on the petiole of a *Solanum quitoense* leaf. This *Solanum quitoense* petiole trichomes image has been obtained by the author(s) from the Wikimedia website where it was made available under a CC BY-SA 3.0 licence. It is included within this book on that basis. It is attributed to Veryhuman.

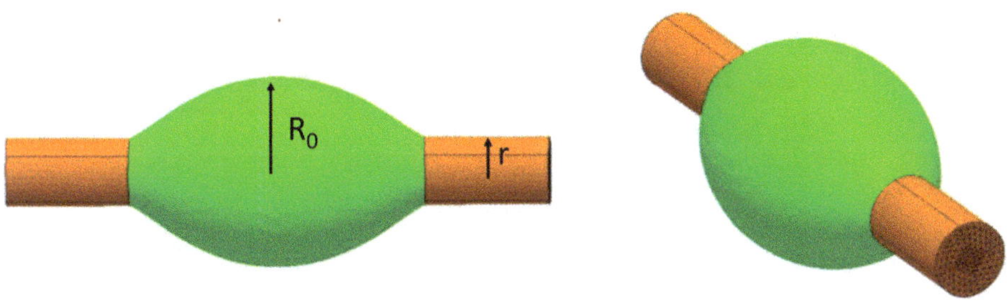

Figure 9.5. A view of a droplet sitting on a horizontal fiber; calculation performed using the software Surface Evolver ($\theta = 25°$, $r = 100$ μm) [7, 8].

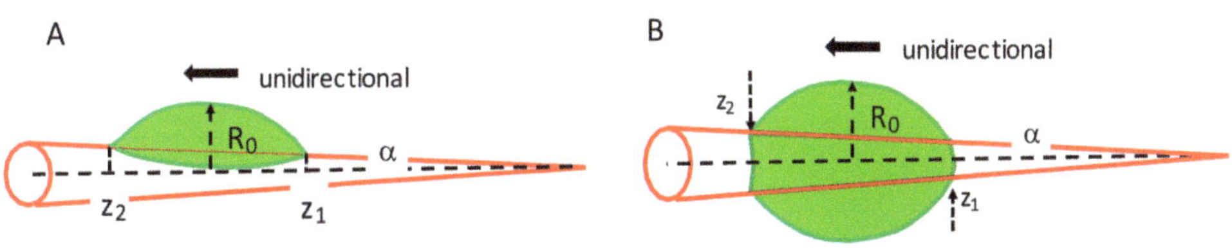

Figure 9.6. A sketch of the cone and the droplet; (A) a small droplet on the cone moving far from the cone apex; (B) a larger droplet englobing the cone, also moving away from the apex.

Hence, as remarked by Liwen Zhang *et al* [10], the resulting imbalance in the Laplace pressure is

$$\Delta P = -\int_{z_1}^{z_2} \frac{2\gamma}{(r + R_0)^2} \sin \alpha dz = -\int_{z_1}^{z_2} \frac{2\gamma}{(z \tan \alpha + R_0)^2} \sin \alpha dz \\ \approx -2\gamma\alpha \int_{z_1}^{z_2} \frac{dz}{(\alpha\, z + R_0)^2}. \tag{9.3}$$

The drop moves away from the apex under the action of a driving force which is the gradient of the Laplace pressure.

Let us now consider the case of *Sarracenia*, which are carnivorous plants that can be found in North America [10]. In addition to the conical shape described above,

Sarracenia trichomes are covered with microgrooves, like striations, parallel to the cone axis, as sketched in figure 9.7. Liquid fingers form in these microgrooves, preceding the droplet. Note that capillary motion along the trichome is not a case of spontaneous capillary flow (SCF), because of the additional Laplace pressure gradient (see chapter 4).

The motion of liquid on the trichome can be decomposed into two regimes: a dry regime and a wet regime. The dry regime corresponds to the capillary motion of a droplet on the dry trichome (figure 9.8). This droplet leaves liquid behind in the microgrooves. When a second droplet forms, it moves along the wet trichome—this is the wet regime (figure 9.9). It was remarked that the flow velocity in the second

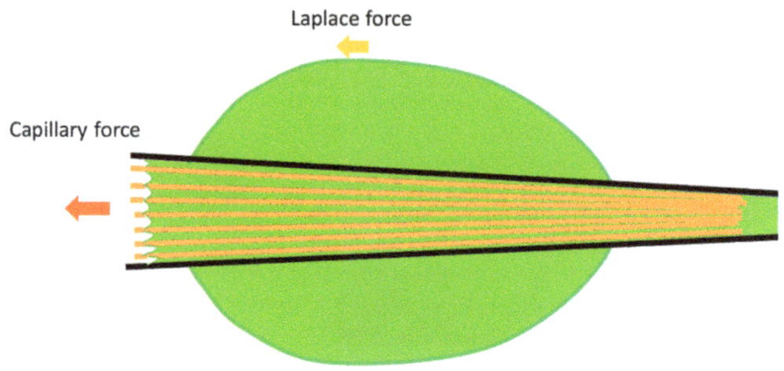

Figure 9.7. A sketch of the droplet with liquid fingers progressing ahead.

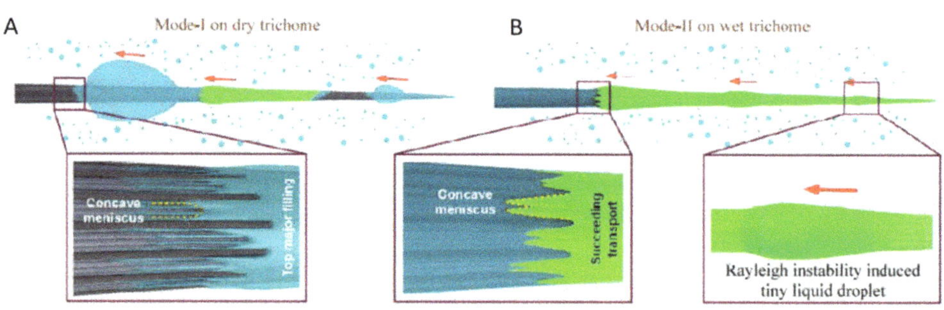

Figure 9.8. The two modes of liquid motion along a *Sarracenia* trichome: (a) a droplet progressing on dry microgrooves; (b) a droplet progressing on wet microgrooves. Reprinted from [11], Copyright (2018), with permission from Springer.

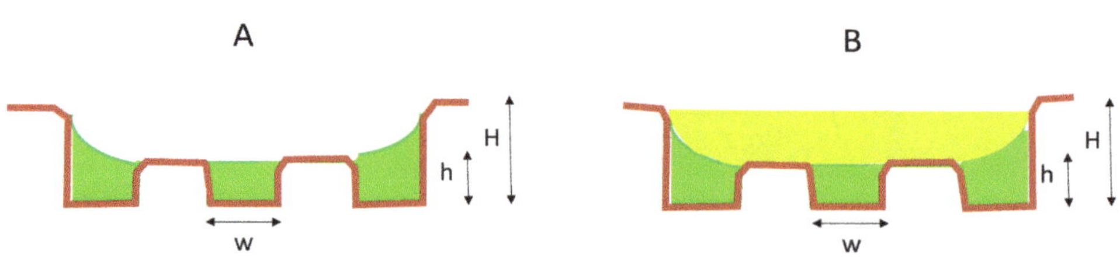

Figure 9.9. The two modes of flow along the trichome: (A) in the initially dry mode, only the minor grooves are filled; (B) in the wet mode, the second capillary flow moves on top of the already-filled grooves.

mode was very fast, approximately 20 times faster than motion in the first mode [10, 11]. In the second mode, as in the first, the driving force is the Laplace pressure of the droplet; however, in the second mode, the friction is small due to the presence of liquid left behind by the first mode.

9.4 Directional spreading on natural surfaces

Conventional forced microfluidic flows usually have no inertia; hence, they obey the Stokes equation. This equation implies that the flow pattern—streamlines—is the same if the flow direction is reversed [12, 13]. The pressure drop associated with wall friction is the same in both directions. In other words, there is no preferred direction. However, capillary-driven microflows are an exception and may be directional due to the particular phenomenon of pinning. They obey Stokes' law except when pinning occurs. In this section, it is shown that directional capillary-driven flows exist in nature and that topological diodes can be derived by observing nature [14, 15]. Interesting reviews of directional capillary-driven microflows have been published by Liwen Zhang and coworkers [16] and Yifan Si and coworkers [17].

9.4.1 Directional spreading on cilia—a pinning–spreading story

Chunhui Zhang and coworkers demonstrated that a directional flow behavior can be obtained for the spread of a liquid on a surface covered by inclined cilia of different elastic moduli [18]. In the case of a soft cilium, the bending of the cilium under the weight of the droplet results in motion in the direction of the cilium. In the case of a rigid cilium, pinning occurs on the tip of the inclined cilium, resulting in motion in the opposite direction to that of the row of cilia (figure 9.10).

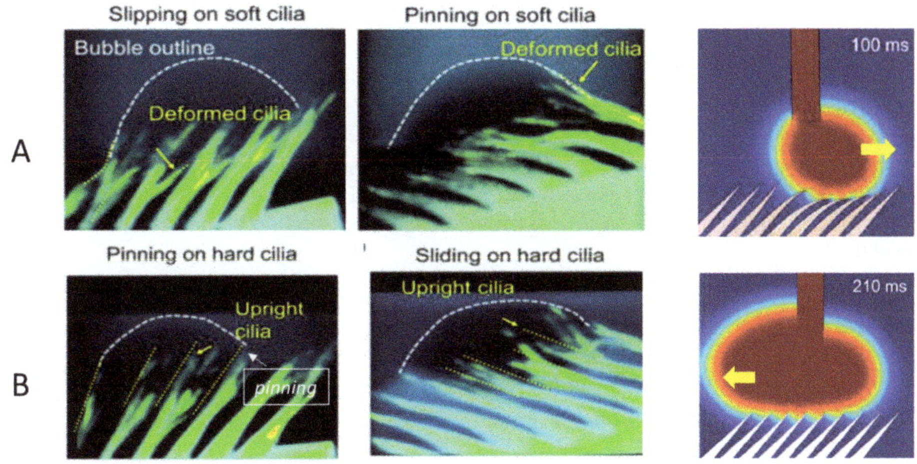

Figure 9.10. The flow directions are not the same on soft (A-bending) and hard (B-rigid) cilia. Panels (A) and (B) reprinted with permission from [10]. Copyright (2022) American Chemical Society.

9.4.2 Anisotropic microfluidics bioinspired by *Morpho menelaus*

The wings of *Morpho menelaus* (and *Morpho peleides*) have special features. *M. menelaus* is a butterfly that resides in the rain forests of Central and South America and is characterized by large iridescent blue wings. The surface of its wings is covered by microstructures that have particular properties that quickly drain liquid away toward the edges of the wings (figure 9.11) [18–20].

Shile Feng and coworkers [21] have shown that in the case of ethanol solutions, the microwings (fins) increase or decrease the capillary flow velocity depending on their orientation (figure 9.12). We shall see that the lateral spacing of the microwings is also of importance.

Let us analyze the motion on a flat plate. The principle of the flow motion is shown in figure 9.13. The liquid injected by the pipette preferentially moves to the right because of the larger capillary force exerted by the microwings. In order to

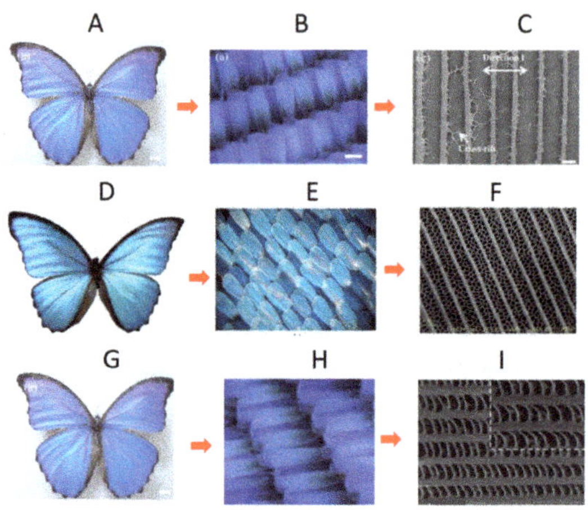

Figure 9.11. Three similar investigations of the structures on the wing surface of an *M. menelaus* butterfly. (A) A photograph of an *M. menelaus*; (B) details of the wing surface; (C) SEM micrography of the wing surface; (D) a photograph of an *M. peleides*; (E) details of the wing surface; (F) SEM micrography of the wing surface; (G), (H) and (I): the same observations for the *M. peleides* butterfly. In all cases, the microstructures show parallel grooves, but with different shapes of cross-ribs. Photographs (A), (B) and (C) reprinted from [18], Copyright (2010), with permission from Elsevier. Photographs (D), (E), and (F) reprinted from [19] with permission. © 2017 ICE Publishing. All rights reserved. Photographs (G), (H), and (I) reprinted from [20], CC BY 3.0.

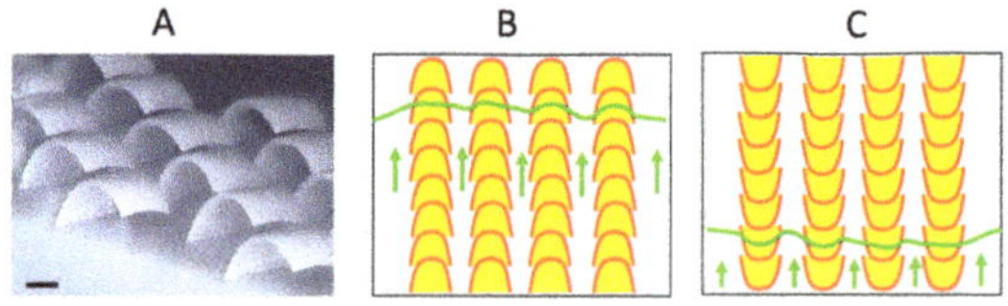

Figure 9.12. (A) 3D printed wings that replicate the details on the surface of the wings of *M. menelaus*; (B) a vertical flow of ethanol enhanced by a microwing array oriented upwards; (C) a vertical flow of ethanol diminished by a microwing array oriented downwards.

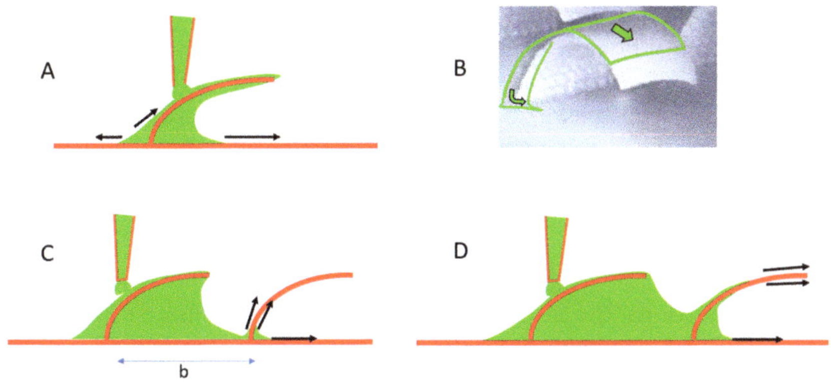

Figure 9.13. A sketch of the progression of a liquid oriented by the curved microwings. (A) The capillary force exerted by the microwing drags the fluid to the right; (B) liquid climbs up the wing and flows around the base of the wing; (C) the flow contacts the next wing; (D) the same as (A).

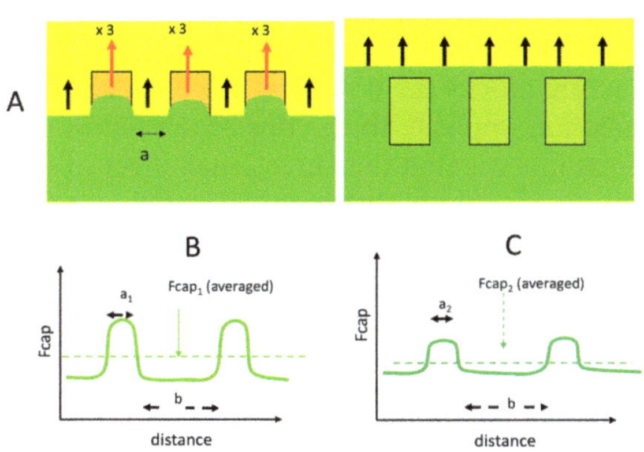

Figure 9.14. (A) A sketch of the capillary forces, showing that the capillary force exerted by the microwing is three times greater than the capillary force on the flat plate. (B) Capillary force for a distance a_1 between two microwings as a function of the travel distance of the fluid; (C) the same as (B) but for a larger value a_2 of the distance between two microwings; the average capillary force is less for a larger gap. The larger a becomes, the more the presence of the cilia is reduced.

continue the rightwards motion, the axial distance between two consecutive microwings should not be too large.

The average capillary force depends on the arrangement of the microwings, as shown in figure 9.14.

When the triple line has passed the microwing row, the capillary force is

$$F_{\text{cap}} = w\,\gamma\cos\theta = [n\,b + (n-1)a]\,\gamma\cos\theta, \tag{9.4}$$

where w is the width occupied by the contact line, n the number of gaps, b the width of a gap, and a the width of a microwing. When the triple line is located in an area where microwings are present, the capillary force is

$$F_{\text{cap}} = [n\,b + 3(n-1)\,a]\,\gamma\cos\theta. \tag{9.5}$$

At the level of the microwings, an additional capillary force $F_{cap} = 2(n - 1)\, a\, \gamma \cos \theta$ promotes the capillary flow in the direction of the wings. Hence, the capillary force considerably increases when the triple line reaches the level of the microwing. Moreover, an increase in the gap *b*—leading to a decrease in the relative number of cilia or wings—decreases the average capillary force at the level of the cilia, as indicated by (9.5).

9.4.3 The particular structure of the leaves of *Nepenthes alata*

In the preceding sections, the capillary flow behavior on the raised structures of some plants and butterflies has been presented. In this section, the behavior of capillary flow in reentrant structures is presented.

The leaves of *Nepenthes alata*, a carnivorous plant native to the Philippines, present very particular structures (figure 9.15) [22–24]. The leaves have reentrant arc-shaped microstructures. These horseshoe structures—or flares—inserted between linear ridges promote a capillary flow in the direction opposed to the tip of the horseshoe flares. The pinning of liquid on the ridges in the direction of the tip of the horseshoe structure blocks the flow. The capillary flow is then unidirectional, and moves in the direction opposite to the tips of the flares.

Chen and coworkers investigated this directional effect of the structure of the *N. alata* leaf using the different microfabricated devices shown in figure 9.16 [24]. As expected, the spread in a smooth channel was symmetrical: the distances over which spreading took place were the same in the two opposing directions (figure 9.16(A)). However, inserting asymmetric surface reliefs resulted in an asymmetric spread of liquid. The asymmetry of the spread increased with the use of pitted grooves (reentrant reliefs) and arc-shaped reliefs, as shown in figures 9.16(B)–(D). The last figure 9.16(D) is very similar to the relief of the leaf of *N. alata* and forced a nearly unidirectional capillary flow. The asymmetric spread was caused by the partial pinning of the liquid on the edges of the reentrant structures.

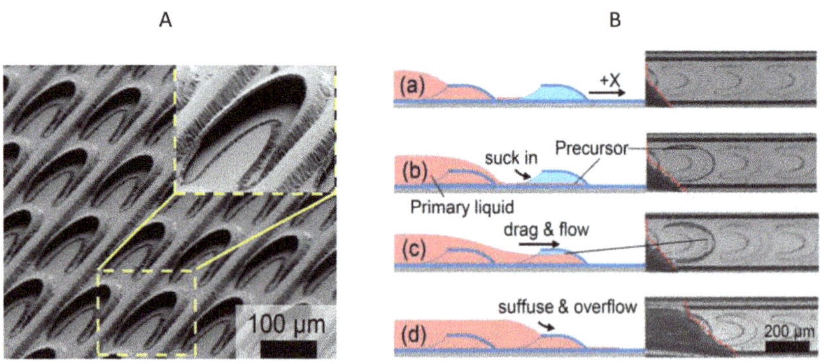

Figure 9.15. (A) A view of the surface of *N. alata* leaves; (B) the mechanism by which liquid spreads on a leaf of *N. alata*. Reproduced from [23] with permission of The Royal Society of Chemistry.

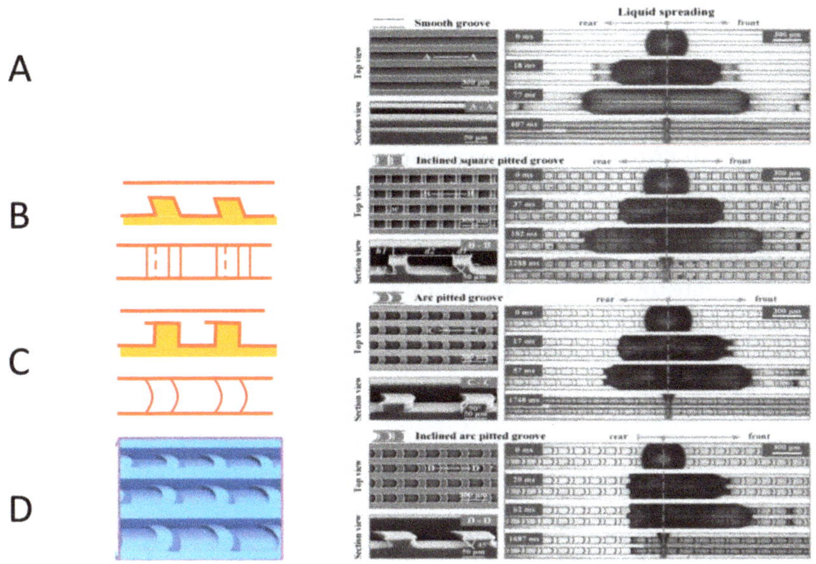

Figure 9.16. Water spreading in different open channels; (A) smooth grooves; (B) inclined square pitted grooves; (C) arc pitted grooves; (D) inclined arc pitted grooves [24] John Wiley & Sons. © 2016 The Authors. Published by WILEY-VCH Verlag GmbH & Co. KGaA, Weinheim.

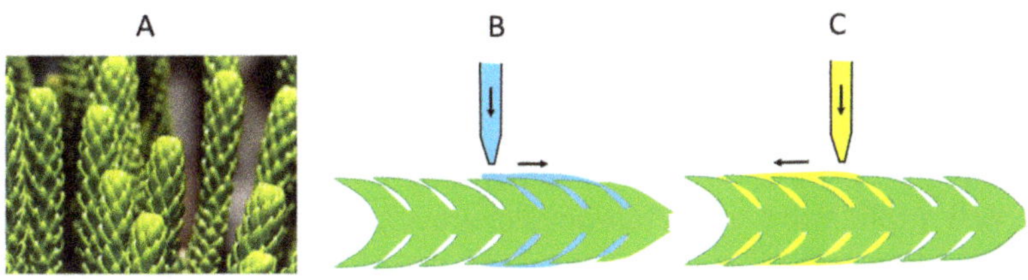

Figure 9.17. (A) A photograph of an *Araucaria* leaf (courtesy of the University of Florida); (B) water being dropped on an *Araucaria* leaf; the direction of the flow is backwards (opposite to the leaf tip) and the water contact angle is approximately 78°; (C) ethanol being dropped on an *Araucaria* leaf, the direction of the flow is downwards (aligned with the leaf tip) and the the ethanol contact angle is approximately 32°.

9.4.4 The pinning paradox: the case of *Araucaria* leaves

The conventional understanding is that a liquid deposited on a surface tends to move in directions that reduce surface energy. Its transport directions are determined mainly by the surface structure and have nothing to do with the liquid's properties, such as surface tension. This conventional idea fails in the case of pinning of the flow. Observing liquid spreading on *Araucaria* leaves in a Hong Kong park, a scientist had the idea that the flow direction might depend on the liquid surface tension—specifically, the contact angle. Indeed, this was confirmed by successively pipetting ethanol and water onto an *Araucaria* leaf (figure 9.17). An *Araucaria* leaf consists of periodically arranged ratchets tilting towards the leaf tip. Water flows in the direction opposite to the ratchets, while ethanol flows in the direction of the leaf tip [21]. In the case of water, the pinning on the extremities of the leaves and the large contact angle (less than 90°)—associated with the high surface tension—blocks the flow in the direction of the leaf tip. In the case of a small contact angle—as associated with low-surface-tension liquids, such as ethanol—the liquid penetrates the interior of the leaf structure and spreads in the direction of the leaf tip. The surface tensions of ethanol–water solutions are given in SI units in this chapter.

Let us recall that the contact angle θ depends on the surface tension according to Zisman's relation (chapter 1)—which is the important parameter here. We recall that the approximate Zisman's relation [25] is

$$\cos\theta = 1 + \frac{\gamma_c - \gamma}{a\,\gamma_c}, \tag{9.6}$$

where γ_c is the critical surface tension and a is a constant. Shile Feng and coworkers used mixtures of water and ethanol. The surface tension of these mixtures decreases with the ethanol concentration (table SI.1), hence their contact angle decreases with the ethanol concentration, according to (9.6), from 78° to 32° for ethanol concentrations increasing from 0% to 40%.

The physical mechanism results from a complicated pinning competition. There are pinning locations at the roots of the fins and at the top of the fins. For a liquid with a low contact angle, the liquid breaks the pinning at the bottom of a fin and propagates from bottom to top. By contrast, a liquid with a high contact angle breaks the pinning at the top of the fins and propagates downwards. Hence the motion occurs in different directions, depending on the contact angle.

A proposed application of such a device concerns the separation of water and oil. Water flows backwards on such a device, while oil flows oppositely downwards.

9.5 Conclusions

Analysis of the spread of liquids on plant leaves or on butterfly wings has led to improvements in our conception of open microflows. Directional or anisotropic capillary-driven microflows can be achieved by special surface reliefs that promote the flow in one direction but pin the liquid in the opposite direction [24]. This observation contradicts the conventional understanding that a liquid deposited on a surface tends to move in directions that reduce surface energy. This conventional idea fails in the case of pinning of the flow, because depinning requires a jump (discontinuity) in the surface energy.

Supplementary information

The values of the surface tension for different concentrations of ethanol are given in table SI.1 [26].

Table SI.1. The surface tensions of ethanol–water mixtures in units of mN m^{-1}. Reprinted with permission from [26]. Copyright (1995) American Chemical Society.

Mass %	x_1	20 °C	25 °C	30 °C	35 °C	40 °C	45 °C	50 °C
0	0.000	72.75	72.01	71.21	70.42	69.52	68.84	67.92
5	0.020	56.41	55.73	55.04	54.36	53.63	52.96	52.16
10	0.042	48.14	47.53	46.88	46.24	45.58	44.97	44.26
15	0.065	42.72	42.08	41.49	40.88	40.27	39.64	38.96
20	0.089	38.56	37.97	37.38	36.83	36.28	35.71	35.12
25	0.115	36.09	35.51	34.96	34.41	33.86	33.31	32.76
30	0.144	33.53	32.98	32.43	31.94	31.42	30.89	30.34
40	0.207	30.69	30.16	29.68	29.27	28.77	28.28	27.82
50	0.281	28.51	27.96	27.53	27.11	26.64	26.21	25.78
60	0.370	26.72	26.23	25.81	25.43	24.97	24.54	24.11
70	0.477	25.48	25.01	24.60	24.21	23.76	23.33	22.92
80	0.610	24.32	23.82	23.39	23.01	22.54	22.12	21.71
90	0.779	23.23	22.72	22.32	21.94	21.53	21.13	20.71
100	1.000	22.31	21.82	21.41	21.04	20.62	20.22	19.82

References

[1] Schmitt O H 1986 *Emerging System Designs for Implementing Computer-Assisted Health Care in the Home* (Chicago, IL: Association for the Advancement of Medical Instrumentation)

[2] Benyus J M 2002 *Biomimicry: Innovation Inspired by Nature* (Boston, MA: Mariner Books) https://www.harpercollins.com/products/biomimicry-janine-m-benyus?variant=32117835366434

[3] Sharma V, Balaji R and Krishnan V 2018 Fog-harvesting properties of *Dryopteris marginata*: role of inter-scalar microchannels in water-channeling *Biomimetics* **3** 7

[4] Lee U N *et al* 2021 Miniaturizing wet scrubbers for aerosolized droplet capture *Anal. Chem.* **93** 11433–41

[5] Carroll B J 1976 The accurate measurement of contact angle, phase contact areas, drop volume, and Laplace excess pressure in drop on fiber system *J. Colloid Interface Sci.* **57** 488–95

[6] McHale G, Kab N A, Newton M I and Rowan S M 1997 Wetting of a high-energy fiber surface *J. Colloid Interface Sci.* **186** 453–61

[7] Brakke K A 1992 The surface evolver *Exp. Math.* **1** 141–65

[8] Berthier J and Brakke K A 2012 *The Physics of Microdroplets* (Beverly, MA: Scrivener Publishing LLC)

[9] Lorenceau E and Quéré D 2004 Drops on a conical wire *J. Fluid Mech.* **510** 29–45

[10] Zhang C *et al* 2022 Bioinspired anisotropic slippery cilia for stiffness controllable bubble transport *ACS Nano* **16** 9348–58

[11] Chen H, Ran T, Gan Y, Zhou J, Zhang Y, Zhang L, Zhang D and Jiang L 2018 Ultrafast water harvesting and transport in hierarchical microchannels *Nat. Mater.* **17** 935–42

[12] Tabeling P 2023 *Introduction to Microfluidics* (Oxford: Oxford University Press) 2nd edn

[13] Berthier J and Silberzan P 2010 *Microfluidics for Biotechnology* (Boston, MA: Artech House) 2nd edn https://uk.artechhouse.com/Microfluidics-for-Biotechnology-Second-Edition-P1267.aspx

[14] Li J, Zhou X, Li J, Che L, Yao J, McHale G, Chaudhury M K and Wang Z 2017 Topological liquid diode *Sci. Adv.* **3** eaao3530

[15] Feng J and Rothstein J P 2013 One-way wicking in open micro-channels controlled by channel topography *J. Colloid Interface Sci.* **404** 169–78

[16] Zhang L, Liu G, Chen H, Liu X, Ran T, Zhang Y, Gan Y and Zhang D 2021 Bioinspired unidirectional liquid transport micro-nano structures: a review *J. Bionic Eng.* **18** 1–29

[17] Si Y, Li C, Hu J, Zhang C and Dong Z 2023 Bioinspired superwetting open microfluidics: from concepts, phenomena to applications *Adv. Funct. Mater.* **33** 2301017

[18] Liu F, Liu Y, Huang L, Hu X, Dong B, Shi W, Xie Y and Ye X 2011 Replication of homologous optical and hydrophobic features by templating wings of butterflies *Morpho menelaus Opt. Commun.* **284** 2376–81

[19] Das S, Shanmugam N, Kumar A and Jose S 2017 Review: potential of biomimicry in the field of textile technology *Bioinspired Biomim. Nanobiomaterials* **6** 224–35

[20] Ahmed R, Ji X, Atta R M H, Rifat A A and Butt H 2018 Morpho butterfly-inspired optical diffraction, diffusion, and bio-chemical sensing *RSC Adv.* **8** 27111–8

[21] Feng S, Zhu P, Zheng H, Zhan H, Chen C, Li J, Wang L, Yao X, Liu Y and Wang Z 2021 Three-directional capillary ratchet-induced liquid directional steering *Science* **373** 1344–8

[22] Chen H, Zhang P, Zhang L, Liu H, Jiang Y, Zhang D and Jiang L 2016 Continuous directional water transport on the peristome surface of *Nepenthes alata Nature* **532** 85–9

[23] Sun W, Tang L, Hong W, Zhan Y, Yang B and Liu J 2019 A novel microstructure inspired from *Nepenthes alata* and lizard skin and its enhanced unidirectional liquid spreading property *RSC Adv.* **9** 7842

[24] Chen H, Zhang L, Zhang P, Zhang D, Han Z and Jiang L 2017 A Novel bioinspired continuous unidirectional liquid spreading surface structure from the peristome surface of *Nepenthes alata Small* **13** 1601676

[25] Zisman W A 1941 The spreading of oils on water. Part III: spreading pressures and the Gibbs adsorption relation *J. Chem. Phys.* **9** 789–93

[26] Vázquez G, Alvarez E and Navaza J M 1995 Surface tension of alcohol + water from 20 to 50 °C *J. Chem. Eng. Data* **40** 611–4

Chapter 10

Two-phase open-channel capillary flows

10.1 Introduction

The study of two-phase open-channel microflows is very recent and has as yet seen only a few experimental developments [1–5]. In this chapter, the first experiments are related and analyzed. Two different approaches have been used in order to investigate two-phase open microflows: (1) the insertion of plugs and drops during or before the open flow, using pipettes; (2) the emission of droplets from a reservoir using capillary forces alone, i.e. without the use of any other energy source.

The first approach consists in dropping immiscible liquids (droplets or plugs) onto an existing open microflow or inserting the immiscible liquid before the arrival of the open microflow (figure 10.1(A)); the second approach (figure 10.1(B)) consists in emitting droplets across a neck using capillary forces and Laplace pressure.

We first analyze the transport of plugs and droplets by an open-channel microflow and present the different modes—or regimes—that occur.

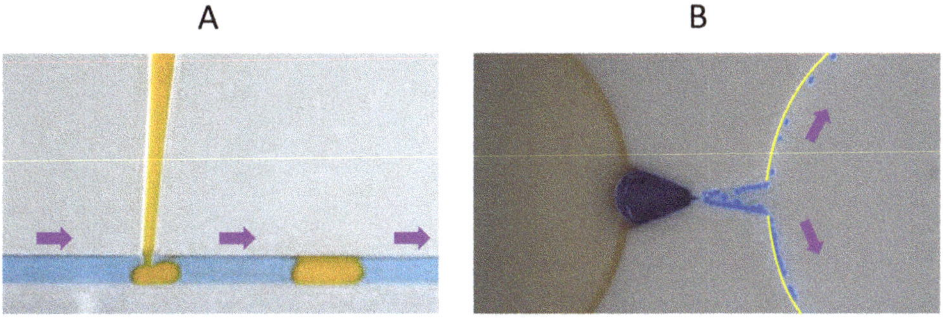

Figure 10.1. (A) The injection of an immiscible liquid (yellow-tinted water) into an open-capillary flow of nonanol; (B) the emission of blue-tinted water across a neck under the Laplace pressure produced by a reservoir of hexafluoroethane (HFE). Photographs courtesy of Jing J Lee and Jian-Wei Khor, University of Washington.

doi:10.1088/978-0-7503-5507-0ch10

10.2 Part 1: plugs and droplets in open channels of uniform cross-section

In this section, we investigate the behavior of plugs and drops in open channels of uniform cross-section, first from a quasi steady-state point of view (small velocities), then from a dynamic point of view by analyzing experimental observations. In particular, it is shown that plugs/drops—even those that do not wet walls—placed in an open flow do not stop the capillary flow. Moreover, it is shown that plugs/drops adopt a few—only three—typical morphologies.

We shall call the liquid filling the inlet reservoir and flowing in the channel the primary or carrier liquid, while the secondary liquid or plug is the liquid droplet pipetted in front or on top of the carrier flow. The two liquids are immiscible.

10.2.1 The quasi steady-state approach: the spontaneous capillary flow condition in the presence of plugs

The software Surface Evolver [6] keeps track of an interface by relocalizing it stepwise in order to minimize the surface energy. Hence, Surface Evolver is well adapted to predict capillary flows, under the condition that there are no inertial effects; in other words, Evolver does not account for the dynamics of the flow. Numerical simulations using Evolver have shown that two-phase capillary flow occurs in rectangular (and suspended) open channels when the spontaneous capillary flow (SCF) condition for the primary (main) fluid is satisfied [1, 3, 4] and that a plug placed in the open channel cannot block the capillary flow, regardless of the wetting properties of the plug.

For example, figure 10.2 shows an Evolver simulation of SCF in suspended channels in the case of a wetting or non-wetting plug, i.e. the contact angle of the

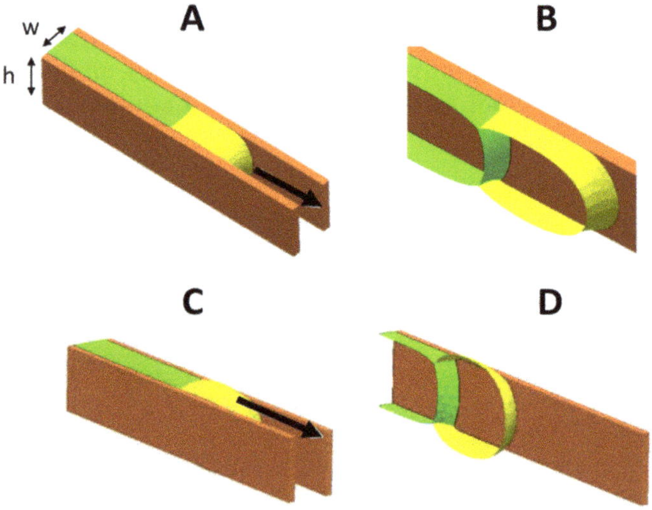

Figure 10.2. A two-phase suspended capillary flow: (A) a lyophilic plug (yellow) and (B) a cutaway view of the channel along the median plane; (C) a lyophobic plug (yellow) and (D) a cutaway view of the channel along the median plane. Figures obtained using Surface Evolver [6].

liquid of the plug is smaller or larger than 90°. Of course, the main fluid satisfies the SCF condition

$$\cos\theta > \frac{w}{h}, \tag{10.1}$$

where w is the distance between the two plates, h the width of the plates, and θ the contact angle of the primary liquid with the plates. It is found that, in both cases, the capillary flow is not interrupted by the presence of the plug.

Let us examine the general case: determining whether SCF occurs involves a static or quasi-static analysis. Inertial and viscous effects can be overlooked in this problem because we are looking for the conditions at the onset of the capillary flow. Consider a plug at the front end of a liquid segment in an open channel (figure 10.3). Under the assumption of a quasi steady state, the contact angles are the Young contact angles.

Let the indices S, A, L_1, and L_2 respectively denote the solid wall, the air, the carrier liquid, and the plug. We will show that the balances of the capillary forces are the same in the two configurations of figures 10.3(A) and (B).

Let us analyze the capillary forces on the floor of the channel. The resultant of the capillary forces is the sum of the capillary forces at the first and second interfaces (figure 10.3(A)):

$$R = \gamma_{\rm L1L2}\cos\theta_{12} + \gamma_{\rm L2A}\cos\theta_2. \tag{10.2}$$

The relation between the different Young contact angles can be deduced by applying Young's law to each of the three triple lines of figure 10.3 [7]. In the case of panel B of the figure, Young's law yields

$$\gamma_{\rm L1A}\cos\theta_1 = \gamma_{\rm SA} - \gamma_{\rm SL1}, \tag{10.3}$$

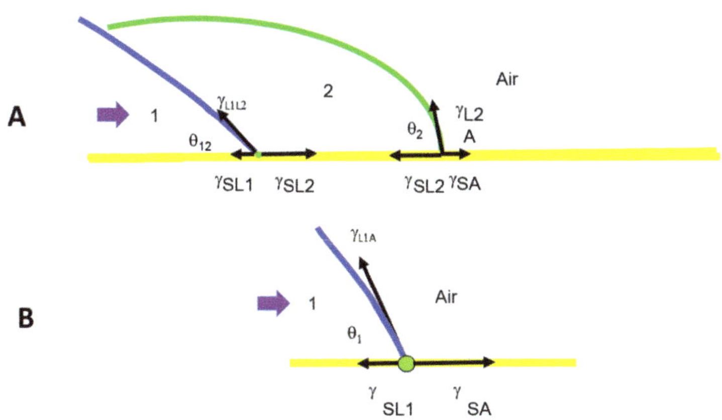

Figure 10.3. Capillary forces acting on a plug placed in front of an SCF. (A) Capillary forces on the triple lines of the plug and the carrier fluid with the solid wall (the plug—index 1—is delimited by the green line, the carrier liquid—index 2—is depicted in blue, and the wall is colored yellow); (B) capillary forces in the absence of the plug. Note that the contact angles are not the same in (A) and (B), i.e. $\theta_{12} \neq \theta_1$.

For the contact between liquid #2 and the wall (the right-hand part of figure 10.3(A)),

$$\gamma_{L2A} \cos \theta_2 = \gamma_{SA} - \gamma_{SL2}, \tag{10.4}$$

and for the contact between liquid #1 and liquid #2 (the left-hand part of Figure 10.3(A)),

$$\gamma_{L1L2} \cos \theta_{12} = \gamma_{SL2} - \gamma_{SL1}. \tag{10.5}$$

Subtracting (10.4) from (10.3) and substituting the result into (10.5) yields

$$\gamma_{L1L2} \cos \theta_{12} = \gamma_{L1A} \cos \theta_1 - \gamma_{L2A} \cos \theta_2. \tag{10.6}$$

Substituting (10.6) into (10.2) yields

$$R = \gamma_{L1A} \cos \theta_1. \tag{10.7}$$

which is simply the capillary force for liquid 1 alone (figure 10.3(B)). Hence, the two-phase configuration A is equivalent to the single-phase configuration B.

The reasoning still applies for a plug embedded in the carrier fluid, as shown in figure 10.4. In this case, because the contact angles are static ones, the resultant of the forces in case A is

$$R = \gamma_{L1L2} \cos \theta_{12} - \gamma_{L1L2} \cos \theta_{12} + \gamma_{L1A} \cos \theta_1 = \gamma_{L1A} \cos \theta_1. \tag{10.8}$$

which is the same as the resultant in case B.

In conclusion, in the case of a channel of constant cross-section, the presence of the plug does not affect the occurrence of SCF. However, we shall see in the following section that it affects the dynamics of the flow. We verify this conclusion using a numerical approach with the Evolver software for rectangular open-channel geometry [4, 6].

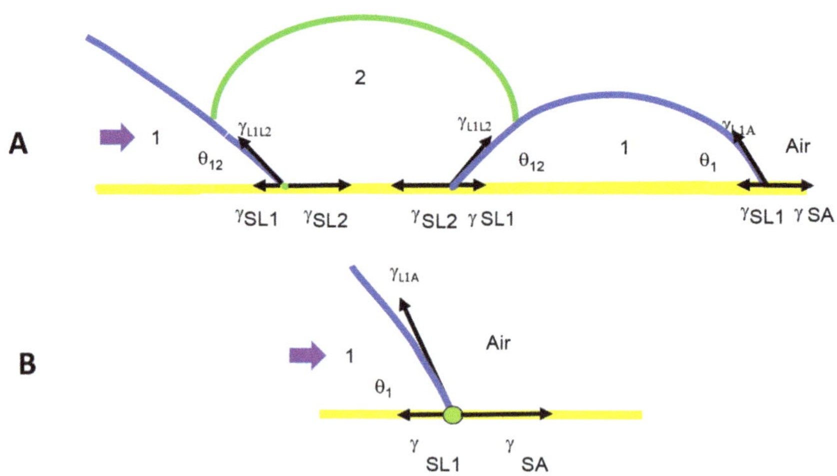

Figure 10.4. A sketch of the capillary forces: (A) the plug is delimited by the green line and the carrier liquid is depicted in blue; (B) a sketch of the SCF of the primary liquid. The yellow line represents the solid wall.

The parameters of the problem are five: two contact angles (that of the primary liquid with the wall, denoted by θ_1, and that of the secondary liquid with the wall, denoted by θ_2) and three surface tensions (the primary liquid surface tension, the secondary liquid surface tension, and the primary–secondary surface tension, respectively denoted by γ_1, γ_2, and γ_{12}). The contact angle between the interface between the two liquids and the wall is deduced from the five preceding parameters using relation (6.3). Evolver calculations have been performed using many different combinations of the parameters $\theta_1 \in \{5°, 50°\}$, $\theta_2 \in \{15°, 90°\}$, $\gamma_1 \in \{20, 50\}$ mN m^{-1}, $\gamma_2 \in \{20, 65\}$ mN m^{-1}, and $\gamma_{12} \in \{10, 55\}$ mN m^{-1}.

Now let us consider a rectangular open channel of uniform cross-section. Depending on the different properties of the fluids, three different plug morphologies—that we shall call modes—are observed: (1) a shift mode, in which the plug is horizontally displaced and translates in front of the flow; (2) a lift mode, in which the plug is lifted and the main liquid flows underneath; (3) a bridge mode in which the plug sticks to the floor of the channel and the flow of the main liquid passes above it. In the following, we detail each of these modes.

a. Shift mode

The shift mode occurs when the plug is translated down the channel by the action of the carrier fluid. Upon initial contact, the plug shape changes. After the contact, the shape of the plug remains unchanged. Figure 10.5 shows a typical shift mode (calculated with Evolver) and figure 10.1(A) an experimentally observed shift mode. This mode principally occurs when the two liquids wet the walls, but this is not a necessary condition: in figures 10.5(A)–(C), the contact angle θ_2 is 90°.

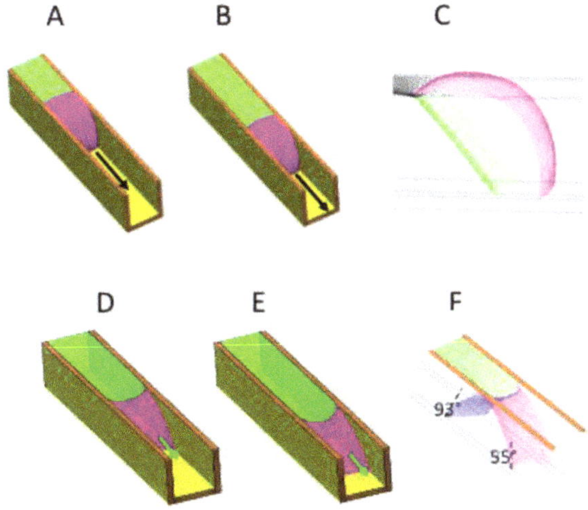

Figure 10.5. Shift mode: (A) and (B) the shift motion of the plug upon arrival of the carrier capillary flow; (C) detail of the shape of the plug; the interfaces have been 'dematerialized' (case obtained with $\gamma_{L1A} = 20$ mN m^{-1}, $\gamma_{L2A} = 72$ mN m^{-1}, $\gamma_{L1L2} = 30$ mN m^{-1}, $\theta_1 = 50°$, $\theta_2 = 90°$). (D) and (E) the shift motion of the plug; (F) detail of the shape of the plug ($\gamma_{L1A} = 30$ mN m^{-1}, $\gamma_{L2A} = 42$ mN m^{-1}, $\gamma_{L1L2} = 16$ mN m^{-1}, $\theta_1 = 40°$, $\theta_2 = 55°$). All figures obtained using Surface Evolver.

b. Lift mode

In the case where the liquid–liquid contact angle θ_{12} is small, the carrier liquid slips below the plug and progressively lifts the plug. The primary liquid SCF continues with the plug of secondary liquid on top of it (figure 10.6).

c. Bridge mode

If the plug makes a small contact angle with the walls and its surface tension is sufficiently high, the plug forms a bridge and the carrier fluid flows above this bridge (figure 10.7). In the Evolver quasi-static approach, the bridge remains immobile. However, experiments show that the bridge may or may not move, depending on the plug–carrier liquid surface tension and the velocity of the flow (inertial effect, not taken into account in the quasi-static approach).

The Evolver results show that three modes exist for the plug under quasi-static conditions and confirm the theoretical result that the SCF is never stopped by a plug in an open channel of uniform cross-section. The parameters corresponding to the different modes obtained in the Evolver calculations are listed in table 10.1. At the time of writing, no theoretical investigations had been performed to characterize the occurrence of each mode. But the results listed in table 10.1 indicate that high

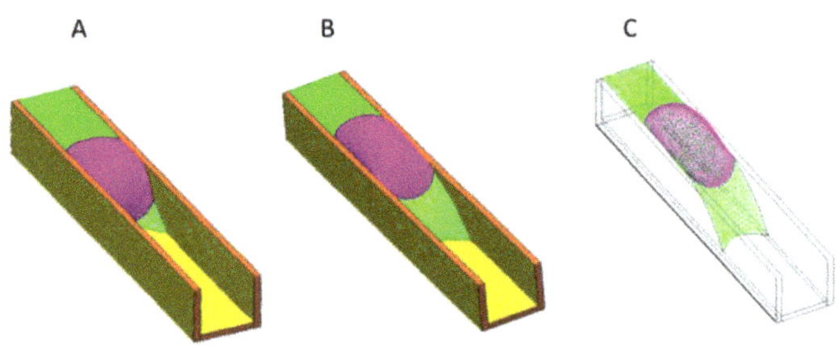

Figure 10.6. Lift mode: (A) and (B) the main liquid uplifts the plug and continues to flow beneath it ($\gamma_{L1A} = 72$ mN m^{-1}, $\gamma_{L2A} = 36$ mN m^{-1}, $\gamma_{L1L2} = 60$ mN m^{-1}, $\theta_1 = 50°$, $\theta_2 = 110°$). Using Young's law (10.6), $\theta_{12} = 12.4°$, and the carrier liquid slips underneath the droplet. All figures obtained using Surface Evolver.

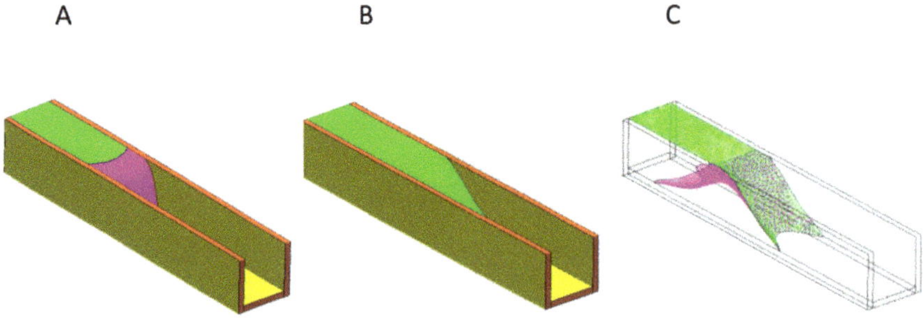

Figure 10.7. Bridge mode: (A) and (B) the primary liquid ($\gamma_{L1A} = 22$ mN m^{-1}, $\gamma_{L2A} = 65$ mN m^{-1}, $\gamma_{L1L2} = 30$ mN m^{-1}, $\theta_1 = 15°$, $\theta_2 = 80°$). Using Young's law (10.6), $\theta_{12} = 112°$. All figures obtained using Surface Evolver.

Table 10.1. The parameters used for the numerical Evolver simulations: high plug surface tension (cells in yellow) corresponds to bridge mode, while a small liquid–liquid–wall contact angle (triggered by a high plug–wall contact angle) corresponds to lift mode (cells in light blue). Note that the lift mode coincides with the establishment of capillary filaments.

Mode	$\theta 1$	$\theta 2$	$\gamma 1$	$\gamma 2$	$\gamma 12$	cos($\theta 12$)	$\theta 12$
Bridge	15	80	22	65	30	0.3	70.6
Bridge	15	80	22	65	30	0.3	70.6
Bridge	15	60	22	65	30	−0.4	112.0
Bridge	15	30	22	65	30	−1.0	180.0
Bridge	15	40	22	65	35	−0.8	144.6
Shift	15	40	22	45	35	−0.4	112.2
Shift	15	30	22	45	35	−0.5	120.4
Shift	10	90	22	45	35	−0.6	126.1
Shift	10	90	22	45	35	0.6	51.8
Shift	30	90	22	55	35	0.5	57.0
Shift	30	70	42	35	55	0.7	48.6
Shift	50	70	42	35	55	0.3	74.1
Shift	50	80	40	35	40	0.3	69.9
Shift	10	90	40	35	40	0.8	33.6
Lift	10	90	45	30	40	1.0	0.0
Lift	10	80	45	20	30	1.0	0.0
Lift	10	80	45	20	30	1.0	0.0
Lift	10	60	45	20	30	1.0	0.0
Shift	10	25	35	50	30	−0.4	111.2
Shift	5	15	35	50	30	−0.4	116.6
Bridge	5	15	35	50	10	−1.0	180.0
Lift	5	15	50	35	10	1.0	0.0

plug surface tension (γ_2) favors the bridge mode and a large plug contact angle with the walls (θ_2)—corresponding to a very small liquid–liquid–wall contact angle (θ_{12})—favors the lift mode. Looking at the table, we note that in the case of the bridge mode, Neumann's rule—Neumann's triangle presented in chapter 1 [8, 9]—is not satisfied because $\gamma_{L2A} > \gamma_{L1A} + \gamma_{L1L2}$, and liquid #1 forms a film at its contact with liquid #2. We shall verify this observation later in this chapter.

Note that the same modes take place in triangular channels, as shown in figure 10.8.

d. Summary

In summary, only three modes for the plug–droplet behavior have been found using a numerical approach. These modes are plotted in figure 10.9. In the next section, the dynamics of the two-phase flow is investigated.

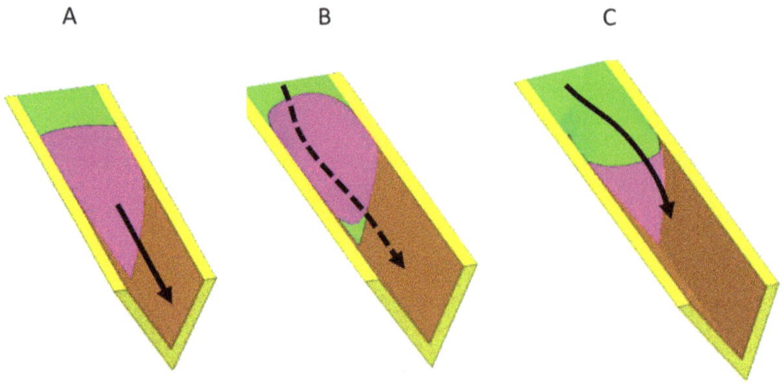

Figure 10.8. The same modes take place in open triangular channels: (A), shift; (B), lift; (C), bridge. All figures obtained with Surface Evolver.

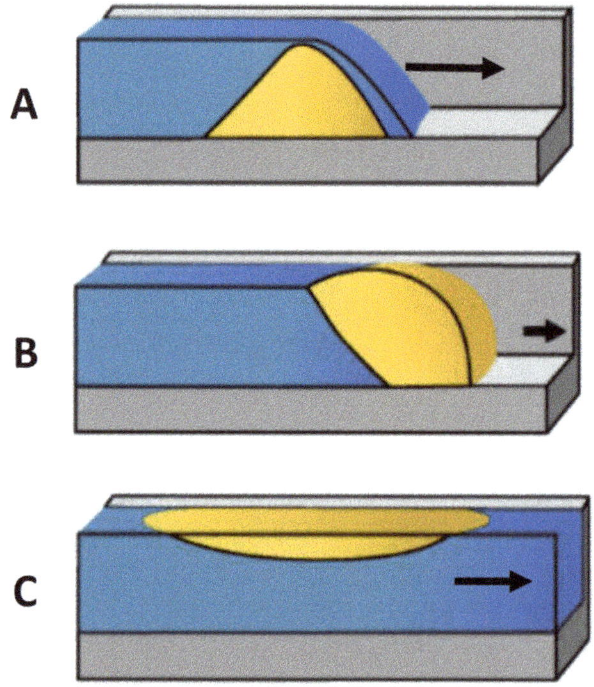

Figure 10.9. The three modes predicted by the numerical approach: (A) bridge; (B) shift; (C) lift. Reprinted with permission from [4]. Copyright 2018 American Chemical Society.

10.2.2 Plug dynamics in open-channel capillary flows—an experimental approach

At the time of publishing, only few experiments had been reported [2–5]. The device considered in this study is shown in figure 10.10. It consists of a large cylindrical inlet reservoir, a long meandering channel etched in poly(methyl methacrylate) (PMMA), and a cylindrical outlet port. Two types of channels are milled in the PMMA: a

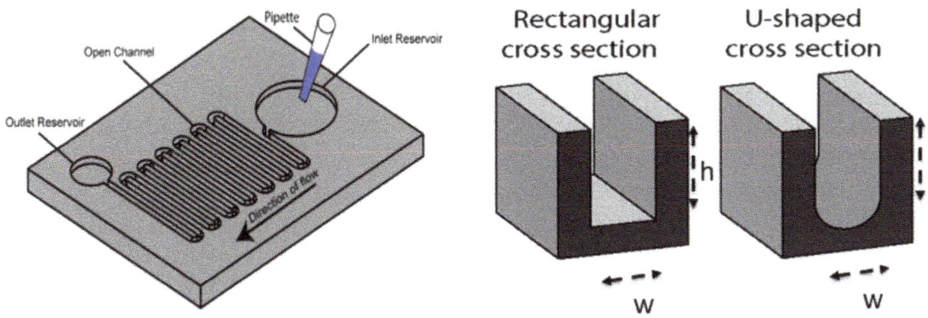

Figure 10.10. Left: a schematic of the open-channel device; right: the two types of channel cross-sections.

Table 10.2. Properties of the organic fluids used as primary liquids and the modes observed.

Liquid	Density (kg m^{-3})	Viscosity (mPa s)	Surface tension w. air (mN m^{-1})	Surface tension w. water (mN m^{-1})	Contact angle with PMMA (°)	Modes observed
1-Pentanol	814	3.50	25.4	4.4	12.5	Shift
1-Nonanol	827	11.7	26.4	8.5	17.0	Shift
Chloroform	1490	0.60	27.5	32.8	17.3	Bridge
Toluene	867	0.55	28.4	36.1	7.8	Bridge
FC40	1850	2.30	16.0	52.1	12.7	Bridge
n-Dodecane	750	1.25	25.4	52.8	26.7	Bridge

rectangular channel (w = 0.8 mm, h = 1.1 mm) and a U-shaped rounded channel (w = 0.8 mm, h = 1.1 mm). The main difference between the two types of channels is the bottom interior corners, which either allow or forbid the presence of capillary filaments [10–12]. In the rectangular channel, capillary filaments precede the bulk flow; this is not the case for the rounded channels, where no filaments exist.

The different organic liquids used in the study and their physical properties are listed in table 10.2.

It has been shown (chapter 4) that the velocity of a capillary flow decreases according to $1/\sqrt{t}$ or $1/z$. It has also been deduced that the capillary number is

$$Ca = \bar{\lambda} \cos\theta / z, \tag{10.9}$$

where $\bar{\lambda}$ is the average friction length. Hence, the dynamic effects decrease quickly with the travel distance. The behavior of a plug placed at the channel entrance is not the same as that of a plug deposited further down the channel. We shall analyze both cases, considering first a plug deposited far from the channel entrance, followed by one near the channel entrance.

a. A plug initially placed far from the channel entrance—small velocities

Let us first remark that $\bar{\lambda} \sim D_H$ (D_H being the hydraulic diameter); then $Ca \sim D_H/z$, and a plug placed far from the channel entrance should not be subject to any dynamic effect. Its behavior should be similar to the quasi-static case presented in section 10.2.1. And indeed, the behavior of the immiscible plug observed experimentally is consistent with the numerical approach: shift and bridge modes are observed (table 10.2). The lift mode—predicted by the numerical approach for a rectangular channel—was, however, not observed, perhaps because the conditions for lift mode were not met by the liquids that we used.

The experimentally observed shift and bridge modes are shown in figure 10.11. The side view of figure 10.11(A) shows that the plug is merely moved without deformation, as in a translation. The side view of figure 10.11(B) shows that the plug is progressively covered by a film of carrier liquid, then the carrier fluid re-forms behind the plug as an open-capillary flow in the channel.

As mentioned earlier in this chapter, the bridge mode does not stop the carrier flow motion. When the carrier fluid meets the plug, it stops, or its velocity decreases, then a film of carrier fluid builds over the immobilized plug and progresses on the other side of the bridge, then carrier flow motion restarts. Figure 10.12 shows a capillary flow of dodecane passing above three successive aqueous bridges. The travel distance of the carrier fluid as a function of time is shown in figure 10.13. The time taken for the dodecane to pass each immobilized bridge increases for each successive bridge (figure 10.13(A)). Between two bridges, the travel distance of the dodecane approximately follows the relation $z \sim \sqrt{t}$. The velocity of the flow of dodecane decreases at each passage over successive bridges (figure 10.13(B)), and the velocity recovery after the bridge is reduced for each successive bridge.

b. Plug initially placed at the channel entrance—high velocities

At the channel entrance, dynamics effects are non-negligible. Not only is the capillary velocity high during the early inertial regime but also at the beginning of the viscous regime. Velocities in the range of 10–30 mm s^{-1} are frequent. From a general standpoint, the Weber number characterizes the ratio between inertial and surface tension forces [13, 14]:

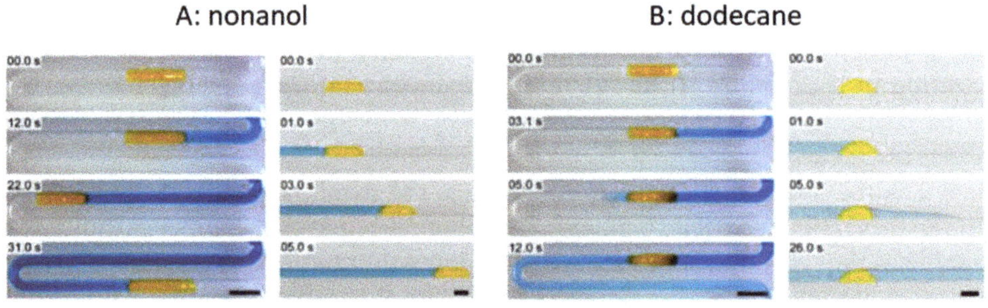

Figure 10.11. (A) An experimental visualization of shift mode using an aqueous plug (yellow) and 1-nonanol as the carrier phase (blue); the left photograph corresponds to the view from the top and the right photograph to the view from the side. (B) The same as (A) but for *n*-dodecane, showing a bridge mode. The scale bar is 2 mm. Reprinted with permission from [4]. Copyright 2018 American Chemical Society.

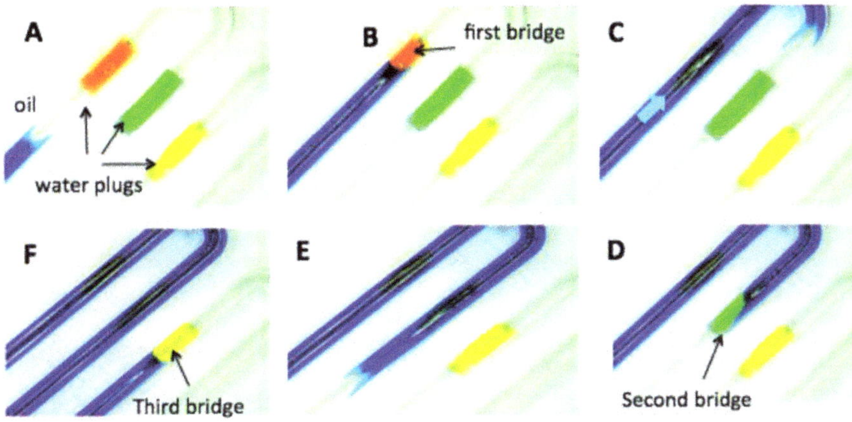

Figure 10.12. A capillary flow of dodecane moving over three aqueous plugs in a PMMA channel: the plugs stay immobile while dodecane continues to flow. Photo courtesy of Jing Lee and Jean Berthier, University of Washington.

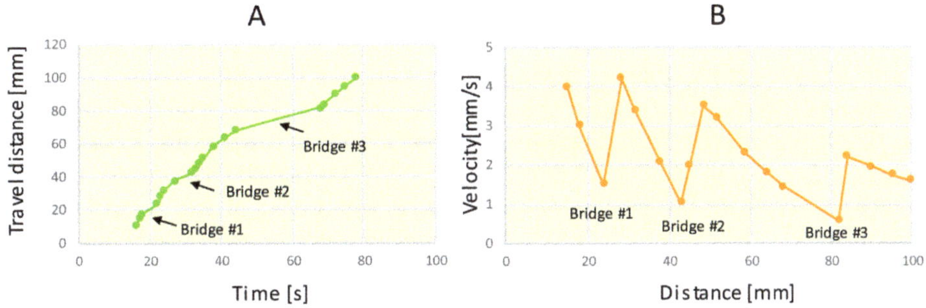

Figure 10.13. (A) The travel distance of dodecane moving past three similar immobile water bridges: each bridge increases the friction of the dodecane flow; an approximate relation $z \sim \sqrt{t}$ is obtained for the dodecane travel distance between the bridges; (B) the dodecane velocity vs. the distance in the channel. At each successive bridge crossing, the velocity becomes smaller, and the velocity recovery is also reduced after the dodecane passes over each bridge.

$$We = \frac{\rho\, V^2}{\gamma / w} = \frac{\rho\, V^2\, w}{\gamma} = \frac{\rho\, V\, w}{\mu}\frac{V\,\mu}{\gamma} = \mathrm{Re}\; Ca, \tag{10.10}$$

where ρ is the density of the liquid, V the velocity of the fluid, μ the fluid viscosity, and Re the Reynolds number characterizing the ratio between inertial and viscous forces. For a two-phase flow, the ratio between the inertial forces of the carrier liquid and the surface tension between the two fluids (γ_{12}) is the Weber number, defined by

$$We_{12} = \frac{\rho\, V^2\, w}{\gamma_{12}} = \frac{\rho\, V\, w}{\mu}\frac{V\,\mu}{\gamma_{12}} = \mathrm{Re}\; Ca\,\frac{\gamma}{\gamma_{12}}. \tag{10.11}$$

Note that γ is the surface tension of the carrier (with the surrounding air), while γ_{12} is the surface tension between the two liquids. In [4], it is shown that in our typical

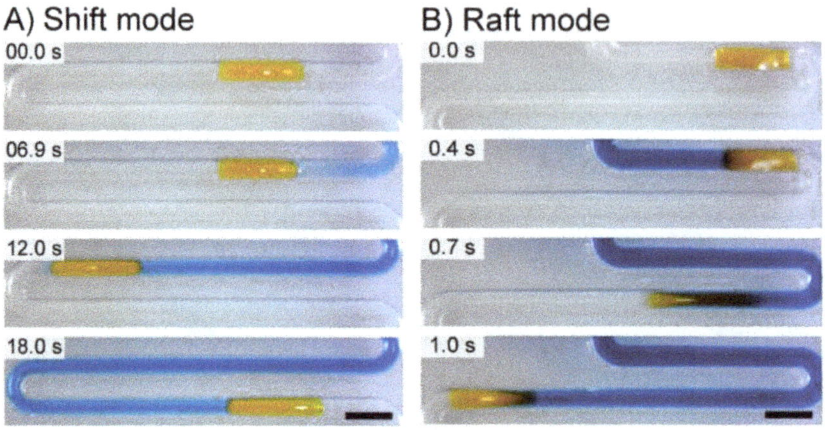

Figure 10.14. An experiment with pentanol as the carrier fluid (blue) and an aqueous plug (yellow) at two different locations in the channel. (A) The plug is in a low-velocity region, showing shift mode. (B) The plug is in a high-velocity region, showing raft mode. Images are representative of three replicated experiments, and the scale bar is 2 mm. Photograph courtesy of Jing J Lee and Jean Berthier, University of Washington.

channels, the Weber number for a plug at a distance of $100\bar{\lambda}$ falls between 0.17 for *n*-dodecane and 3.3 for chloroform, showing that some dynamic effects are present. In figure 10.14, we display top view photographs showing transitions in behavior for pentanol and a water plug as dynamic effects are increased. The quasi-static shift mode becomes a 'raft' mode in which the plug is lifted up and entrained by the flow of pentanol.

Similarly, the bridge mode can be altered by dynamic effects when the flow velocity is sufficient. Figure 10.15 first shows the bridge mode occurring at low velocities, then a 'bridge–translate' mode at higher velocities, and finally a raft mode at high velocities.

10.2.3 Summary

Three modes (shift, bridge, and lift) are characteristic of static and quasi-static conditions. However, these interfacial tension-dominated modes are affected by the dynamic forces of the carrier fluid, resulting in the emergence of mixed modes, such as the bridge–translate mode—in which the bridge is progressively shifted downstream—and raft mode—a mixed mode between lift and shift, at higher velocities.

10.2.4 Injecting a droplet/plug into an open flow

Plugs and droplets may be directly injected into an open-capillary flow, using a pipette, as shown in figure 10.16, where plugs are transported by the carrier fluid in the 'shift' mode. The injection of the plug was numerically simulated using Evolver and the results are shown in figure 10.17.

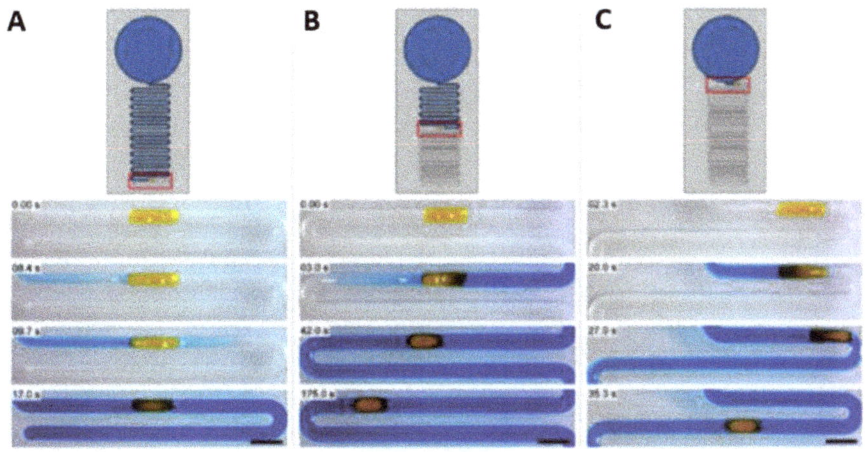

Figure 10.15. An experiment with toluene as the carrier fluid (blue) and an aqueous plug (yellow) at three different locations in the channel. (A) The plug is in a low-velocity region, showing bridge mode. (B) The plug is in a mid-velocity region, showing the bridge–translate mode. (C) The plug is in a high-velocity region, showing the raft mode. The scale bar is 2 mm. Reprinted with permission from [4]. Copyright 2018 American Chemical Society.

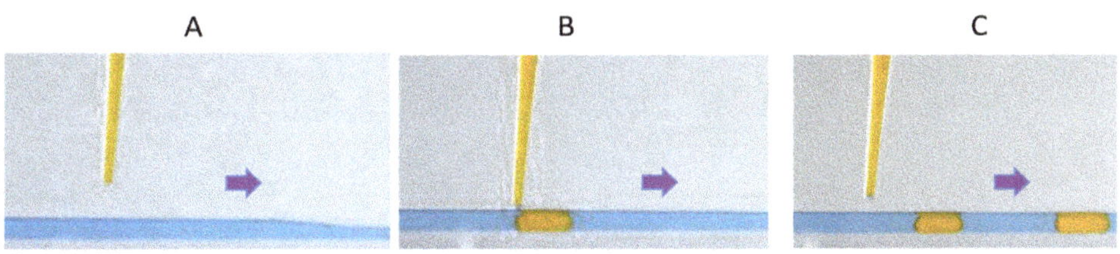

Figure 10.16. (A) The injection of the first plug; (B) the injection of the second plug; (C) the injection of the third plug. Blue-tinted nonanol is the carrier liquid, and yellow-tinted water is the discontinuous phase. Photograph courtesy of Jing J Lee, University of Washington.

10.2.5 Capillary wagons

As shown in figure 10.16, trains of capillary wagons can easily be formed in an open-channel capillary flow. This property has been numerically checked using Surface Evolver (figure 10.18).

Figure 10.19 shows the formation of capillary wagons in the shift mode in a rounded rectangular open channel. It demonstrates the ability to add fluid to a plug using a simple pipette. Every other yellow water droplet is selected, and a bolus of red-colored water is added.

10.2.6 The case of capillary filaments

In square open channels—or in any channel with sharp corners—capillary filaments form if the Concus–Finn condition [10–12] is met, i.e. $\theta < \frac{\pi}{4}$ for a square open

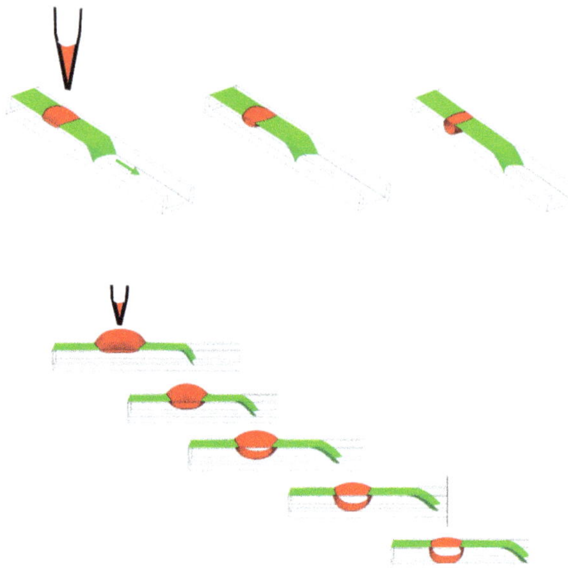

Figure 10.17. (A) An Evolver calculation of the injection of a plug into a low-velocity open microflow (dynamic effects are neglected) in the shift mode; (B) the same view as (A) but with dematerialized walls. Calculation data: $\gamma_{L1A} = 20$ mN m^{-1}, $\gamma_{L2A} = 72$ mN m^{-1}, $\gamma_{L1L2} = 30$ mN m^{-1}, $\theta_1 = 50°$, $\theta_2 = 90°$.

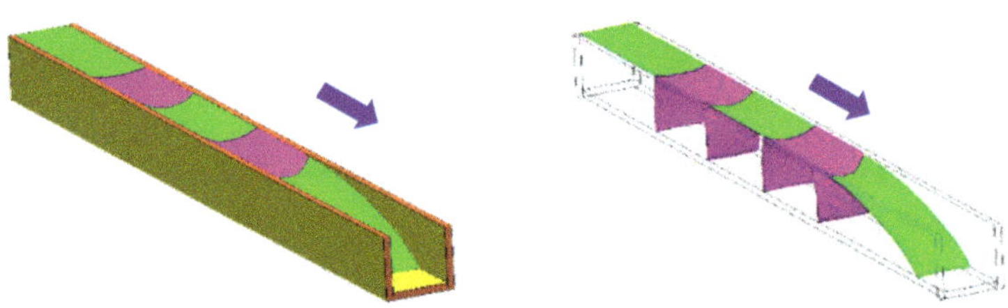

Figure 10.18. (A) An Evolver calculation of two plugs transported by a carrier flow at low velocity (inertia is neglected) in the shift mode; (B) the same view as (A) but with dematerialized walls. Calculation obtained using $\gamma_{L1A} = 20$ mN m^{-1}, $\gamma_{L2A} = 72$ mN m^{-1}, $\gamma_{L1L2} = 30$ mN m^{-1}, $\theta_1 = 50°$, $\theta_2 = 90°$.

channel or, more generally, $\theta < \frac{\pi}{2} - \alpha$ for an interior sharp corner of semi-angle α (see chapter 7).

Figure 10.20 shows that capillary filaments have an important role in the transport of plugs due to the high shear rate that they exert on plugs.

In the following section, we analyze how capillary filaments may transport plugs.

a. Plugs moved by capillary filaments

Capillary filaments may extend for a relatively long distance ahead of the bulk flow. In fact, they grow faster than the bulk flows and the length of the filaments increases during the flow (figure 10.21).

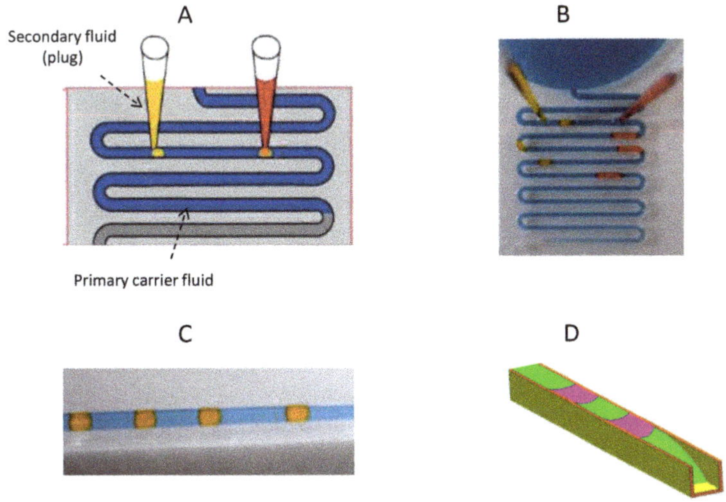

Figure 10.19. (A) A sketch showing the primary carrier fluid and two secondary liquids injected into the flow using pipettes; (B) sequential plug formation in an open microfluidic channel. The carrier solvent is 1-pentanol with blue dye; the droplets are water-based with yellow or red dye; (C) a side view of the channel with the water 'wagons'; (D) an Evolver calculation of the capillary wagons. Panel (B) reprinted with permission from [4], Copyright 2018 American Chemical Society. Panel (C) photograph courtesy of Jing J Lee, University of Washington.

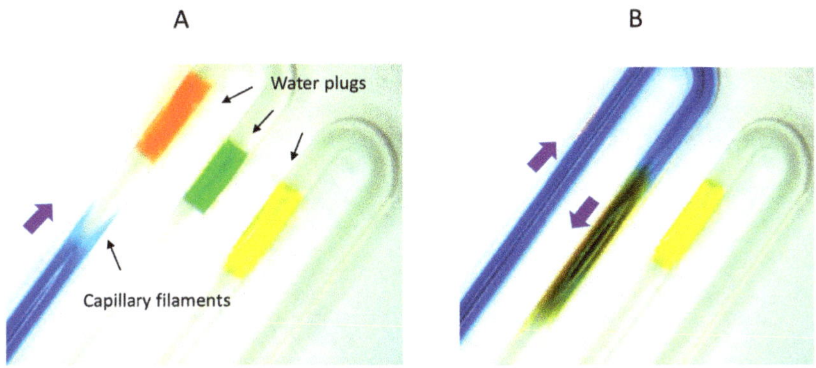

Figure 10.20. A capillary flow of nonanol (tinted blue) successively moving water plugs (tinted red, green, and yellow). Note the fast mixing of the plugs. Photographs courtesy of Jing J Lee and Jean Berthier, University of Washington.

When a plug is placed in the channel, the two capillary filaments are the first to contact the plug. If the filaments are long, they can move the plug downstream, as shown in figure 10.22. This mode can be categorized as the 'raft' mode.

The motion of a plug moved by capillary filaments has a characteristic dynamic (figure 10.23). Although no study has interpreted this motion, it seems that the plug —after an initial acceleration—has a constant velocity (yellow curve in figure 10.23). In fact, the plug translates along the channel. It has a constant friction length with

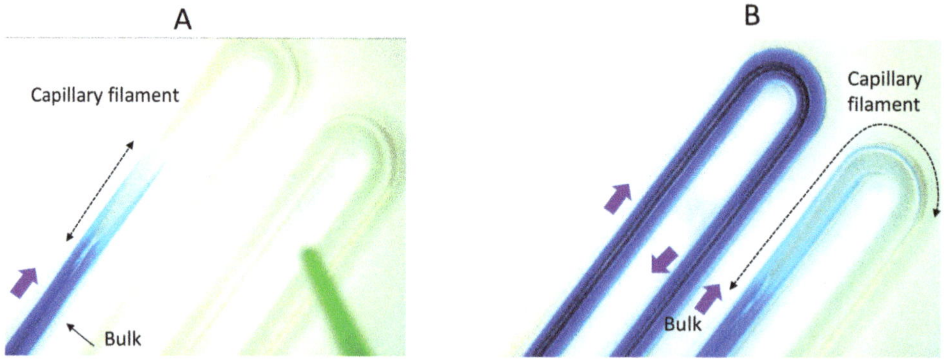

Figure 10.21. A capillary flow of toluene (tinted blue) in a winding rectangular open channel with sharp corners, preceded by long capillary filaments. The filaments grow during the flow (from A to B). Photographs courtesy of Jing J Lee and Jean Berthier, University of Washington.

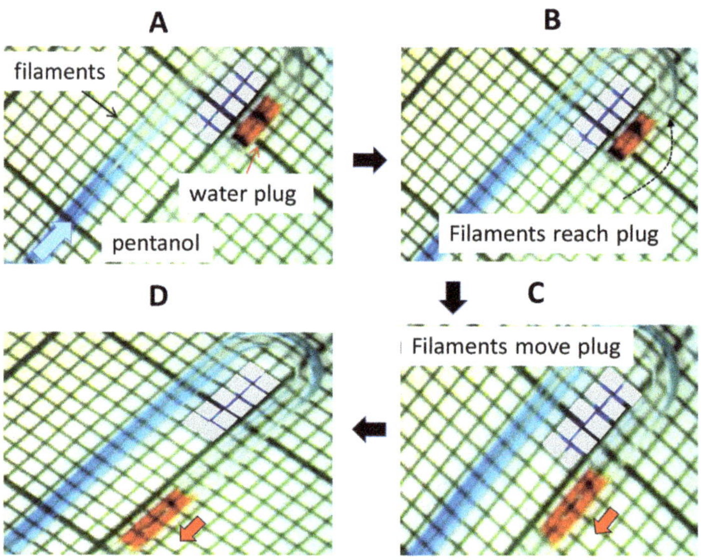

Figure 10.22. Capillary filaments of pentanol (tinted blue) moving a water plug (tinted red): the plug moves under the action of the filaments without any contact with the bulk of the flow. Photographs courtesy of Jing Lee and Jean Berthier, University of Washington.

the walls and a constant capillary force is exerted by the filaments; hence, the constant velocity.

10.2.7 Bifurcations and bypasses

This section first presents the behavior of a plug transported by a carrier capillary flow and arriving at a bifurcation; then, the focus moves to bypasses, which are used to maintain a separation between two plugs.

a. Bifurcations

Single-phase capillary flow across a bifurcation has been investigated and reported in the literature by Mehrabian *et al* [15]. Open-channel two-phase flow in bifurcations is still underdeveloped. It has been shown that plugs can be divided in two by a symmetrical bifurcation [16]. Figure 10.24 shows a water plug

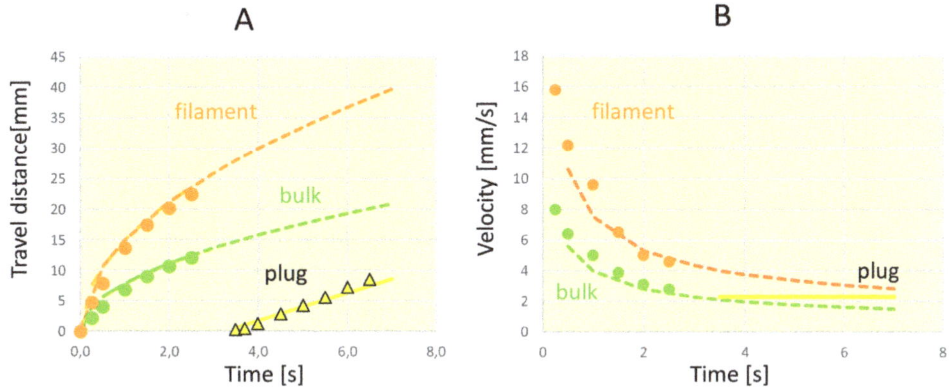

Figure 10.23. (A) Travel distance vs. time for dodecane filaments (green triangles), bulk dodecane (green circles), and water plugs (yellow squares). The yellow squares are aligned along a straight line, showing a constant-velocity translation of the plug. (B) Velocity vs. time using the same colors as (A). As long as the filament velocity is greater than the plug velocity (yellow), the plug has a constant velocity.

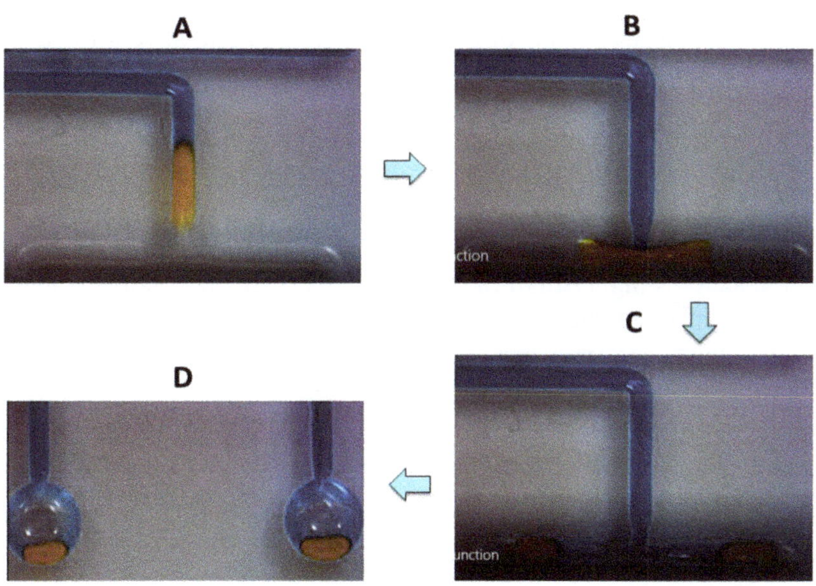

Figure 10.24. A water plug (tinted yellow) transported by a capillary flow of pentanol (tinted blue) separates into two at a symmetrical T-junction. (A) The arrival of the plug at the junction; (B) the plug separates; (C) the primary flow divides evenly at the junction; (D) the two plugs arrive at the outlet ports. Photographs courtesy of Samuel Barnes Berry, University of Washington.

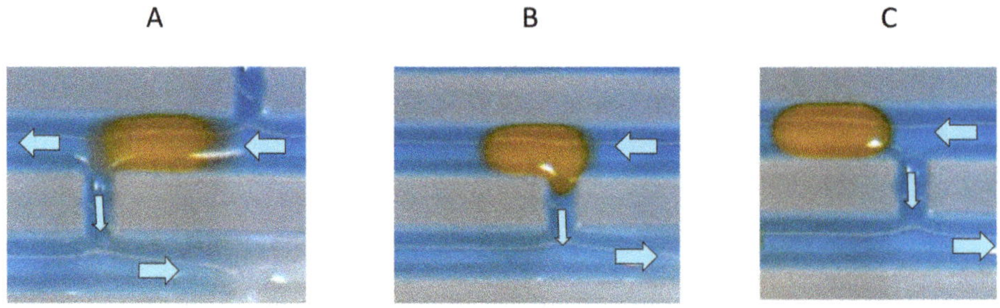

Figure 10.25. A water plug resists division in a T-junction (in the case of a rounded rectangular open channel) due to the high Laplace pressure imposed by the narrow channel. Photographs courtesy of Samuel Barnes Berry and Jing J Lee, University of Washington.

transported by a capillary flow of pentanol in an open rounded channel made of PMMA. The water contact angle with PMMA is approximately 80°, the pentanol contact angle with PMMA is 13°, and the surface tension of pentanol is 25 mN m^{-1}. When the plug reaches the bifurcation, the pentanol spreads along the walls and then pinches the water plug, leading to its symmetrical division.

Note that a plug does divide when the two daughter channels have different cross-sections, as shown in figure 10.25. The Laplace pressure in the narrow channel opposes the carrier fluid entrainment and the plug keeps moving in the larger channel.

b. Bypasses

It has been observed that trains of plugs in the 'raft' mode have uneven velocities, resulting in the merging of the plugs. This phenomenon has been experimentally observed but not yet analyzed (figure 10.26).

Plugs can be kept separate using bypasses, as shown in figure 10.27. The flow divides at the junction with a bypass, creating a separation downstream.

10.2.8 Capillary filaments and bifurcations, networks and bypasses

Since their velocity is higher than the bulk velocity, capillary filaments may play an unexpected role in two-phase flows.

a. A comparison between rounded and sharp corners

We first compare the effect of narrow bypasses on the two-phase flow in square and rounded open channels. The different behaviors encountered are shown to be due to the velocity difference inside the bypasses.

We have seen previously that these filaments have a high velocity, especially at the beginning of their formation. Hence, such filaments invade short bypasses very quickly. Depending on the geometry, they exit the bypass before the arrival of the main flow (figure 10.28(A)).

Conversely, in rounded open channels, there are no capillary filaments. The fraction of the liquid penetrating the bypass flows at a velocity determined by the

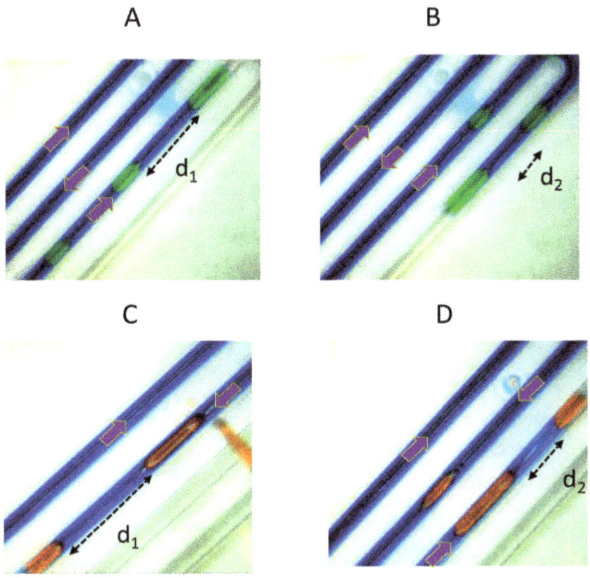

Figure 10.26. Water plugs tend to get closer and even merge in a capillary-driven open microflow in rectangular winding channels; (A) and (B) nonanol (blue) and water plugs; (C) and (D) dodecane and water plugs. Photographs courtesy of Jean Berthier and Jing J Lee, University of Washington.

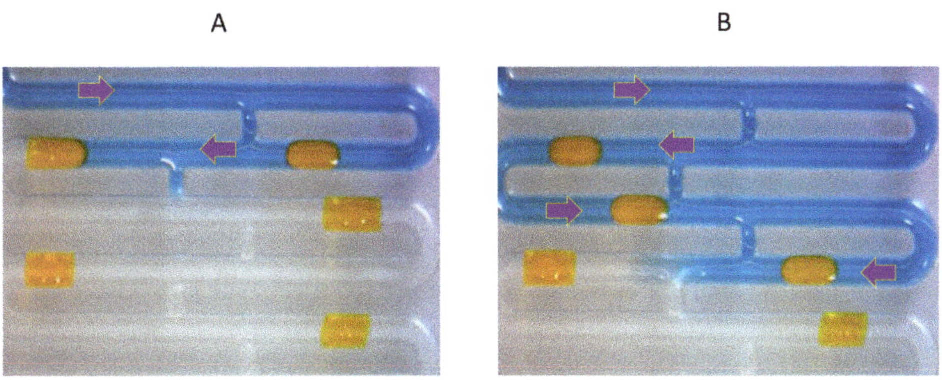

Figure 10.27. Water plugs stay separated due to bypasses in a capillary-driven open microflow in a rectangular winding channel. Reproduced from [16] with permission of The Royal Society of Chemistry.

ratio of the capillary and frictional forces, as presented in chapter 4. In the case of figure 10.28(B), the bypasses fill later than the main channel.

b. A plug temporarily blocked by reverse capillary filaments

In the case of bifurcations or junctions in an open-channel network, capillary filaments may play an unexpected role. This is illustrated by the network shown in figure 10.29, where a water plug is temporarily blocked by reverse capillary filaments. In the trifurcation at the right-hand side of figure 10.29(A), the capillary

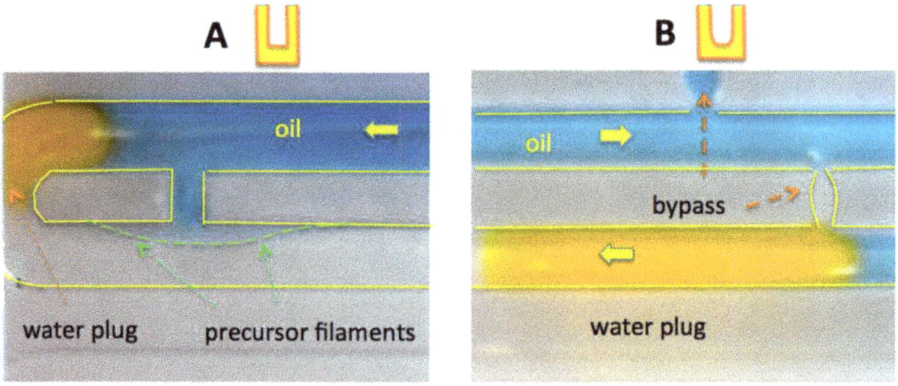

Figure 10.28. Different effects of a bypass depending on the presence or absence of capillary filaments: (A) in square channels, capillary filaments drive the fluid into the bypass very quickly and the fluid exits in front of the plug; (B) in rounded channels, the filaments are absent, bypasses fill slowly and the plug stays in the front of the capillary flow. Photos courtesy of Jing J Lee and Samuel Barnes Berry, University of Washington.

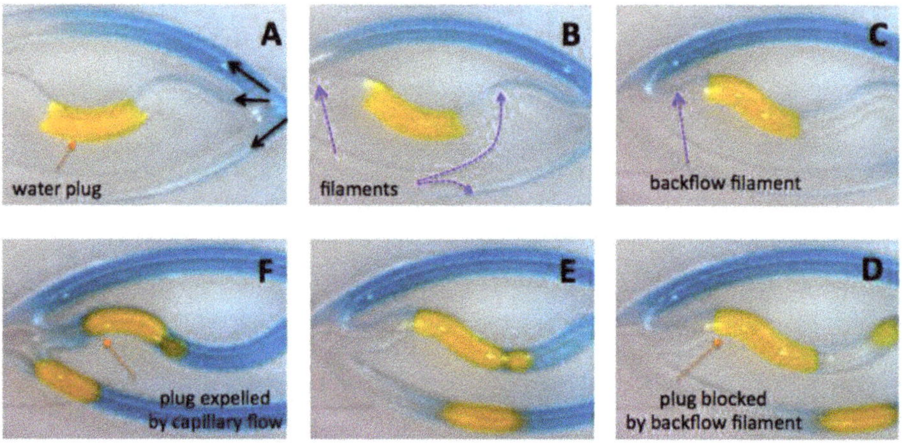

Figure 10.29. A backflow capillary filament blocks a plug temporarily: (A) and (B) the water plug (yellow) at the center is moved to the left by an incoming capillary filament of dodecane (blue); (C), (D), and (E) the plug is stopped by the reverse filament coming from the downstream junction; (F) the plug is finally expelled from the median channel by the bulk capillary flow. Photos courtesy of Jing J Lee and Samuel Barnes Berry, University of Washington.

filaments extend for different lengths depending on the curvature of the root channel. The capillary filaments progress quickly and move back into the middle channel, temporarily blocking the plug. The plug is 'released' when all the channels are completely filled with liquid.

c. Uneven division of a plug due to the action of capillary filaments

Capillary filaments may play an important role at a bifurcation. Figure 10.30 shows a trifurcation in a rectangular open channel. Because of the small contact angle of

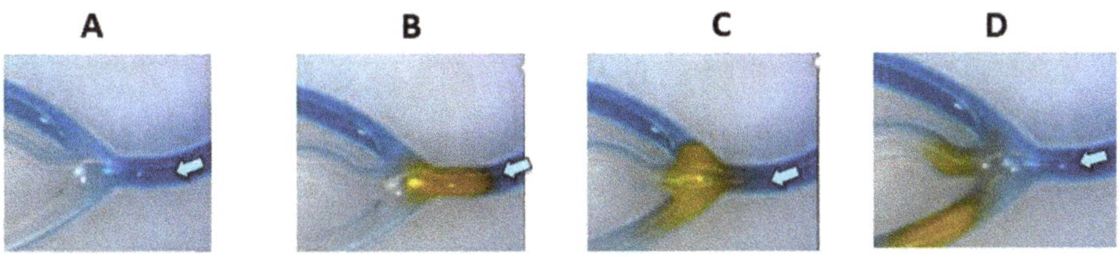

Figure 10.30. The division of a water plug (yellow) at a triple fork in a rectangular open channel. (A) The carrier fluid (pentanol, tinted blue) divides unevenly in the three branches due to uneven capillary filaments; (B) the plug reaches the fork; (C) the plug starts stretching in the three channels; (D) the plug is pulled by the capillary filaments in the two lower channels. Note that the duration of the plug division is less than 0.2 s. Photographs courtesy of Jing J Lee and Samuel Barnes Berry, University of Washington.

pentanol with PMMA ($\theta \sim 10°–15°$), capillary filaments reach the trifurcation first. They first extend into the two lateral channels. An unsymmetrical situation results, and the transported plug arriving at the trifurcation is pushed aside towards the two channels which are not yet filled.

10.2.9 An introduction to two-phase microflows in nonuniform open channels

Open capillary-driven microflows in nonuniform channels have not been investigated to date. The first numerical approaches used in Surface Evolver resulted in a very different situation than in the case of the uniform cross-section. Considering the case of a cylindrical enlargement of a rectangular open channel, the SCF of the continuous phase—carrier fluid—depends on the different contact angles. Figure 10.31 shows different flow behaviors. Note that the approach taken by Surface Evolver is a quasi-static approach and does not account for dynamic effects.

10.3 Part 2: the production and manipulation of droplets

In this section, we show that droplets can be emitted using only capillary forces in open channels and that these droplets can easily be manipulated using pipettes.

10.3.1 Droplet emission

Conventionally, droplets are produced by devices such as flow focusing devices (FFDs) [17–20]. A sketch showing the principle of a FFD is shown in figure 10.32. In such devices, a high shear is obtained at the neck due to the two lateral flows driven by pumps.

In open-channel flows, pressures and velocities are small. The pinching effect described above is then much less than for a forced flow. However, droplets can still be emitted if the surface tension between the two immiscible liquids is small. Let us consider the device shown in figure 10.33 [21].

Other conditions for obtaining aqueous droplets by capillary means only are high hydrophobicity of the solid bottom and walls and high wetting of the walls by the continuous phase. In our case, polytetrafluoroethylene (Teflon) [22] is used for the solid walls and HFE with a surfactant is used as the continuous liquid. Under these

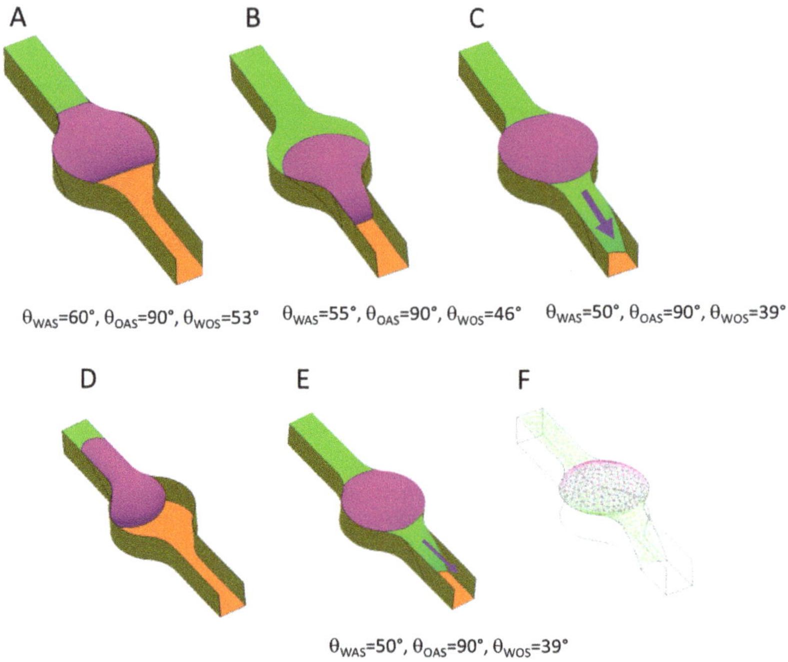

Figure 10.31. A plug immobilized in a cylindrical enlargement. (A) The flow stops in the enlargement; (B) the flow is blocked at the enlargement outlet; (C) the plug is immobilized in the enlargement and the carrier fluid flows beneath; (D), (E) and (F) details of case (C). Figures obtained using Surface Evolver.

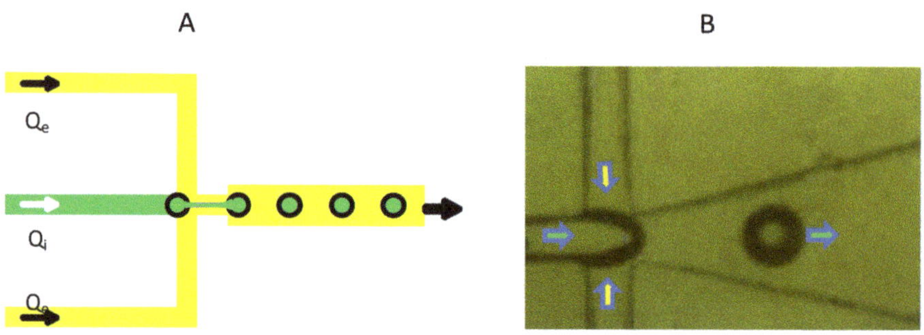

Figure 10.32. (A) The principle of an FFD: the discontinuous liquid (green) is laterally pinched by two microflows of immiscible liquid and droplets are emitted at the outlet of the neck (narrow cross-section). (B) An image of an FFD emitting alginate droplets in a silicon oil flow. Photographs courtesy of Jean Berthier and Sophie Le Vot, CEA Leti.

conditions, HFE forms filaments in the bottom corners, according to the second Concus–Finn condition [9]:

$$\theta > \frac{\pi}{2} + \alpha, \tag{10.12}$$

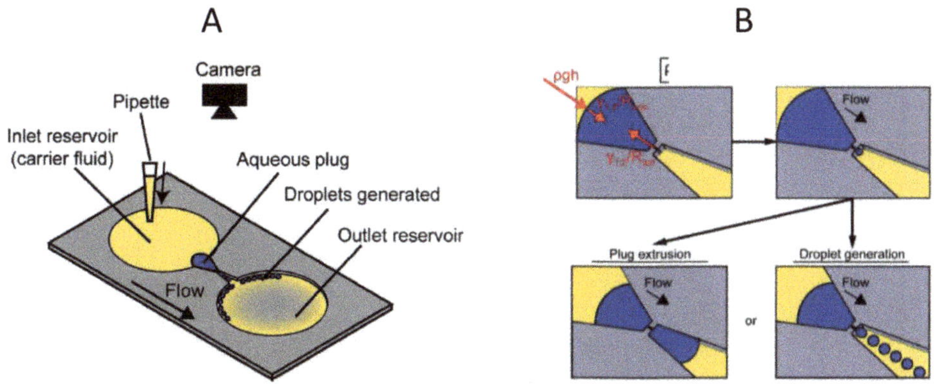

Figure 10.33. (A) A sketch of the device. The inlet reservoir is filled by the carrier fluid and the plug (blue) placed in the convergence before the neck. (B) The liquid of the plug slowly flows through the neck (the narrow section between the two reservoirs) under the action of the Laplace pressure and hydrostatic pressure. If the neck is narrow, droplets are emitted in the outlet reservoir, otherwise the plug is extruded through the neck. Reproduced from [21]. CC BY 4.0.

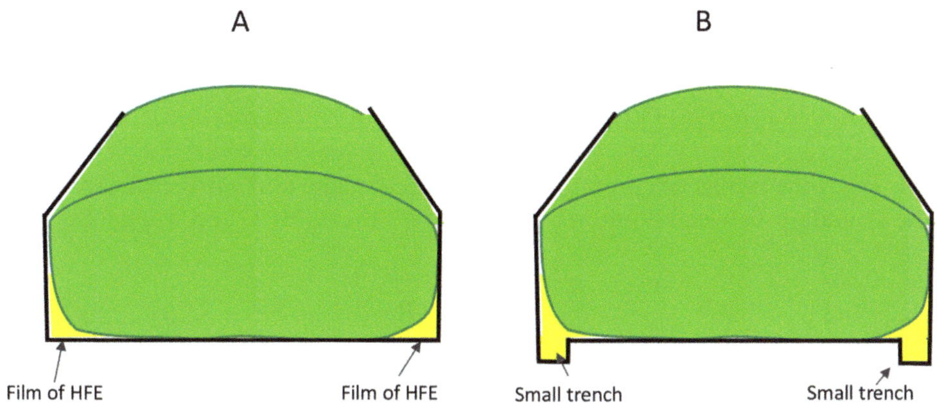

Figure 10.34. (A) A sketch of the films of HFE in the bottom corners below the aqueous plug; (B) an additional passage for HFE provided by small trenches in the bottom corners.

where θ is the contact angle between HFE, Teflon, and air (here, $\theta \geqslant 120°$) and α is the corner angle (here, $\alpha = 90°$). Moreover, a film of HFE forms in the corners beneath the water plug, as shown in figure 10.34(A). The surface area of the film is still increased by additional trenches in the corners (figure 10.34(B)). Hence a small flow of HFE occurs below the plug and wets the other side of the aqueous plug in the outlet reservoir. This wetting of the other interface is fundamental, as we shall see next.

Let us consider the schematization of figure 10.35(A), in which a plug of liquid #1 (green in the figure—aqueous liquid in the experiment) is placed in a constriction formed from another immiscible liquid #2 (yellow in the figure—HFE in the

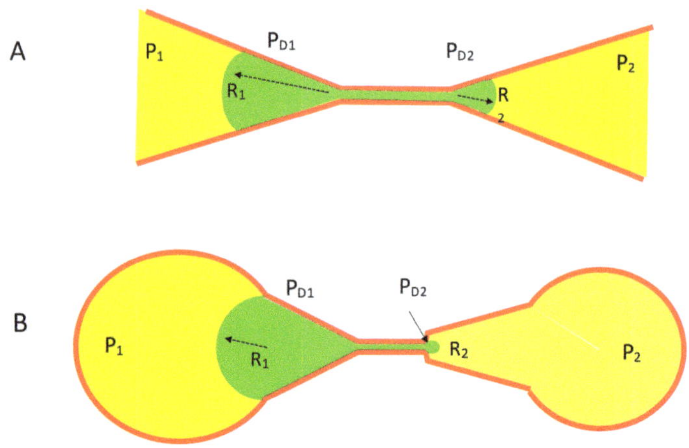

Figure 10.35. (A) A sketch of a plug placed in a constriction; pressures P_1 and P_2 are applied at the sides; (B) a sketch of the plug at the very beginning of the droplet formation.

experiment). The pressures on the two sides in liquid #2 are supposed to be different: $P_1 > P_2$.

Let us write the Laplace pressure for each end of the droplet:

$$P_{D1} - P_1 = \gamma\left(\frac{1}{R_1} + \frac{1}{R_{v\,1}}\right), \tag{10.13}$$

where R_{v1} is the radius of curvature in the vertical plane and γ is the surface tension at the boundary between liquid #1 and liquid #2 (here, HFE and water). Similarly,

$$P_{D2} - P_2 = \gamma\left(\frac{1}{R_2} + \frac{1}{R_{v\,2}}\right), \tag{10.14}$$

where γ is also the surface tension at the boundary between liquid #1 and liquid # 2 (because of the flow of liquid #2 below the plug). Subtracting (10.14) from (10.13) yields

$$P_{D1} - P_{D2} = (P_1 - P_2) + \gamma\left(\frac{1}{R_1} - \frac{1}{R_2} + \frac{1}{R_{v\,1}} - \frac{1}{R_{v\,2}}\right). \tag{10.15}$$

If we make the assumption that $R_{v\,1} \approx R_{v\,2}$—which is approximately valid because the contact with the bottom plate is the same—we are left with

$$P_{D1} - P_{D2} \cong (P_1 - P_2) + \gamma\left(\frac{1}{R_1} - \frac{1}{R_2}\right). \tag{10.16}$$

The pressure in the inlet reservoir P_1 (on the left in figure 10.35(B)) is the hydrostatic pressure, while that in the outlet (nearly empty) reservoir is $P_2 \sim 0$. Then relation (10.16) becomes

$$P_{D1} - P_{D2} \cong \rho g h + \gamma \left(\frac{1}{R_1} - \frac{1}{R_2} \right). \tag{10.17}$$

The condition for liquid #1 (green in figure 10.35(B)) to move towards the right-hand side to form droplets is $P_{D1} > P_{D2}$, or

$$\rho g h + \gamma \left(\frac{1}{R_1} - \frac{1}{R_2} \right) \gtrsim 0. \tag{10.18}$$

At the beginning of droplet formation, the volume of aqueous (green) liquid #1 is large and the curvature radius R_1 is also large. Thus $1/R_1$ is negligible compared to $1/R_2$. Therefore, the condition for droplet formation is the approximately

$$\rho g h > \frac{\gamma}{R_2}. \tag{10.19}$$

When a droplet forms (figure 10.35(B)), the initial curvature radius R_2 is approximately [2]

$$R_2 \approx \frac{w}{2 \mid \cos \theta \mid}. \tag{10.20}$$

The condition for droplet emission is then

$$\rho g h > \frac{2\gamma \mid \cos \theta \mid}{w}. \tag{10.21}$$

Relation (10.21) shows that the constriction width w chosen cannot be too small: $w > 2\gamma \mid \cos \theta \mid / \rho g h$. It also shows that the surface tension γ liquid #1/liquid #2 must be small. For this reason, surfactants were added to the HFE.

Once droplet formation is initiated, the radius of curvature R_1 decreases, and the contribution of the term γ / R_1 increases. We then must consider the relation

$$\begin{aligned} (P_{D1} - P_{D2}) &\cong \rho g h + \gamma \left(\frac{1}{R_1} - \frac{1}{R_2} \right) \\ &\cong (P_{D1} - P_{D2})_{\text{initial}} + \frac{\gamma}{R_1} > (P_{D1} - P_{D2})_{\text{initial}}. \end{aligned} \tag{10.22}$$

Hence, once the emission process has started, it continues as long as the outlet pressure stays small and the inlet pressure stays constant. Figure 10.36 shows the

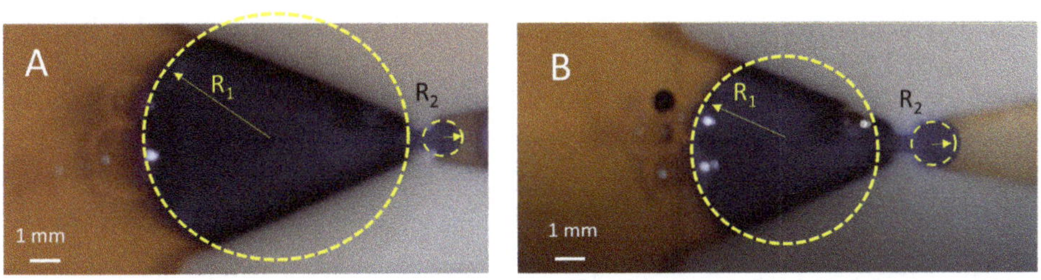

Figure 10.36. The formation of droplets/plugs in the 0.4 mm constriction. (A) The beginning of the droplet emission process; (B) during droplet emission (R_1 decreases). Photograph courtesy of Jian-Wei Khor, University of Washington.

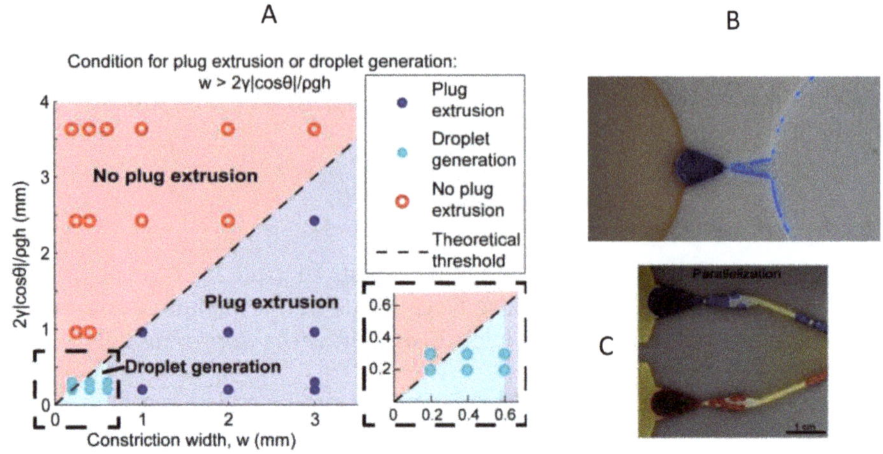

Figure 10.37. (A) A diagram of droplet emission, showing that condition (10.21) approximately predicts the emission process; (B) droplets emitted and carried by the filaments of HFE in the outlet reservoir; (C) the device can be parallelized by inserting two plugs into two converging channels—two populations of droplets are then produced. Reproduced from [21]. CC BY 4.0.

droplet emission at the start (figure 10.36(A)) and later (figure 10.36(B)) in the emission process.

The width of the narrow channel (neck) conditions the droplet emission, as stated by inequality (10.21). If the width is too small, droplet emission is impossible. If the width is too large, the whole plug is extruded (figure 10.33(B)). A comparison between the experimental results and the simplified model is shown in figure 10.37(A). The domain of droplet emission is rather small, showing that capillary-driven droplet emission is not straightforward.

10.3.2 Droplet manipulation

Accessibility is a key advantage of open microfluidics (OM). The emitted droplets in the outlet reservoir are easily accessible and can be manipulated using tweezers, styluses, or pipettes.

Tweezers are small hand tools used for grasping objects too small to be easily handled with the human fingers. The principle of tweezers is very old. There are drawings of Egyptian craftsmen holding hot pots over ovens with a double bow-shaped tool. Asiatic tweezers, consisting of two strips of metal brazed together, were commonly used in Mesopotamia and India from about 3000 BC. The word tweezer comes from 'etwee,' which describes a small case that people would use to carry small objects with them. Etwee takes its origin from French *étui*, 'small case,' from the Old French verb *estuier*, 'to hold or keep safe.' Tweezers are often made of superalloy (CX) or high-alloy stainless steel (DX). Tweezers with ultrafine tips are ideal for handling extremely minute materials and grids under a microscope. Figure 10.38(A) shows the use of tweezers for transferring and patterning droplets.

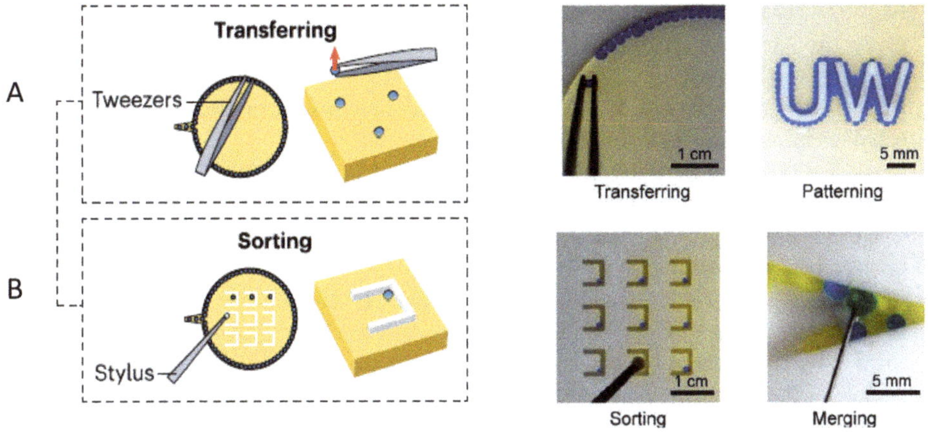

Figure 10.38. (A) Operations performed using tweezers—transferring and patterning; (B) operations performed using a stylus—sorting and merging. Reprinted from [23], Copyright (2023), with permission from Springer.

The name stylus comes from the Greek '*stylos,*' which means pillar and from the similar Latin word '*stilus,*' which designates a long, sharply pointed piece of metal. In French, the Greek stylos has been adapted to designate a 'stylo,' which is 'pen' in English. In today's biology, a stylus is a thin, sharp, small stick that can move liquid or cells using adhesion forces. Figure 10.38(B) shows how a stylus can be used for droplet sorting and merging.

10.4 Conclusions

In this chapter, the different shapes of plugs in an open channel were investigated. Under static or quasi-static conditions—corresponding to a plug at some distance from the channel entrance—three plug morphologies (modes) are possible: shift mode (translation of the plug), bridge mode (immobilization of the plug in the bottom of the channel permitting the flow to pass over it) and lift mode (the plug floats on the flow).

Plug morphologies change at sufficiently high velocities—corresponding, for example, to a plug located at the channel entrance—and a raft mode dominates, which is an intermediate between the shift mode and the lift mode. In some cases, when the liquid properties are suitable, the bridge mode can still persist.

We have also shown that, using the proper mode, plugs can be moved by open-channel capillary flows, and capillary wagons can be transported by the capillary flow. Plugs can be also be divided in bifurcations.

Today, the study of two-phase capillary flows is still in its infancy, and many observed phenomena are still unexplained.

References

[1] Berthier J, Brakke K A and Berthier E 2016 *Open Microfluidics* (New York: Wiley)

[2] Li C, Boban M and Tuteja A 2017 Open-channel, water-in-oil emulsification in paper-based microfluidic devices *Lab Chip* **17** 1436–41

[3] Berthier J, Brakke K, Lee J, Gosselin D, Navarro F, Theberge A and Berthier E 2017 Two-phase capillary flow in open uniform cross section microchannels and the transport of capillary wagons *Proc. of the 2017 Nanotech. Conf. (Washington, DC, May 13–17)*

[4] Lee J J, Berthier J, Brakke K A, Dostie A M, Theberge A B and Berthier E 2018 Droplet behavior in open biphasic microfluidics *Langmuir* **34** 5358–66

[5] Berry S B, Lee J J, Berthier J, Berthier E and Theberge A B 2018 Open channel droplet-based microfluidics *Proc. of the 2018 MicroTas Conf. (Kaohsiung, Taiwan)* (San Diego, CA: Chemical and Biological Microsystems Society) 1093–6 https://cbmsociety.org/conference/proceedings_archive.html

[6] Brakke K A 1992 The Surface Evolver *Exp. Math.* **1** 141–65

[7] Li W, Nie Z, Zhang H, Paquet C, Seo M, Garstecki P and Kumacheva E 2007 Screening of the effect of surface energy of microchannels on microfluidic emulsification *Langmuir* **23** 8010–4

[8] Neumann F 1894 *Vorlesungen über die Theorie der Capillarität* ed B G Teubner (Leipzig)

[9] Berthier J and Brakke K A 2014 *The Physics of Microdroplets* (New York: Wiley)

[10] Concus P and Finn R 1969 On the behavior of a capillary surface in a wedge *Proc. Natl Acad. Sci.* **63** 292–9

[11] Concus P and Finn R 1994 Capillary surfaces in a wedge—differing contact angles *Microgravity Sci. Technol.* **7** 152–5

[12] Berthier J, A. Brakke K, Gosselin D, Huet M and Berthier E 2014 Metastable capillary filaments in rectangular cross-section open microchannels *AIMS Biophys.* **1** 31–48

[13] Berthier J and Silberzan P 2009 *Microfluidics for Biotechnology* (Norwood, MA: Artech House) 2nd edn https://us.artechhouse.com/Microfluidics-for-Biotechnology-Second-Edition-P1328.aspx

[14] Soh G Y, Yeoh G H and Timchenko V 2016 Numerical investigation of the velocity fields during droplet formation in a microfluidic T-junction *Chem. Eng. Sci.* **139** 99–108

[15] Mehrabian H, Gao P and Feng J J 2011 Wicking flow through microchannels *Phys. Fluids* **23** 122108

[16] Berry S B, Lee J J, Berthier J, Berthier E and Theberge A B 2019 Droplet incubation and splitting in open microfluidic channels *Anal. Methods* **11** 4528

[17] Cristini V and Tan Y-C 2004 Theory and numerical simulation of droplet dynamics in complex flows—a review *Lab Chip* **4** 257–64

[18] Christopher G F and Anna S L 2007 Microfluidic methods for generating continuous droplet streams *J. Phys.* D **40** R319–36

[19] Utada A S 2005 Monodisperse double emulsions generated from a microcapillary device *Science* **308** 537–41

[20] Berthier J and Wagh J 2012 A novel micro flow-focusing device for the control of droplet contents *J. Basic Appl. Phys.* **1** 89–92

[21] Khor J W, Lee U N, Berthier J, Berthier E and Theberge A B 2023 Interfacial tension driven open droplet microfluidics *Adv. Mater. Interfaces* **10** 2202234

[22] Manko D, Zdziennicka A and Janczuk B 2017 Surface tension of polytetrafluoroethylene and its wetting by aqueous solution of some surfactants and their mixtures *Appl. Surf. Sci.* **392** 117–25

[23] Zeng Y, Khor J W, van Neel T L, Tu W-C, Berthier J, Thongpang S, Berthier E and Theberge A B 2023 Miniaturizing chemistry and biology using droplets in open systems *Nat. Rev.* **7** 439–55

IOP Publishing

Open-Channel Microfluidics (Second Edition)
Fundamentals and applications
Jean Berthier, Ashleigh B Theberge and Erwin Berthier

Chapter 11

Applications

11.1 Introduction

The development of open-channel microfluidics has been rapid during the last few years. The number of publications on the topic of open-channel microfluidics has soared. Some interesting applications have already appeared, new devices have been designed, and new perspectives are unfolding.

In the following, some typical examples of applications of open-channel microfluidics are given that pertain to different scientific domains, such as materials, microfluidic systems, chemistry, biology, and medicine. These examples illustrate the wide range of applications of open-channel microfluidics. A whole chapter dedicated to its applications in space is presented after this chapter.

11.2 Materials and fabrication

Open-channel microfluidics has many different applications in the domain of materials and fabrication. One of the first applications reported by Rye, Mann, Romero, and Yost—dating from 1995 and 1996—was the fabrication of micro solder paths [1–4] using open-capillary flows in V-grooves etched in copper (figure 11.1(A)). More recently, open microfluidics was used to conveniently coat microfluidic channels and build objects such as microelectrodes.

As a continuation of the preceding work, still using open-capillary flows in V-grooves, Shao and coworkers developed electrodes made of carbon nanotube ink [5]. They showed that a capillary carrier ink containing carbon nanotubes (CNTs) could flow in V-grooves and form continuous electrodes upon drying (figure 11.1(B)). The aim if this approach was to develop low-cost pliable electronics.

Our third example has the same inspiration as the preceding example: Zang and coworkers [6] reported the utilization of capillary microfluidics to rapidly create nanostructure-patterned electrodes for energy storage applications. Using patterned photoresist as a substrate for open-channel capillary microfluidics, *tobacco mosaic* virus (TMV) bio-nanoscaffolds suspended in solution were autonomously delivered

doi:10.1088/978-0-7503-5507-0ch11

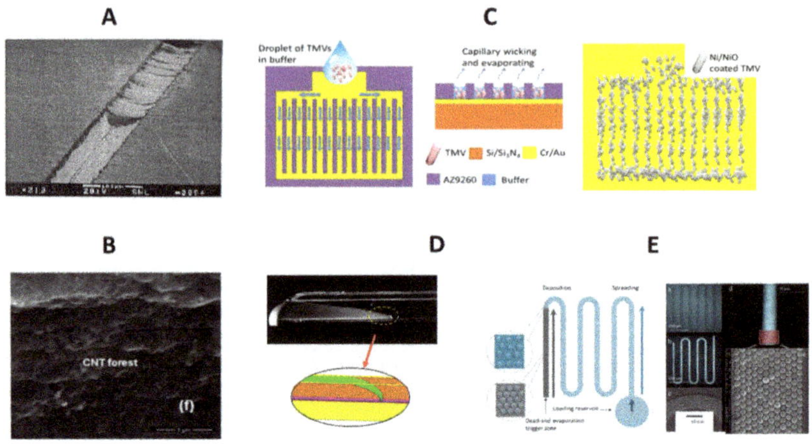

Figure 11.1. (A) A solder capillary flow in a V-groove; (B) dry carbon nanotube (CNT) ink in a V-groove, forming a conductive electrode; (C) the construction of a Ni/NiO electrode using tobacco mosaic virus (TMV) and an open-capillary flow; (D) an anomalous capillary flow in a V-groove induced by the deposition of polymers; (E) ordered deposition of particles in an evaporating capillary flow. Panel (A) reprinted with permission from [2], Copyright (1996) American Chemical Society. Panel (B) reprinted from [5], Copyright (2011), with permission from Elsevier. Panel (C) reproduced from [6], © IOP Publishing Ltd, CC BY 3.0. Panel (E) reprinted with permission from [9], Copyright (2017) American Chemical Society.

onto planar gold electrodes over a 1 cm^2 area (figure 11.1(C)). The TMVs assembled on the electrode and formed a dense bio-nanoscaffold layer due to the enhanced evaporation-assisted assembly provided by the open-channel capillary microfluidic device within an hour. The TMV structures were then coated with Ni/NiO through electroless plating and thermal oxidation to form supercapacitor electrodes. The galvanostatic charge/discharge cycle showed a 3.6-fold increase in areal capacitance for the nanostructured electrodes compared to planar structures.

Note that the capillary flow of heavily loaded carrier fluids, such as those of the two preceding examples, might be subject to a brutal stop due to pinning triggered by evaporation. This was observed by Lade and coworkers for the flow of regular two-dimensional wires of different polymers, such as poly (vinyl alcohol), poly(3,4-ethylenedioxythiophene) (PEDOT) ink, and graphene ink, in open channels [7]. Note also that deposition in an open channel—due to evaporation or sedimentation or any other cause—modifies the flow pattern and may even block the flow [8] (figure 11.1(D)). In biotechnology, this can be the case for the flow of whole blood.

Our fourth example is the fabrication of regular two-dimensional wires using open-channel microflows. Lone and coworkers used open-capillary flows to deposit micro- and nanoparticles in grooves via evaporative lithography, akin to the coffee-ring effect [9]. Hydrophilic conditions ensured the rapid spread of the suspension from the loading reservoir to fill the entire channel length. Evaporation during the spreading process and after the channel was full increased the particle concentration toward the end of the channel. This evaporation-induced convective transport brought particles from the loading reservoir to the channel end, where this flow deposited a continuous multilayered particle structure (figure 11.1(E)). The particle deposition front propagated backward over the entire channel length. The final dry deposit of the particles was thereby much thicker than the initial volume fraction of the suspension. The aim of this approach was to build low-cost and large-scale thin optoelectronic films with variable thicknesses and interspacing distances.

11.3 Microfluidic channels

New types of channels or structures for guiding open microflows have been developed. We present some of them in this section.

11.3.1 Capillary channels on paper

Paper-based microfluidic devices and paper strips are becoming very popular devices as biosensors in medicine for point-of-care use or self-testing. Such devices are simple to use and inexpensive. However, these systems present limitations such as retention in the paper matrix, evaporation, and low liquid flow rates. Microfluidic channels built in paper matrices have been developed to palliate these drawbacks. Different approaches have been followed: (1) etching a channel in a paper substrate [10], (2) embossing a paper substrate [11–13], (3) building rails between a paper substrate and a cover plate [14].

The embossing of paper is inexpensive and highly compatible with industrial processes. This approach was first demonstrated by Thuo and coworkers [13]. More recently, Gosselin and coworkers developed a paper substrate treated with styrene butadiene rubber (SBR) to make it impermeable to aqueous liquids and coated with reticulated polyvinyl acetate (PVA) to give its surface hydrophilic properties [11, 12]. Channels were formed by embossing these paper substrates at a pressure of 40 bars and a temperature of 90°. It was demonstrated that such channels were compatible with open-capillary flows (figure 11.2(A)) and that glucose tests were successful in such systems.

Another approach that has been used to facilitate the capillary flow in a paper-based system was the fabrication of rails proposed by Songok and colleagues [15]. They developed a paper-based capillary-driven flow system that sped up flow rates by utilizing a narrow gap geometry between two parallel surfaces separated by a spacer. The top surface was hydrophobic, while the bottom surface was a hydrophobic paper substrate with a microfluidic channel defined by a hydrophilic pathway, leaving the sides of the channel open to the air (figure 11.2(D)). The liquid flowed on the hydrophilic path in the gap without spreading onto the hydrophobic regions, which is typically what we have defined as 'rail microfluidics' (see chapter 4). Note that this geometry reduces liquid drying and evaporation.

11.3.2 Smart textiles

Capillary flows in textiles fall in the category of open microfluidics because these flows are in contact with the surrounding air. They have the advantage of being able to transport liquid via capillary wicking without the need for an external driving force; they are lightweight and flexible and they can fit the human body comfortably. Textile-based microfluidics is a promising way to realize the concept of portable,

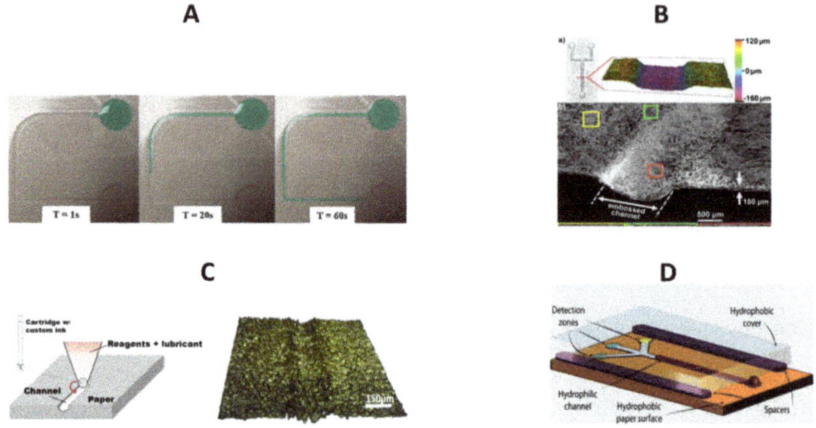

Figure 11.2. Open channels in a paper matrix: (A) capillary flow in embossed paper [12]; (B) the profile of an embossed channel; (C) a channel obtained by asymmetric mechanical modification of a paper surface using a rolling ball (a ball-point pen); (D) open rail geometry on paper. Panel (A) reprinted from [12], Copyright (2017), with permission from Elsevier. Panel (B) reprinted with permission from [13], Copyright (2014) American Chemical Society. Panel (C) reprinted from [10], with the permission of AIP Publishing. Panel (D) reprinted with permission from [15], Copyright (2016) American Chemical Society.

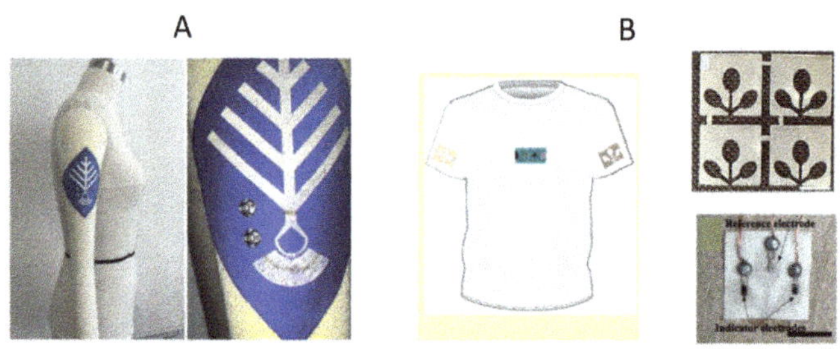

Figure 11.3. Electrochemical biosensors embedded in textiles: (A) an electrochemical biosensor as a patch on the arm. Reproduced from [19] with permission of The Royal Society of Chemistry. (B) A biosensor for sweat analysis, reprinted with permission from [16], Copyright (2022) American Chemical Society.

low-cost, and field-based diagnostic technology [16–19]. Hence, so-called smart textiles—textiles with imbedded sensors—have been developed in the last ten years.

In order to guide the fluids in the textile structure, different approaches have been followed. The first one consists in creating patterned wettability by selective hydrophobic modification of the porous substrate to manipulate capillary flow (fibers, which are often composed of cellulose, are usually hydrophilic) [16, 19]. Embedded in clothes, such microfluidic systems are used as biosensors for sweat monitoring, for example (figure 11.3).

Another approach is that of *in situ* laser engraving, which enables directional spreading of liquids [20]. The laser beam can heat up the surface and subsequently melt or vaporize the materials, thus allowing structural inhomogeneities to be created on the surface. The permeability of selected bundles of yarns is locally

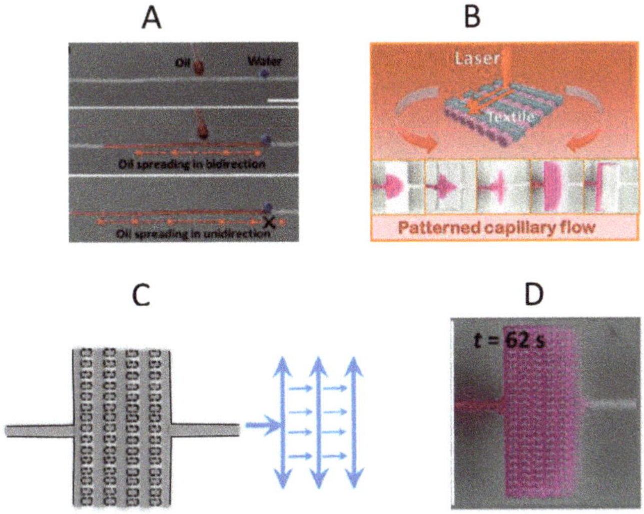

Figure 11.4. (A) Directional wicking in a Tex-50 hydrophilic thread using a water barrier; (B) the principle of applying a laser treatment to the weft and warp to create selected pathways; (C) an example of the use of a weft and warp arrangement to obtain selective wicking; (D) oil spreading in the Tex-50 textile design of figure (C). Reprinted with permission from [20]. Copyright (2020) American Chemical Society.

increased, leading to the creation of directional paths on the textile (figure 11.4). The laser engraving strategy has the advantages of simplicity, full scalability, and time rapidity. It was confirmed by Li *et al* that the wicking length z is given by a Lucas–Washburn law with a variable wicking rate constant

$$z^2 = K\ t, \tag{11.1}$$

where K is a wicking rate constant that depends on the textile structure and t is time.

The concept of smart bandages is associated with smart textiles. The aim of smart bandages is to improve the healing of superficial wounds. They facilitate the monitoring of wound healing and they incorporate wound dressings. The subject of smart bandages is too wide to be treated in this book and the reader is referred to the book '*Advanced Textiles for Wound Care* (2nd edition)' [21]. Textiles are also often used for cell culture (figure 11.5). The principle is to associate regulators—such as growth factors, cytokines, etc.—and cells in nonwoven fibers (figure 11.5(A)).

11.3.3 3D-printed capillary structures

Midway between fiber-based devices and open channels, new rigid structures have been developed to guide open microflows. The concept stems from the observation that conventional closed microfluidic devices have limited uses in multiphase process engineering. Open microsystems are more useful because they allow direct access to the fluid. Dudukovic *et al* [24] developed 3D-printed structures for what they call 'cellular fluidics,' which enabled cellular culture, transport, and reaction processes (figure 11.6). These structures were built by adding millimeter-scale cubes known as

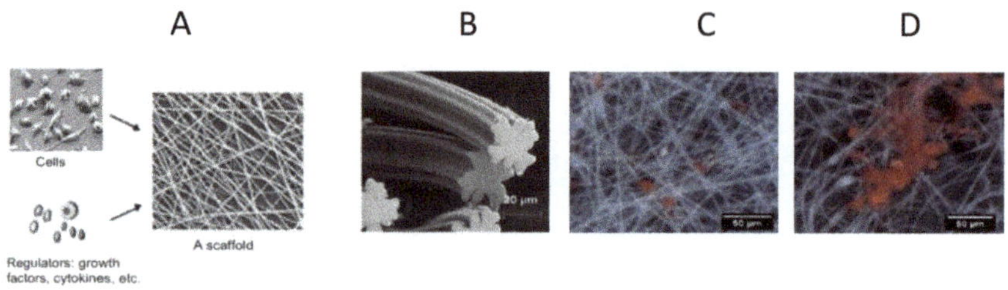

Figure 11.5. (A) Elements of tissues in smart textiles: cells, scaffold/matrix, and regulators; (B) an example of fibers tailored for tissue engineering: cross-sectional features of polyvinylidene fluoride fibers; (C) culture of human osteoblast-like cell line MG63 (ECACC), day 1; (D) same culture, day 7. Panel (B) reproduced from [22], CC BY 4.0. Panels (C) and (D) reprinted with permission from [23], Copyright (2018) American Chemical Society.

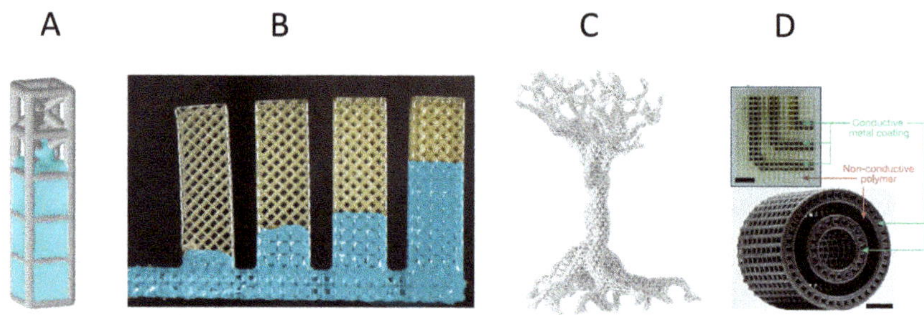

Figure 11.6. Capillary flow and a transpiration-like process in cellular fluidic systems. Dudukovic *et al* reported a technological platform called cellular fluidics.: (A) sketch of millimeter-scale cubes (unit cells) which have internal architectures that draw up fluids through capillary action.; (B) the combined capillary actionof the stacked unit cells produces a vertical flow of fluid; (C) a view of a tree-like structure built from the unit cells that continuously delivers liquid from a reservoir to the tips of the branches, where the liquid evaporates—a process that mimics transpiration in natural trees; (D) Selected preferred flow pathways in the 3D-printed structure obtained by tayloring the wettability of the structure. Reprinted from [24], Copyright (2021), with permission from Springer.

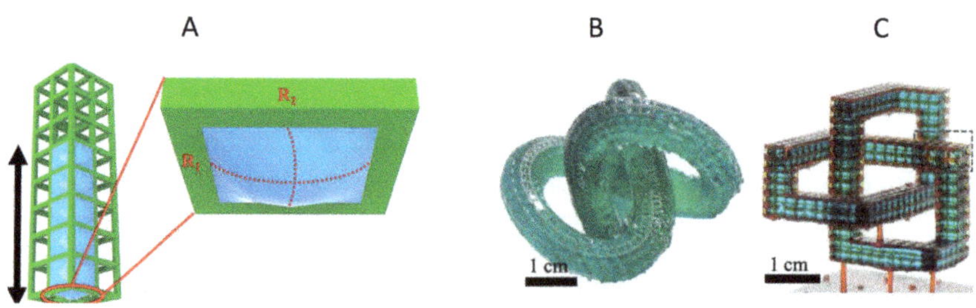

Figure 11.7. (A) Capillary flow in 3D-printed unit cells; (B) and (C) the assembly of unit cells to design ad-hoc capillary channels. Reproduced from [25]. CC BY 4.0.

unit cells to build up the desired object. The motion of the fluid was driven by the capillary pressure in the unit cells.

The same idea was developed by the team of Noo Li Jeon *et al*, who also used unit cells to design new capillary 'channels' (figure 11.7) [25].

11.3.4 Directional steering of liquids

Unidirectional liquid transport on solid surfaces without an energy input would advance a variety of applications, such as biofluidic devices, self-lubrication, and high-resolution printing. We have already seen that wicking can be oriented by microstructures cleverly placed in an open channel or on a plane (Feng and Rothstein [26], chapter 7). Chapter 9 presented the particular structure of the leaves of *Nepenthes alata*, a carnivorous plant native to the Philippines. Inspired by the structure of the surface, Chuxin Li and coworkers [27] fabricated a peristome-mimicking surface through high-resolution stereolithography. They demonstrated that such a surface produces a unidirectional capillary liquid flow. Figure 11.8 shows that the horseshoe structures promote capillary flow in the direction opposed to the tips of the horseshoe flares. Pinning of the liquid on the ridges in the direction of the tip of the horseshoe structure blocks the flow.

11.3.5 Evaporative capillary pumping

Capillary-driven microflows have a duration limited by the volume of the outlet reservoir or the pump at the end of the channel. Capillary-driven microflows can be made to work on longer timescales using evaporative capillary pumps. At the scale of the microsystem, one can speak of continuous capillary microflow [28, 29]. In the

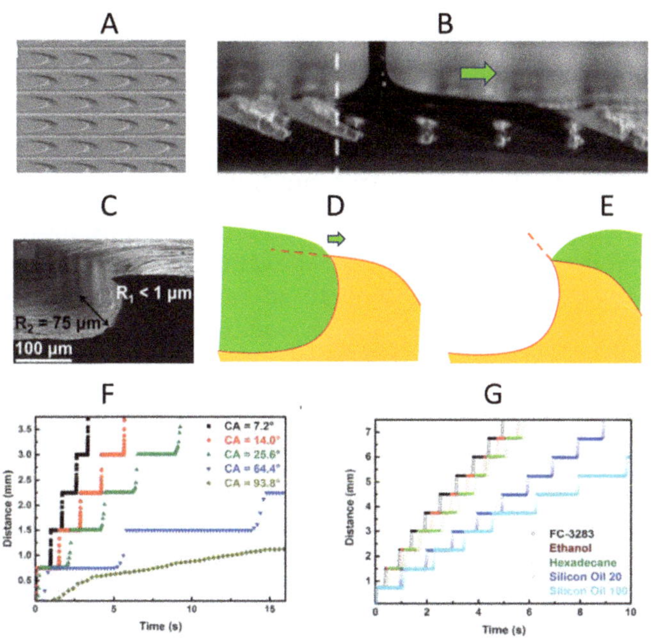

Figure 11.8. Directional capillary flow: (A) the top view of the surface; (B) the directional spread of the liquid; (C) a cross-sectional view of the surface, showing the 'beak'; (D) the bulging liquid spreads above the 'beak'; (E) the liquid on top of the scale stays pinned; (F) the observed distance vs. time for different contact angles; (G) the observed distances vs. time for different liquids; [27] John Wiley & Sons. © 2016 Wiley-VCH Verlag GmbH & Co. KGaA, Weinheim.

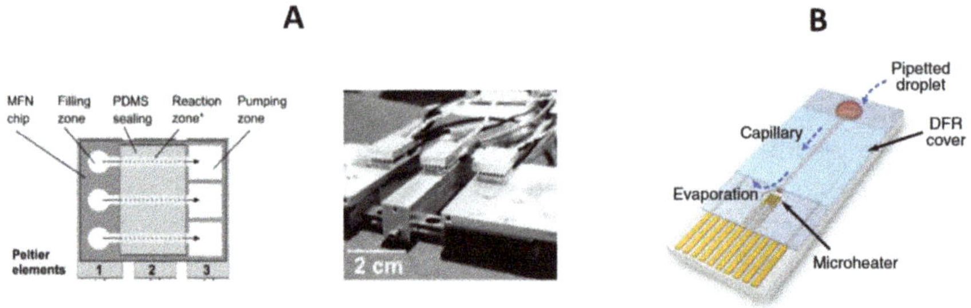

Figure 11.9. (A) A sketch and a photograph of the continuous capillary system produced by Zimmermann *et al.* Reproduced from [30] with permission of The Royal Society of Chemistry. (B) A schematic of the evaporation capillary pump produced by Temiz *et al.* Reproduced from [28] © 2014 SPIE.

system proposed by Zimmermann and coworkers [30], a microfluidic network includes open-capillary pumps (CPs) at its end. A Peltier element underneath the CPs heats the pumps, thereby inducing evaporation in the CPs and determining the flow rate in the microchannels (figure 11.11(A)). The authors claim that this method achieves flow rates in the microchannels ranging from ~1.2 nl s^{-1} to ~30 pl s^{-1} and is able to keep 90% of a 0.6 ml solution placed in an open filling port for 60 min. Note that the device does not strictly fall within the category of open microfluidics, since the 'transport channels' are closed (covered by a seal).

Another example, very similar to the preceding one, was proposed by Temiz and coworkers [28]. The heater facilitating evaporation is no longer a bulky Peltier but a microheater embedded in the substrate (figure 11.9(B)). Again, note that the system is not completely open but only partially open.

11.4 Biology, biotechnology, and medicine

Open-channel microfluidics is of primary importance in biotechnology, biology, and medicine. In the following sections, we show examples of the use of open microflows for gel electrophoresis, which is very useful in biotechnology for the separation of biologic objects, for cell culture in biology, and in *in vivo* sensors in medicine.

11.4.1 Microdots for cell studies

Casavant and coworkers used suspended microfluidics to create arrays of collagen membranes, known as microdots (μDots), in a horizontal plane separating two fluidic chambers [31] (figure 11.10(B)). Using this technique, a transwell platform able to discern collective or individual cellular invasion was fabricated. μDots can also be used as a simple multiplexed 3D cellular growth platform. Based on the same fluidic principles, a suspended microfluidic metabolite extraction platform was built using a multilayer biphasic system to leverage the accessibility of open microchannels to easily retrieve steroids and other metabolites from cell culture. In conclusion, suspended microfluidics brings the high degree of fluidic control and unique functionality of closed microfluidics into the highly accessible and robust platform of open microfluidics.

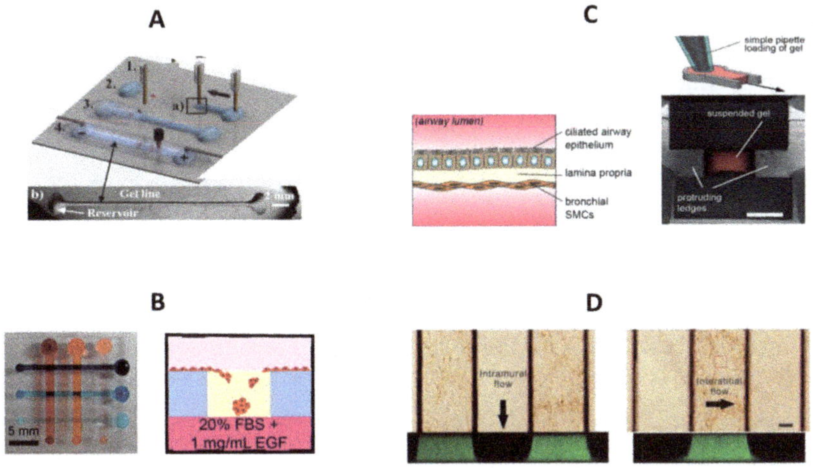

Figure 11.10. (A) A gel electrophoresis device; (B) suspended μDots made of gel used to study the migration of cells; (C) suspended gel used as a substrate to grow cells in order to mimic the lung epithelium; (D) the fabrication of collagen porous walls for the study of intramural and interstitial flows. Panel (A) reproduced from [41] John Wiley & Sons. © 2017 WILEY-VCH Verlag GmbH & Co. KGaA, Weinheim. Panel (B) reproduced from [31], © 2013 PNAS. Panel (C) reproduced from [32] with permission of The Royal Society of Chemistry. Panel (D) reproduced from [36] with permission of The Royal Society of Chemistry.

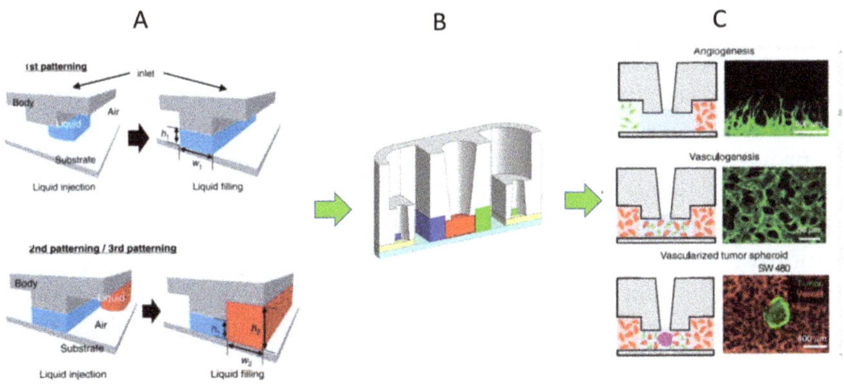

Figure 11.11. (A) First patterning in a rail geometry, followed by a second patterning; (B) a view of the device with the hydrogels; (C) the cell culture step. Reproduced from [38]. CC BY 4.0.

11.4.2 Mimicking the lungs

Suspended microfluidics has been revealed to be very useful in the development of organs on chips, for example to mimic the complex interactions between the many different cell types residing in lung airway tissues [32] (figure 11.11(C)). The authors of this study claim that, in order to advance our understanding of lung pathophysiology—especially chronic lung disease (CLD)—and accelerate drug development processes, new and innovative *in vitro* tissue models are needed that can reconstitute the complex *in vivo* microenvironment of human lung tissues. Hence, they designed an *in vitro* component that enables the study of the interactions between airway epithelial cells (ECs) and airway smooth muscle cells (SMCs), which have been shown in part to play major roles in the pathogenesis of CLDs. The design is based on a suspended hydrogel with upper and lower interfaces for EC and SMC culture, respectively. The hydrogel is first flowed in as a suspended liquid, then gelled to obtain a thin membrane. ECs and SMCs are then cultivated on each side of the membrane.

11.4.3 Cellular microfluidics

Open microfluidics has been proved to be very useful for the study of cell culture, proliferation, and crosstalk. In this section, we present some examples of these studies.

A. Cell culture

Open-channel microfluidics marries well with cell studies. The reason is accessibility: on one hand, it is easy to add and remove cells using pipettes; on the other hand, nutrients and oxygen are easily available for the cell culture.

Hydrogels are the most used host materials for cell culture due to their biocompatibility, 3D scaffolds, and liquid-to-gel transformation [33–35].

One of the first applications was that of Tung and coworkers, who developed a method for building collagen walls using enhanced capillary flow between solid ridges [36] (figure 11.10(D)). The device was used to study the effect of the microenvironment on cellular behavior, using controlled interstitial and intramural flows. In this approach, two liquid flows of collagen, actuated by capillary forces and pressure, filled the space delimited by four parallel—linear or curved—ridges. The collagen progressively gelled, and a cover plate was adjusted on top. A cell culture was introduced into the channel delimited by the collagen walls, and the effect of controlled transverse (interstitial) or axial (intramural) flow on cell growth and proliferation was monitored.

Bouaidat and coworkers [37] developed a simple microfluidic system consisting of two planar parallel surfaces separated by spacers (rail microfluidics). One of the surfaces was entirely hydrophobic, the other was mainly hydrophobic but had hydrophilic pathways defined on it by photolithographic means. By controlling the wetting properties of the surfaces in this manner, the liquid was confined to certain areas defined by the hydrophilic pathways. The capillarity-driven microflow system was used to pattern and immobilize biological material on planar substrates. Well-defined strips of human cells (HeLa) 200 μm wide were cultivated using the system.

B. Tissue engineering and organs on chips

More recently, it was found that open microfluidics is a very useful tool for tissue engineering and organ-on-a-chip studies [38–40]. We recall that organ-on-a-chip platforms recapitulate the physiological and pathological conditions *in vivo* (e.g. lung, brain, and liver) and are used to gain a better understanding of disease mechanisms and allow more efficient drug development. There is a need to develop reliable and reproducible platforms that precisely mimic aspects of the human body. The concept is based on using open channels, such as rails, to flow liquid hydrogels in place before gelling them. Different layers of hydrogel walls can be successively deposited, as shown in figure 11.11, providing a suitable environment for cell culture. From a physical point of view, this method consists in flowing a first hydrogel structure, then gelling it, and then using this gelled hydrogel as a wall for the next capillary flow.

11.5 Biosensors

11.5.1 Gel electrophoresis

Gel electrophoresis is one of the most applied and standardized tools for the separation and analysis of macromolecules and their fragments. Gel electrophoresis entails passing an electrical charge through a porous material after adding a solution containing the targeted substance. DNA, RNA, and proteins require different extraction, gel makeup, and buffering methods. Electrophoresis exploits the fact that shorter strands travel through a gel medium at higher speeds, thus separating out different molecules. Staining picks out the individual strands for identification. Gutzweiler and coworkers [41] (figure 11.10(A)) developed an approach in which the gel filled an open channel. Their approach combines the advantages of slab gel, capillary, and chip-based methods and offers low consumable costs, the circumvention of cost-intensive microfluidic chip fabrication, short process times (5 min per analysis), and high sensitivity combined with reasonable resolution (17 bases).

11.5.2 *In vivo* sensors

Threads have recently emerged as a promising material for the creation of tissue constructs and biomedical implants for organ replacement and repair. The wicking property and flexibility of threads also make them promising candidates for the creation of 3D microfluidic circuits. Mostafalu and coworkers have demonstrated that thread-based microfluidic networks can interface intimately with biological tissues in three dimensions [42] (figures 11.12(A) and (B)). A thread implanted in the human body and connected to a diagnostic device placed on the skin can retrieve information about an internal wound by transporting liquids from the wound to the surface.

11.5.3 Open-channel microfluidics for whole blood analysis

A. Blood serum analysis

Kimura and coworkers have developed [43] an open-type capillary-assembled microchip (CAs-CHIP) which is very easy to fabricate and use, for the rapid and simultaneous sensing of different serum components; viz. glucose, cholesterol, and alkaline phosphatase (ALP) and pH (figures 11.12(C) and (D)). The device included multiple square glass capillary sensors (with an outer diameter of 30 mm^2) embedded into a black polydimethylsiloxane (PDMS) microchannel array (30 mm wide and deep) forming an open-type CAs-CHIP into which samples were

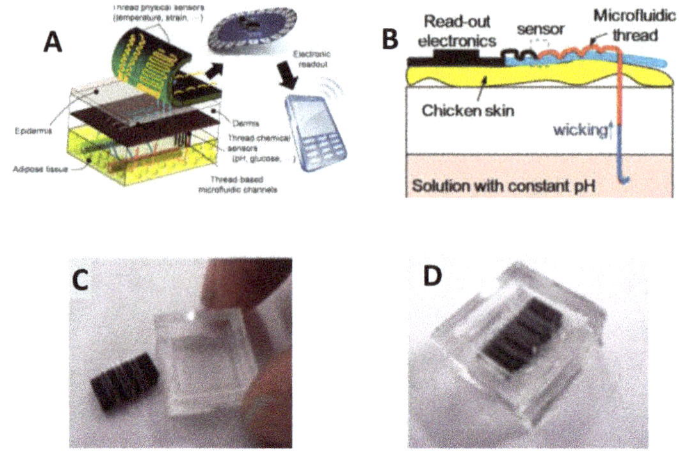

Figure 11.12. Top (A) and (B): a thread-based system used to extract biologic information from under the skin, reproduced from [42], CC BY 4.0; (C) and (D) an open diagnostic device for the analysis of serum samples, reproduced from [43] with permission of The Royal Society of Chemistry.

introduced by capillary action. The open-type CAs-CHIP was embedded in a PDMS reservoir chip where PDMS oil was dropped onto both ends of the capillary to avoid the evaporation of the nanoliter sample inside the capillaries. Fluorescein-based probes achieved simultaneous multicomponent sensing using a single fluorescence filter and accurate measurements were obtained with acceptable everyday precision (5.3%–8.8%).

B. Blood sampling and blood analysis

Prothrombin time (PT) and the associated international normalized ratio (INR) are routinely tested to assess the risk of bleeding or thrombosis and to monitor patient response to anticoagulant therapy. Hence, a portable device aimed at monitoring PT and INR would be a useful tool for patients [44, 45]. As whole blood has been shown to flow easily in a sharp V-groove [46], such a groove was used for the fast transport of whole blood from a finger prick to a reservoir where speckle analysis was done to measure the prothrombin time (figure 11.13(A)). The reader may recall that speckle analysis is an optical method used to follow alterations in the viscoelastic properties of a drop of whole blood following the activation of coagulation by thromboplastin. In other words, the random motion of the RBCs (red blood cells)—which is not Brownian motion but a chemically driven motion—decreases with coagulation, until immobilization takes place.

The analysis of human blood is of utmost importance in medicine. Blood sampling is conventionally performed using syringes. However, this method requires the presence of a nurse or doctor and usually transport to a laboratory. Recently, a new method was developed by Berthier and Casavant [47]. The new device extracts blood from the forearm with no pain. It comprises open channels which are used to collect clinical-grade samples directly from the human body without a phlebotomist.

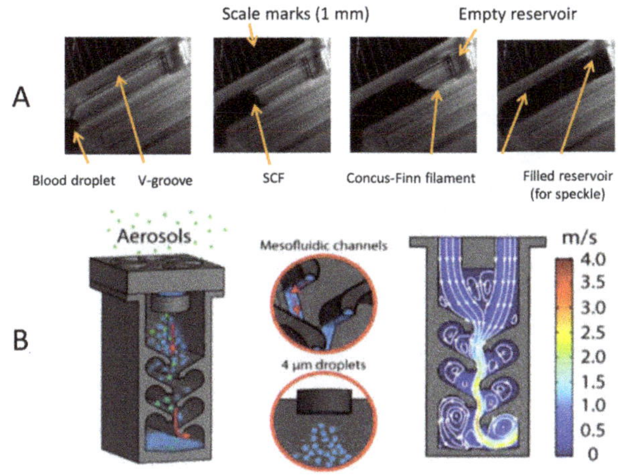

Figure 11.13. (A) The flow of whole blood in a narrow V-groove towards the reservoir where the speckle motion is analyzed; (B) an aerosol sampling device for the capture of volatile organic compounds (VOCs). Panel (A) photographs courtesy of Patrick Pouteau, Avalun SA. Panel (B) reprinted with permission from [49], Copyright (2021) American Chemical Society.

In other words, open microfluidics is a technique that can obtain blood samples without the need for venipuncture.

11.5.4 Biochemistry: liquid–liquid extraction

Open-channel biphasic systems have recently been used for liquid–liquid extraction. In traditional liquid–liquid extraction of metabolites from cell culture, numerous flasks and large volumes of solvent are used; the ability to evaporate solvent from the extracted metabolites is also a limiting factor. Open microfluidic devices have been used to study the culture conditions of multiple cell types, including mammalian [31], fungal, and bacterial cultures. Careful control of immiscible fluids via open device features generates a stable biphasic interface, enabling the extraction of metabolites by diffusion and direct access for solvent collection via a pipette or automatic liquid handler. The open microfluidic devices designed by Barkal *et al* and Casavant *et al* [48] give researchers the ability to: use solvents not commonly used for liquid–liquid extraction; co-culture different cell types that are connected by diffusion; screen for novel metabolites; and reduce the experimental time and materials used.

11.5.5 Aerosol sampling

Open microfluidics forms the basis of the development of a portable aerosol sampling device aimed at detecting the concentrations of airborne particles (dust particles, ultrafine particulate matter (PM), biologically derived aerosols, and potentially infectious bioaerosols). These gases are termed volatile organic compounds (VOCs), referring to gases released into the atmosphere by products or processes. The molecules or nanoparticles transported by the ambient air are

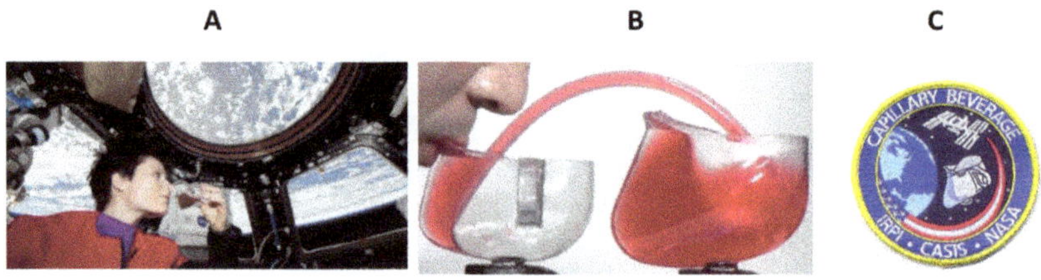

Figure 11.14. The space cup: (A) the space cup is used to drink coffee in the space station; (B) in a low-gravity environment, liquid can be confined in cleverly designed wetting corners; (C) the logo of the laboratory designing the space cup. Photographs courtesy Mark Weislogel, IRPI and Portland University.

responsible for environmentally induced respiratory illnesses. Figure 11.13(B) shows the principle of the portable device developed by Lee and coworkers [49]. A piezo crystal diffuses a dense cloud of microdroplets of phosphate-buffered saline (PBS) solution into an airflow activated by a fan. Airborne particulate aerosols are captured by the droplets and fall onto a plate with striations. Because of the open geometry, no clogging occurs and the droplets fall into a well that is used for detection.

11.6 A space-based application—the space cup

The following chapter, written by Mark Weislogel, is dedicated to the use of open-capillary fluidics aboard spacecraft. To cite Mark Weislogel: 'almost no conventional fluidic systems that we have on earth will work in space.' We just add here an interesting application of open microfluidics—more precisely, capillary filaments. In space, capillary forces are important even at a scale larger than the microscale. This is due to the weak gravitational force. The Bond number characterizing the ratio between the gravitational force and the surface tension force

$$Bo = \frac{\rho\, g\, d^2}{\gamma}, \tag{11.2}$$

is often small due to the small value of the gravitational constant g. Free liquids in space form deformable floating spheres under the action of surface tension, which imposes the minimization of the free surface. In order to guide liquids, hydrophilic solid structures are needed. We now describe a device invented in Mark Weislogel's lab to guide liquids in space for astronauts to drink. The 'space cup' illustrates the use of open capillarity and capillary filaments [50] (figure 11.14). The wedge shape of the space cup confines the liquid and enables liquid to flow alongside the bottom of the wedge as a capillary filament.

11.7 Conclusions

The emergence of open microfluidics has promoted numerous new applications in many scientific domains. In this chapter, we have presented some interesting applications in the domain of space, biotechnology, and biology. The accessibility to the flow provided by the open geometry is a determining factor for many applications, especially in biology. The ability to flow a liquid polymer using an open channel as a mold and then gel the polymer to form a target structure is also a great advantage of open-channel microfluidics. More applications of open microfluidics are presented in the review paper by Oliveira *et al* [51].

References

[1] Mann J J A, Romero L, Rye R R and Yost F G 1995 Flow of simple liquids down narrow V-grooves *Phys. Rev.* E **52** 3967

[2] Rye R R, Yost F G and Mann J A 1996 Wetting kinetics in surface capillary grooves *Langmuir* **12** 4625–27

[3] Romero L A and Yost F G 1996 Flow in an open channel capillary *J. Fluid Mech.* **322** 109–29

[4] Yost F G, Rye R R and Mann J A 1997 Solder wetting kinetics in narrow V- grooves *Acta Mater.* **45** 5337–45

[5] Shao F, Ng T W, Fu J, Shen W and Ling W Y L 2011 Electrical circuits from capillary flow driven evaporation deposition of carbon nanotube ink in non-porous V-grooves *J. Colloid Interface Sci.* **363** 425–30

[6] Zang F, Chu S, Gerasopoulos K, Culver J N and Ghodssi R 2015 Rapid fabrication of supercapacitor electrodes using bio-nano-scaffolds in capillary microfluidics *J. Phys.: Conf. Ser.* **660** 01201

[7] Lade R K, Jochem K S, Macosko C W and Francis L F 2018 Capillary coatings: flow and drying dynamics in open microchannels *Langmuir* **34** 7624–39

[8] Berthier J, Gosselin D, Huet M, Sabatte G, Pouteau P, Navarro F and Cubizolles M 2016 Capillary flow of whole blood in microsystems: non-Newtonian blood behavior and substrate reagent-coating effect *Proc. of the 2016 Nanotech Conf. (Washington, DC, May 23–25)*

[9] Lone S, Zhang J M, Vakarelski I U, Li E Q and Thoroddsen S T 2017 Evaporative lithography in open microfluidic channel networks *Langmuir* **33** 2861–71

[10] Oyola-Reynoso S, Frankiewicz C, Chang B, Chen J, Bloch J-F and Thuo M M 2017 Paper-based microfluidic devices by asymmetric calendaring *Biomicrofluidics* **11** 014104

[11] Gosselin D, Baumlin J-M, Joyard-Pitiot B, Navarro F, Belgaceml N, Chaussy D and Berthier J 2016 Open microfluidics in embossed paper micro-channels *Proc. of the 2016 MicroTas Conf. (Dublin, Ireland)* (San Diego, CA: Chemical and Biological Microsystems Society) 984–5 https://cbmsociety.org/conference/proceedings_archive.html

[12] Gosselin D, Belgacem M N, Joyard-Pitiot B, Baumlin J M, Navarro F, Chaussy D and Berthier J 2017 Low-cost, embossed-paper micro-channels for spontaneous capillary flow *Sensors Actuators* B **248** 395–401

[13] Thuo M M, Martinez R V, Lan W-J, Liu X, Barber J, Atkinson M B J, Bandarage D, Bloch J-F and Whitesides G M 2014 Fabrication of low-cost paper-based microfluidic devices by embossing or cut-and-stack methods *Chem. Mater.* **26** 4230–7

[14] Yu L and Shia Z Z 2015 Microfluidic paper-based analytical devices fabricated by low-cost photolithography and embossing of parafilm *Lab Chip* **15** 1642–5

[15] Songok J and Toivakka M 2016 Enhancing capillary-driven flow for paper-based microfluidic channels *ACS Appl. Mater. Interfaces* **8** 30523–30

[16] Zhang T, Ratajczak A M, Chen H, Terrell J A and Chen C 2022 A step forward for smart clothes fabric-based microfluidic sensors for wearable health monitoring *ACS Sens.* **7** 3857–66

[17] Kun M, Chan C and Ramakrishna S 2019 Textile materials and structures for topical management of wounds *Advanced Textiles for Wound Care* (Cambridge: Woodhead Publishing) 2nd edn pp 55–104 The Textile Institute Book Series
[18] Nilghaz A, Wicaksono D H B, Gustiono D, Abdul Majid F A, Supriyanto E and Abdul Kadir M R 2012 Flexible microfluidic cloth-based analytical devices using a low-cost wax patterning technique *Lab Chip* **12** 209–18
[19] Yang Y, Xing S, Fang Z, Li R, Koo H and Pan T 2017 Wearable microfluidics: fabric-based digital droplet flowmetry for perspiration analysis *Lab Chip* **17** 926–35
[20] Li Y, Fischer R, Zboray R, Boillat P, Camenzind M, Toncelli C and Rossi R M 2020 Laser-engraved textiles for engineering capillary flow and application in microfluidics *ACS Appl. Mater. Interfaces* **12** 29908–16
[21] Rajendran S 2019 *Advanced Textiles for Wound Care* (Cambridge: Woodhead Publishing)
[22] Schellenberg A, Ross R, Abagnale G, Joussen S, Schuster P, Arshi A, Pallua N, Jockenhoevel S, Gries T and Wagner W 2014 3D non-woven polyvinylidene fluoride scaffolds: fibre cross section and texturizing patterns have impact on growth of mesenchymal stromal cells *PLoS One* **9** e94353
[23] Szewczyk P K, Metwally S, Karbowniczek J E, Marzec M M, Stodolak-Zych E, Gruszczyński A, Bernasik A and Stachewicz U 2019 Surface-potential-controlled cell proliferation and collagen mineralization on electrospun polyvinylidene fluoride (PVDF) fiber scaffolds for bone regeneration *ACS Biomater. Sci. Eng.* **5** 582–93
[24] Dudukovic N A, Fong E J, Gemeda H B, De Otte J R, Ceron M R, Moran B R, Davis J T, Baker S F and Duoss E B 2021 Cellular fluidics *Nature* **595** 58–65
[25] Lee B, Kim S, Ko J, Lee S-R, Kim Y, Park S, Kim J, Hyung S, Kim H-Y and Jeon N L 2022 3D micromesh-based hybrid bioprinting: multidimensional liquid patterning for 3D micro-tissue engineering *Asia Mater.* **14** 6
[26] Feng J and Rothstein J P 2013 One-way wicking in open micro-channels controlled by channel topography *J. Colloid Interface Sci.* **404** 169–78
[27] Li C, Li N, Zhang X, Dong Z, Chen H and Jiang L 2016 Uni-directional transportation on peristome-mimetic surfaces for completely wetting liquids *Angew. Chem. Int. Ed.* **55** 14988–92
[28] Temiz Y, Skorucak J and Delamarche E 2014 Capillary-driven microfluidic chips with evaporation-induced flow control and dielectrophoretic microbead trapping *J. Micro/Nanolith. MEMS MOEMS* **13** 033018
[29] Li J, Lu N and Cong T 2021 Experimental study on evaporation-capillary pumping flow in capillary wick and working fluid system *Thermal Sci.* **25** 367–75
[30] Zimmermann M, Bentley S, Schmid H, Hunziker P and Delamarche E 2005 Continuous flow in open microfluidics using controlled evaporation *Lab Chip* **5** 1355–9
[31] Casavant B P *et al* 2013 Suspended microfluidics *Proc. Natl Acad. Sci.* **110** 10111–6
[32] Humayun M, Chow C-W and Young E W K 2018 Microfluidic lung airway-on-a-chip with arrayable suspended gels for studying epithelial and smooth muscle cell interactions *Lab Chip* **18** 1298–309
[33] Wong A P, Perez-Castillejos R, Christopher Love J and Whitesides G M 2008 Partitioning microfluidic channels with hydrogel to construct tunable 3-D cellular microenvironments *Biomaterials* **29** 1853–61
[34] Su C, Chuah Y J, Ong H B, Tay H M, Dalan R and Hou H W 2021 A facile and scalable hydrogel patterning method for microfluidic 3D cell culture and spheroid-in-gel culture array *Biosensors* **11** 509

[35] Lou J and Mooney D J 2022 Chemical strategies to engineer hydrogels for cell culture *Nat. Rev. Chem.* **6** 726–44

[36] Tung C K, Krupa O, Apaydin E, Liou J-J, Diaz-Santana A, Kima B J and Wu M 2013 A contact line pinning based microfluidic platform for modelling physiological flows *Lab Chip* **13** 3876–388

[37] Bouaidat S, Hansen O, Bruus H, Berendsen C, Bau-Madsen N K, Thomsen P, Wolf A and Jonsmann J 2005 Surface-directed capillary system; theory, experiments and applications *Lab Chip* **5** 827–36

[38] Lee S-R, Kim Y, Kim J, Park S, Rhee S, Park D, Lee B, Baek K, Kim H-Y and Jeon N L 2022 U-IMPACT: a universal 3D microfluidic cell culture platform *Microsyst. Nanoeng.* **8** 126

[39] Ahn J, Sei Y J, Jeon N L and Kim Y T 2017 Tumor microenvironment on a chip: the progress and future perspective *Bioengineering* **4** 64

[40] Yang C, Yu Y, Wang X, Wang Q and Shang L 2021 Cellular fluidic-based vascular networks for tissue engineering *Eng. Regen.* **2** 171–4

[41] Gutzweiler L, Gleichmann T, Tanguy L, Koltay P, Zengerle R and Riegger L 2017 Open microfluidic gel electrophoresis: rapid and low-cost separation and analysis of DNA at the nanoliter scale *Electrophoresis* **38** 1764–70

[42] Mostafalu P, Akbari M, Alberti K A, Xu Q, Khademhosseini A and Sonkusale S R 2016 A toolkit of thread-based microfluidics, sensors, and electronics for 3D tissue embedding for medical diagnostics *Microsyst. Nanoeng.* **2** 16039

[43] Kimura Y, Henares T G, Funano S, Endo T and Hisamoto H 2012 Open-type capillary-assembled microchip for rapid, single-step, simultaneous multi-component analysis of serum sample *RSC Adv.* **2** 9525–30

[44] Tripathi M M, Egawa S, Wirth A G, Tschikudi D M, van Cott E M and Nadkarni S K 2017 Clinical evaluation of whole blood prothrombin time (PT) and international normalized ratio (INR) using a Laser Speckle Rheology sensor *Sci. Rep.* **7** 9169

[45] Faivre M, Peltié P, Planat-Chrétien A, Cosnier M-L, Cubizolles M, Nougier C, Négrier C and Pouteau P 2011 Coagulation dynamics of a blood sample by multiple scattering analysis *J. Biomed. Opt.* **16** 057001

[46] Berthier J, Brakke K A, Furlani E P, Karampelas I H, Poher V, Gosselin D, Cubizolles M and Pouteau P 2015 Whole blood spontaneous capillary flow in narrow V-groove micro-channels *Sensors Actuators* B **206** 258–67

[47] Tasso Inc. www.tassoinc.com

[48] Barkal L J *et al* 2016 Microbial metabolomics in open microscale platforms *Nat. Commun.* **7** 10610

[49] Lee U N, van Neel T L, Lim F Y, Khor J W, He J, Vaddi R S, Berthier J, Novosselov I V, Theberge A B and Berthier E 2021 Miniaturizing wet scrubbers for aerosolized droplet capture *Anal. Chem.* **93** 11433

[50] Concus P 1983 On existence criteria for fluid interfaces in the absence of gravity *Waves on Fluid Interfaces* (New York: Academic) 113–22

[51] Oliveira N M, Vilabril S, Oliveira M B, Reis R L and Mano J F 2019 Recent advances on open fluidic systems for biomedical applications: a review *Mater. Sci. Eng.* C **97** 851–63

IOP Publishing

Open-Channel Microfluidics (Second Edition)
Fundamentals and applications
Jean Berthier, Ashleigh B Theberge and Erwin Berthier

Chapter 12

Open-capillary fluidics aboard spacecraft

This chapter was written by M M Weislogel, PhD, Principal Scientist, IRPI LLC, Wilsonville, Oregon, USA. Even if the applications of open fluidics in space are at a larger scale than that presented above, we found it interesting to present another aspect of capillary flows in open environments.

12.1 Introduction

Open microfluidics applications abound for both forced and spontaneous liquid-handling operations aboard spacecraft. In fact, if the relative impact of local acceleration forces is small in comparison to capillary forces, the topics of this book are equally pertinent to applications aboard spacecraft but over much larger length scales.

The Bond number, as a ratio of local acceleration forces (i.e. gravity) to surface tension forces, is redefined here as

$$Bo \equiv \frac{\Delta\rho a R L}{\gamma} f(\theta,\ R/L), \tag{12.1}$$

where $\Delta\rho$ is the fluid density difference across the liquid free surface, a the scalar local acceleration field strength that has both magnitude and direction (i.e., $a \sim |\boldsymbol{a}|$), R a characteristic capillary length scale, L a characteristic hydrostatic length scale, and γ the surface or interfacial tension. Rarely included in the Bond number is an additional dimensionless function $f(\theta,R/L)$ characterizing the free surface boundary conditions accounting for the often-dramatic impacts of the contact angle θ, the contact-angle hysteresis, pinning conditions, and others, including an additional specific dimensionless geometry represented here by R/L [1]. Accounting for $f(\theta,R/L)$ improves the chances that $Bo \sim O(1)$ demarcates the transition between capillary- and gravity-dominated behavior. The condition $Bo \ll 1$ is defined herein as the low-g state.

Terrestrial ($a = g_o \equiv 9.8$ m s^{-2}) open microfluidic phenomena assume such small length scales that $Bo \ll 1$ is often taken for granted, as gravity plays a negligible role in capillary flows when limited largely to horizontal planes. However, in the nearly weightless environments of orbiting or coasting spacecraft, local accelerations of $a < 10^{-6}\ g_o$ are common and enormous capillary fluid systems on the order of meters in length can result. The phrase 'open-capillary fluidics' provides a slightly broader description of phenomena that include both 'terrestrial open microfluidics' and 'low-g open macrofluidics.' For separated fluid phases without bubbles or droplets, the distinction is moot provided that $Bo \ll 1$. However, the presence of free bubbles and/or droplets in such 'capillary fluidic' systems can lead to a highly divergent outcome when employed in space compared to their outcome on Earth. In both microfluidic or macrofluidic systems in space, bubbles do not rise and droplets do not fall. Such flows are truly unearthly and pose challenges to both closed and open microfluidic (aka microgravity capillary fluidic) solutions when applied aboard spacecraft.

For example, applications in microfluidic processes are found aboard spacecraft such as the International Space Station (ISS), where an increasing number of wet lab activities are necessary for capillary fluidic bio-fluid sample preparation, processing, and analysis with miniaturization and automation in mind, e.g. DNA sequencing [2]. As previously stated, terrestrial microfluidic methods and devices can be unaffected by the low-g environment due to their small capillary length scales on Earth, i.e. the Minion (Minion is a biological sample device for DNA sequencing). As another example, nanopore sequencing has succeeded in microgravity with only slight accommodation, as demonstrated by McIntyre *et al* [3] who made simple pipette adjustments to avoid introducing bubbles into liquid samples. They also confirmed that gravity does not affect nanopore sequencing. In keeping with the simple pipette example, Burton *et al* [4] reported that the ISS crew had no issues pipetting in the microgravity environment, since surface tension 'was a driving factor in liquids staying in their respective tubes.' However, it has since been noted that a closer microgravity 'capillary fluidic' investigation of such pipetting operations revealed the release of micro-satellite droplets following nearly every pipette withdrawal in microgravity [5]. These droplets were too small and too fast to be visible by the crew and thus were not observed and reported. This may pose a safety concern for high-throughput, high-automation applications since such unmitigated droplets may be deposited anywhere. It is conjectured that microbubble ingestions are also likely and that there are low-g complications even with simple commercial terrestrial well plates [6]. Further spacecraft microgravity microfluidics challenges remain aboard the ISS, including difficulties removing air bubbles that form during sample handling, increases in susceptibility to air bubbles introduced into flow cells during loading, rogue bubble damage to flow cell membranes or nanopores, bubble occlusions leading to complete nanopore blockage, and more.

Challenging fluidics problems in space are not only limited to 'microfluidics' problems. Pertinent to this audience are problems that can benefit from current and pending terrestrial developments in open-capillary fluidics technologies extended to the large capillary length scale, low-gravity environment. The list of challenging critical fluidics functions aboard spacecraft is long [5, 7, 8] and includes vital systems for life support including CO_2 removal, trace contaminant controls, particulate filtration, condensing heat exchangers, O_2 recovery from CO_2, O_2 generation, high-pressure O_2 production, water microbial control, urine collection, wastewater processing, urine brine processing, non/metabolic solid waste management, and other systems where spurious bubbles and droplets wreak havoc. Systems where fluidics play a 'second-tier' role include atmosphere, water, and microbial systems and particulate monitoring. Further examples include the large-scale systems for liquid fuels, propellants, cryogens, coolants, food production, and hygiene and the smaller-scale systems for medical fluids, housekeeping, science experiments, and more. The primarily passive solutions of open-capillary fluidics are attractive partly because of their ability to scale up, the challenges being the somewhat nonintuitive considerations for the correct initial conditions, stability, and the increased inertia in open-capillary fluidic solutions aboard spacecraft, which are reviewed in part below.

12.2 Statics: configurations, initial conditions, and stability

The presence of gravity determines the characteristic capillary length for any terrestrial open microfluidic flow, $\lambda \sim (\gamma/\Delta\rho g)^{1/2}$, and in most cases the characteristic channel dimension d is the appropriate length scale. But for $g = 0$, the capillary length scale increases to the system length scale, $l \equiv L \sim R$, which can be enormous ($R \gg d$), challenging the notion of 'open fluidic flow' in many applications since the enclosure length scale has a high chance of being part of the problem. Thus, open-capillary fluidic solutions in space are far more dependent on system-, container-, enclosure-, and tank-level length scales than on Earth. Open-capillary fluidics in low-g environments may be viewed as faster flows based on local length scales within slower flows based on global length scales. For example, the outer length scale may set the critical global boundary conditions (i.e. capillary pressure) for the local open-capillary fluidic flow, which may slowly modulate the global boundary condition, in turn modulating the local flow, and so on. Examples will be provided. In addition, in the absence of up or down, open fluidic solutions must account for any number of metastable interfacial states that depend on the wetting conditions and the system geometry. Knowledge of the stability limits of any capillary fluidic solution is critical to assure local if not global minimum energy states, as opposed to states that are readily destabilized into useless or non-recoverable configurations. Several examples are reviewed here.

Concus *et al* [9] demonstrated the conundrum of low-g interface configurations by testing partially liquid-filled rotationally symmetric containers within which an infinite number of equal-energy rotationally symmetric surface configurations satisfying the contact-angle boundary condition are possible (figure 12.1(A)). In the end, none of the symmetric configurations were preferred in low-g but rather a cascade of lower-energy metastable asymmetric configurations. And these metastable asymmetric surfaces possess differing stability limits, natural frequencies, damping rates, and capillary pressures which might feed into the more local open-capillary fluidic application. Another example by Concus *et al* [10, 11] demonstrates the decay of metastable states due to contact-angle hysteresis; the metastable states are eventually overcome by mechanical perturbations, temperature fluctuations, and

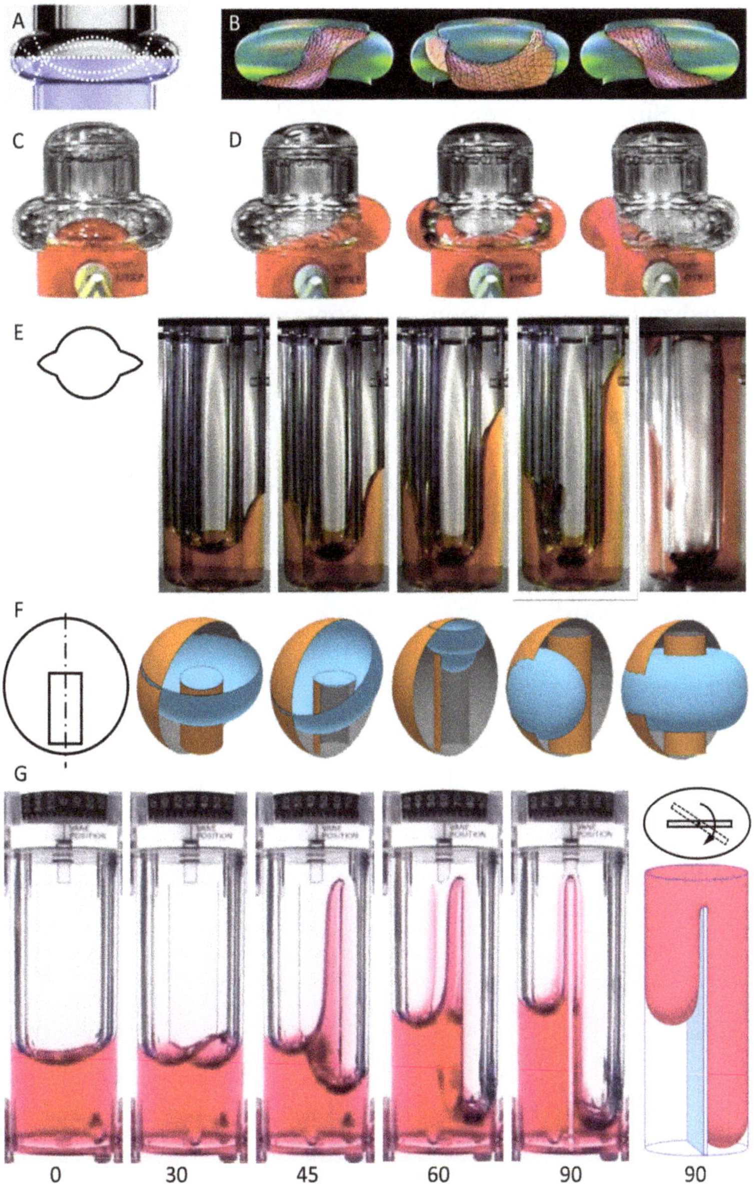

Figure 12.1. Complex metastable low-g interface configurations. (A) Infinite rotationally symmetric equal-energy low-g interfaces satisfying volume and contact-angle conditions; (B) numerical lower-energy asymmetric surfaces; (C) initial configuration in space; (D) metastable perturbed states; (E) metastable perturbed states of liquid in a Proboscis cylinder—last state imaged after several days; (F) a Mercury 07 standpipe tank with Surface Evolver-computed symmetric and asymmetric equilibria; (G) images of the Vane Gap experiment displaying critical geometric wetting as a function of vane angle with a deeply converged Evolver configuration beneath an elliptical cylindrical container section. Panels (A)–(D), reprinted from [9] with permission. © 1999, Cambridge University Press. Panel (E) reprinted from [10], Copyright (2000), with permission from Springer. Panel (F) reprinted from [12], Copyright © 2017 by the American Institute of Aeronautics and Astronautics, Inc. Panel (G) reproduced from [15], Copyright © 2008 by the American Institute of Aeronautics and Astronautics, Inc.

time (figure 12.1(B)). A model for an early spacecraft fuel tank tested on the Mercury 7 mission [12] produced a variety of stable and metastable Surface Evolver-computed interfaces [13, 14] in what appears to be a rather simple axisymmetric tank (figure 12.1(C)). Surface Evolver images were also compared to the $\theta = 0°$ ISS Capillary Flow Experiment-Vane Gap test [15], which demonstrates the coupling between critical geometric wetting on bulk and local menisci (figure 12.1(D)). Somewhat analogously to the Concus–Finn geometric wetting condition [16], in this case two extremely subtle geometric wetting conditions occur simultaneously—one leading to an obvious global bulk shift, another leading to obvious local gap wetting. As a conceptualization, despite being in a container, for application purposes, the latter might be considered to be an open-capillary flow, but it is only the faster, smaller part of a slower, larger capillary flow. These are compound capillary flows, and they are highly exaggerated in scale in low-g environments.

What these low-g investigations demonstrate is that unlike microfluidic droplets on surfaces in terrestrial laboratories, bulk meniscal configurations in space can be numerous, nonintuitive, metastable, and strongly (and sometimes discontinuously) dependent on system geometry and wettability. But such surfaces often serve as the 'reservoir' for an otherwise open-capillary fluidic flow in microgravity. From asymmetric blobs in poorly wetting well plates to asymmetric droplets on pipette cannulas, to large asymmetric wall-bound droplets, the larger liquid reservoir feeding the open-capillary fluidic flow may be a complex capillary flow problem in its own right.

A simple low-g example is shown schematically in figure 12.2, where spontaneous capillary flows are imbibed by parallel perfectly wetted triangular grooves in a 2.1 s drop tower experiment in which gravity is effectively and suddenly turned off when the experiment is released into free fall [17, 18]. The upstream capillary curvature boundary condition for the flows is established by the receding bulk meniscus in the

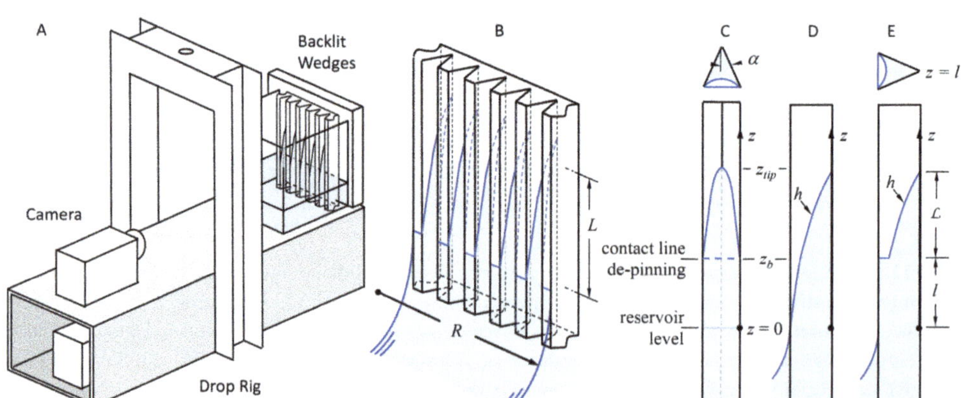

Figure 12.2. Schematics of low-g compound capillary rise along open wedge channels in drop tower tests; (A) a schematic of the drop test rig; (B) a schematic of the parallel wedge channels; (C) a schematic of the camera view; (D) the true profile view; (E) the compound capillary flow model profile. Reprinted from [17] with permission. © 2012, Cambridge University Press.

container (length scale R), which advances the liquid filaments in the channels to the point at which they de-pin from the groove edges (length l), beyond which free contact-line boundary conditions drive the flows ever further toward the zero-height advancing front (length L). All boundary conditions for such compound capillary flows are dynamical and in this case obey the $\sim t^{1/2}$ power law, the former being potentially uniquely low-g dependent, requiring particular low-g attention.

12.3 Dynamics: inertia and bubble separations

For truly microfluidic applications in low-g environments, the dynamics of the flow phenomena are largely unchanged from their terrestrial counterparts. It is true that in the near absence of a 'gravitational' force, significantly longer flows may be established incorporating any number of interesting out-of-plane elements. In fact, in a microgravity environment, a pore size of 1 mm can conduct a capillary flow as far as $\sim O(10^4 \text{ m})$ while maintaining $Bo < 1$, as defined in (12.1), with $f = 1$. However, practical disruptions to the 'low-g' environment are common aboard spacecraft due to orbital maneuvers, docking, thruster firing, spin stabilization, gravity gradients, drag, crew handling, crew kick loads, exercise, and others. Thus, capillary fluidics solutions must tolerate perturbations on the order of $\sim O(10^{-2} g_o)$ or larger, such that, say, pore sizes of 1 mm conduct capillary flows only as far as $< O(1 \text{ m})$. This is still a long capillary length for microfluidics, which only gets longer as pore size gets smaller.

The case has been made for the application of terrestrial microfluidics technologies aboard spacecraft at macrofluidic length scales. The advent of free bubbles and droplets in low-g, however, poses unique challenges with potentially disastrous consequences [19]. The presence of gravity assists greatly in open microfluidic flows in this regard—when bubbles do rise and burst, the microshower of free droplets returns to the substrate due to gravity. Without this passive force, bubbles merely languish, obstruct, and occlude liquid passageways, while droplets merely drift, adhere, and occlude gas passageways. Successful capillary fluidic solutions in low-g must also passively mitigate the often ill effects of bubbles and droplets.

In 2011, the Capillary Channel Flow (CCF) experiment flew aboard the ISS to investigate forced inertial–visco–capillary flows in open channels. The fundamental and applied investigation pursued the flow rate limits in open rectangular slots, open rectangular channels, and open triangular channels [20–22]. The latter devoted a significant effort to the study of bubbly flow phenomena in open wedge channels and how the wedge might be applied to practically and passively separate bubbles. A schematic of the experiment is provided in figure 12.3. Bubbles of a variety of sizes were introduced near the inlet vertex of the wedge at a variety of frequencies, varying liquid flow rates, liquid-to-gas flow rate ratios, and open wedge channel lengths. Bubble confinement mechanisms and imbalanced capillary forces drove the bubbles toward inscribed elevations within the wedge [21–24]. If the bubbles were large enough, or merged to become large enough, and they were not convected downstream fast enough, the bubbles found time to coalesce with the free surface and leave the flow. Thus, passive low-g bubble separation may be achieved using the

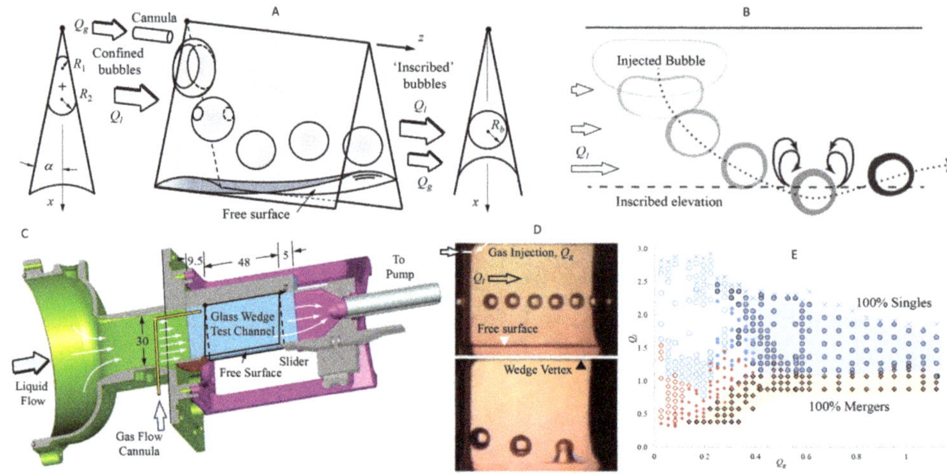

Figure 12.3. (A) The behavior of bubbles vertex injected into a forced open wedge flow; (B) the bubble path with respect to the inscribed elevation; (C) a schematic of the CCF test channel plumbing used during space experiments; (D) flight images of the steady inscribed bubble flow and 100% single bubble separations; (E) a regime map identifying 100% separation conditions. Reproduced from [22]. © 2015 NASA.

open-channel wedge geometry. A sample regime map is provided in figure 12.3(C). It identifies the conditions under which 100% single and/or 100% merged bubble separations occur. Over 20 such maps were constructed (~25 000 data points) for these large-length-scale low-g open inertial–visco–capillary fluidic flows [22]. The CCF-wedge channel dimensions were 30 mm in height and up to 48 mm long, with an included angle of 15.8° for the perfectly wetting dielectric HFE-7500 test fluid. Flow rates were limited to <5 ml s^{-1} and velocities were limited to <10 cm s^{-1}. The open wedge channel has been applied to further spacecraft fluid applications as a result. We note that with each bubble coalescence, satellite droplets from the rupturing thin films are likely to be ejected into the surroundings. Such droplets are not always detected, but for long-term processes, significant fluid volume loss is possible into perhaps undesirable regions within the system. Separate mitigations should be considered if there is cause for alarm at such potentially significant, hazardous, or annoying fluid escape.

In addition to the unearthly large non-buoyant bubbles and droplets arising in low-g capillary fluidics systems, the simultaneous unearthly large viscous length scales $\sim R$ ($\sim 10^3$l) leading to unearthly low viscous resistance $\sim 1/R^2$ permit unique inertial–capillary regimes to occur over longer durations than those experienced on Earth. The Reynolds number is defined as the ratio of inertial to viscous forces given by

$$\mathrm{Re} \equiv \frac{\rho V d}{\mu} \sim \frac{\rho \gamma d}{\mu^2} \equiv Su, \tag{12.2}$$

where V is a characteristic velocity and d the characteristic length ($\sim R \sim L$). But for spontaneous capillary flows, the visco–capillary balance yields the velocity scale $V \sim \gamma/\mu$, which is substituted into equation (12.2) to define the Suratman number—a

Reynolds number based on the visco–capillary velocity scale. In the microgravity environment, d can be 1000 times larger than in a terrestrial environment and still maintain capillary dominance, $Bo \ll 1$ (refer to equation (12.1)). In this way inertial–capillary flows are greatly amplified in low-g capillary fluidics problems and solutions. An identical result is achieved by substituting $V \sim \gamma/\mu$ into the Weber number. For equivalent velocity scales between terrestrial and microgravity environments, the fact that the length scales of the latter can be up to 1000 times larger implies that flow rates $Q \sim Vd^2$ may be up to 10^6 times larger than those observed on the ground. The large underdamped low-g capillary flows and phenomena routinely occurring in low-g are susceptible to perturbations $Bo > 1$, especially at fundamental frequencies $\omega \sim (\gamma/\rho d^3)^{1/2}$. They are capable of auto-ingestions, auto-ejections, and numerous other exaggerated geyser phenomena [25, 26]. Such phenomena occur in terrestrial flows but over significantly shorter inertial and viscous timescales; $\tau_{\text{inert}} \sim (\rho d^3/\gamma)^{1/2}$ ($\sim 3 \times 10^4$ times longer in low-g) and $\tau_{\text{visc}} \sim \rho d^2/\mu$ ($\sim 10^6$ times longer in low-g), respectively. The Suratman number appears in correlations for low-g forced open wedge channel flows, e.g. the CCF-wedge flow rate limit due to gas ingestion [17] is given by

$$Q_{\text{ing}} \sim \frac{\mu L}{2\rho\alpha^2}\left(\left(1 + \frac{4}{3}\text{Su}\left(\frac{H\alpha^2}{L}\right)^2\right)^{1/2} - 1\right), \tag{12.3}$$

where H is the channel height, L the channel length, and α the wedge half-angle, with $Su \equiv \rho\gamma H\alpha/\mu^2 \sim O(10^5)$—two to three orders of magnitude larger than a similar flow on the ground. The uncertainty in (12.3) is ±5% for 15 mm $\leqslant L \leqslant$ 48 mm.

12.4 Applications of open macrofluidics aboard spacecraft

12.4.1 Bubble separation

The passive bubble-separating nature of the open wedge channel aboard spacecraft provides an attractive open-capillary fluidics solution to several critical systems in which the presence of bubbles can have deleterious consequences.

Beginning in 2010, the Capillary Flow Experiment-Interior Corner Flow (CFE-ICF) investigation began on the ISS with a series of simple perfectly wetting passive capillary flow tests within containers of centimetric scale (~40 mm diameter, ~160 mm length) [27]. Wire models and annotated flight images of the partially liquid-filled devices are provided in figure 12.4 representing several families of tapered sections with at least one interior corner (wedge) spanning the length of each container. The tapers of the containers and channels within the containers are such that the bulk menisci curvatures for liquids covering the bases and lids of the containers are different, leading to spontaneous corner flows from the larger base regions to the narrower lid regions, which displace gas ullages from the narrower lid regions toward the larger base regions. In keeping with the arguments of this chapter, in low-g the container diameter may be significantly larger than the interior corner flow section length scale, providing the temptation to categorize the corner flows in the CFE-ICF devices as open-capillary flows. In a way they are, because

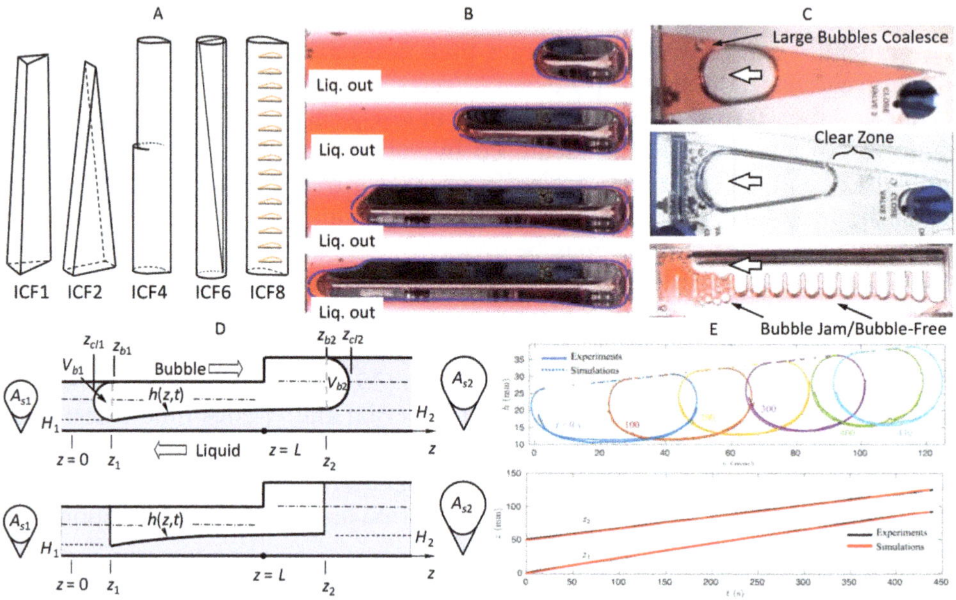

Figure 12.4. (A) A selection of wire models of the CFE-ICF test cells; (B) ullage migration in ICF6 with numerical predictions during draining; (C) large and small bubble separations for ICF2 and bubble jam separations for ICF8; a schematic of the model assumption for ICF4; (E) numerical comparisons with theory for ICF1 (nearly coincident). Panels (A), (C), and (D) reprinted from [28] with permission. © 2011, Cambridge University Press; panels (B) and (E) reprinted from [30], Copyright (2022), with permission from Springer.

they are fast and the gas pressure above them is essentially constant. On a short timescale, the corner flows act like open-capillary flows, but on a longer timescale, the upstream and downstream boundary conditions are only quasi steady and depend on the flow rates of the corner flow itself. These flows have been investigated and benchmarked theoretically [28] and numerically [29, 30]. They serve as pertinent examples of how, in low-g, the container length scale can play a weak but non-negligible role in an otherwise open-capillary fluidics operation.

Numerous bubble separation demonstrations were also performed as part of the CFE-ICF tests (figure 1.4D), providing additional examples of the passive bubble-separating characteristics of the interior corner flow. In these cases, the bubbles are typically larger than the corner flow section length scale; the bubbles are driven into the entrance to the interior corner flow where they become confined, merge, and coalesce through the bulk meniscus and are separated from the flow. Liquid drops ejected from the coalescence are observed rebounding within the ullage until an eventual impact and statistical merger with the liquid take place. If the bubbles are much smaller than the entry corner flow length scale, they are convected into the corner flow, becoming increasingly confined with distance along the corner, where they merge and are driven ever closer to the free surface, where they coalesce and leave the flow. Structures within the bubble flow path serve as dams that back up bubbles and also as small-scale self-filters, such that 100% liquid separation can be achieved despite the presence of myriad bubbles. Nonetheless, over time, the confined bubbles merge, coarsen, and coalesce, leaving the liquid free of bubbles.

12.4.2 CO_2 scrubbing

Life support aboard the confined quarters of spacecraft requires the removal of CO_2 from the cabin air. There are many competing processes that can accomplish this task, but it remains one of the key challenges for advanced exploration in space [7] to do so using a lightweight, compact, low power, and reliable system. One approach employs liquid sorbents, as commonly applied aboard submarines. Such systems have been successful for decades on Earth, but they make tremendous use of falling sprays and films and rising bubbles and hot gases. These are not available in low-g environments that lack the usual impact of gravity. Furthermore, thin planar films are inherently unstable in low-g, where they either break up into separate wall-bound drops in the case of partial wetting, or they wick into some unexpected region within the system in the case of perfect wetting.

However, thin planar films may be constructed of numerous open-capillary fluidic channels with integrated 3D-printed channel manifolds providing inlet and outlet flows, with an open-capillary wedge flow connecting the two. The Capillary Structures for Exploration Life Support (CSELS) technology demonstration on the ISS was perhaps the first to gang open wedge channels into parallelized capillary channel flows that collectively function like a planar 'thin film' [31]. The hardware setup is shown in figure 12.5, which includes a magnified view of the 16-way parallel open wedge channel contactor (and degasser). The channels for the contactor were 10 mm deep and 10 mm long and had a 10° included wedge angle. The channels for the degasser were 10 mm deep and 20 mm long and had a 15° included wedge angle. The objectives of CSELS (which were achieved) were to demonstrate a stable flow between two serially connected parallelized open-capillary fluidic devices using a single pump and to demonstrate the limits of stability in the face of significant crew-induced perturbations. The degasser headspace was sealed such that positive and negative gage pressures could be achieved there. The system naturally separated rogue and injected bubbles, was stable to convective fan-driven air flows over the contactor, could operate in both directions, could operate when both the contactor and the degasser were open to atmospheric pressure, and was capable of stable flow in inertial–capillary flow regimes, $Q_{\mathrm{pump}} \approx 10$ ml s^{-1} with individual channel velocities of $V \approx 5$ cm s^{-1}. Also shown in figure 12.5 are separate CSELS tests with larger open wedge channels conducted to demonstrate the stability limits of such parallel open-capillary flows, the impacts of the number of channels, channel pinning, fluid inertia, fluid viscosity, flow resistance imbalances, and others. The larger open wedge channels (30 mm deep, 100 mm long, and with a 15° included vertex angle) were employed for adequate flow visualization and quantitative interface profile comparisons with theory for steady interface profiles and at specified flow rates [32]:

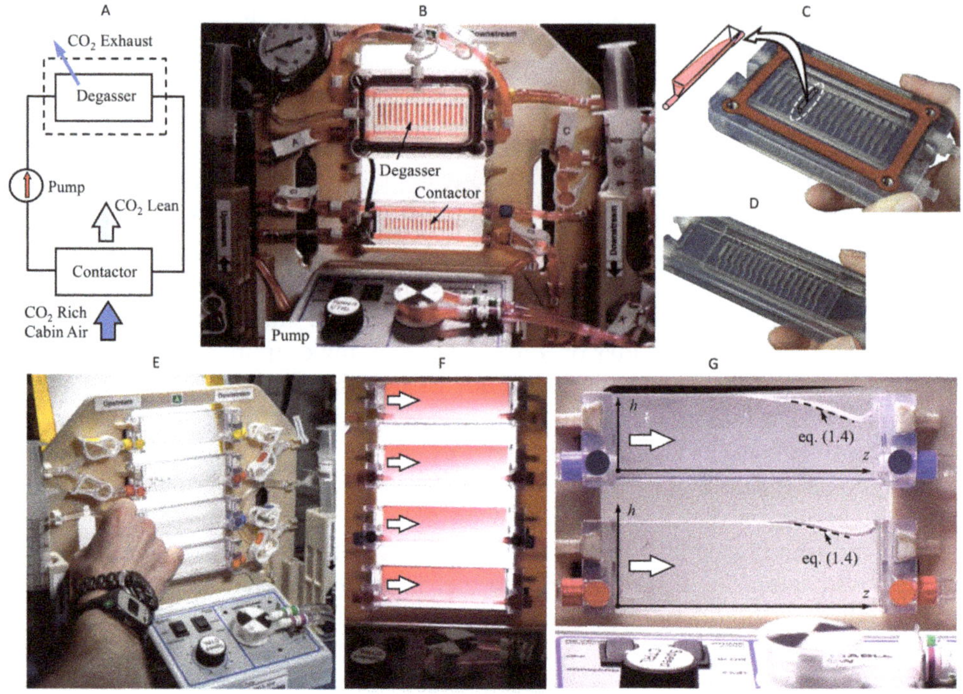

Figure 12.5. (A) A simplified schematic of the CSELS technology demonstration which was analogous to CO_2 scrubbing in low-g; (B) an image of the CSELS-CapiSorb hardware installed for tests on the ISS; (C) a magnified view of the CSELS degasser consisting of 16 parallel open wedge capillary channels with an exaggerated magnified sketch of an individual channel; (D) a magnified view of the smaller 16-channel CSELS contactor; (E) the CSELS four-cell science hardware on the ISS with a larger four-way parallel open wedge flow; (F) a demonstration of a highly stable pinned low flow condition (μ = 1 cP); (G) the CSELS viscous demonstration of a larger two-channel parallel flow configuration to illustrate the impacts of increased viscosity (μ = 45 cP)—equation (12.4) profiles overlaid for comparison. Figure courtesy of M M Weislogel.

$$h = H_1\left(1 - \left(1 - \frac{H_2}{H_1}\right)\frac{z}{L}\right)^{1/3} \text{ and } Q = \frac{\sigma}{3\mu L}\frac{F_A F_i \sin^2\alpha}{f}\left(H_1^3 - H_2^3\right); \qquad (12.4)$$

an example comparison is provided in figure 12.5(G). The four large pinned parallel open wedge channel flows in figure 12.5(F) at a low flow rate $Q = 0.184$ ml s^{-1} were highly stable, as the pinning and low flow allowed for significant capillary pressure accommodation between channels that balanced the flow despite small viscous resistance variations between the different legs of the plumbing. In cases in which contact lines de-pinned, the poor wetting and large contact-angle hysteresis served a similar though weaker role as the pinning edges of the channels. Free contact lines, low contact angles, low contact-angle hysteresis, low liquid inventory, and elevated inertia reduce resistance to hydrodynamics imbalances that can lead to the drying out of a given channel and/or the flooding of another. Thus, massively parallelized networks of open-capillary channels do well to exploit creeping flows of pinned and poorly wetting liquids.

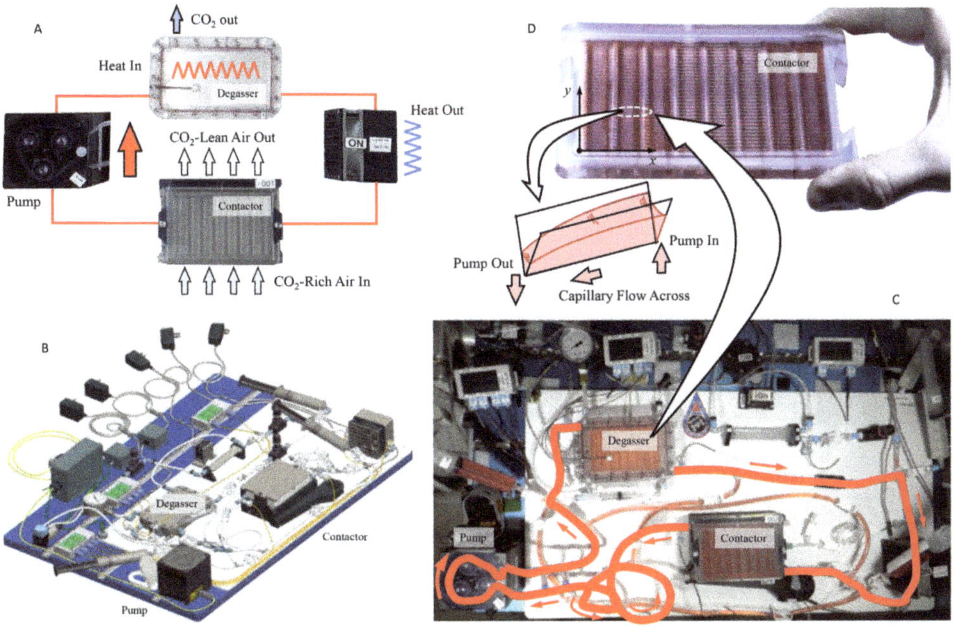

Figure 12.6. (A) A simplified schematic of the analogous low-g CO_2 scrubbing cycle; (B) a solid-model perspective view of the CVS hardware; (C) a top view image of the CVS on the ISS with the flow path and direction highlighted; (D) a magnified image of the 256-channel parallel open wedge contactor with a further magnified schematic of a single open wedge flow. Note that C labels the primary axes of the crew-induced perturbations for dynamic device stability tests. Figure courtesy of M M Weislogel.

The successful CSELS demonstrations on the ISS led to the CapiSorb Visible System (CVS) experiment conducted on ISS in 2023 [33, 34], which employed serial 256-channel parallel open wedge channel contactor and degasser operating with a viscous ammine ersatz liquid. The system included a heated sealed degasser, an open humidified contactor, and a capillary condensing heat exchanger and performed dozens of diagnostic measures: system temperatures, pressures, flow rates, condensation rates, fluid volume distributions, valve configurations, and more (figure 12.6). Over 20 h of flight footage was recorded and $\sim 10^5$ data points were collected. In the end, with the system operating at highest flow rate ($\sim$10 ml min^{-1}), the crew demonstrated the stability of the open wedge channel network by manually imparting perturbations of increasing intensity in the x-, y-, z-, r-, and ϕ-directions, increasing their frequency and amplitude until a single 3 mm droplet detached from the contactor. The deflections were digitized and twice differentiated to find the stability of the open fluidic devices at up to $2g_o$ at 4.6 Hz, which exceeds the standard $\sim 10^{-2}g_o$ requirement by a comfortable margin. For the CVS, the dimensions of the open wedges in the contactor (120 cm^2) and degasser (120 cm^2) are identical: 5.6 mm deep, 16 mm long; the included wedge angle was 24° and the temperature-dependent liquid viscosity was 4 cP $\leqslant \mu \leqslant$ 42 cP. The CVS represented a 1/40th-scale CO_2 scrubber for a crew of four with a margin. This approach is stable, simple, and quiet; it requires low volume, limited controls, and moving parts. The open-capillary fluidic channels establish effective steady-flowing bubble-separating thin films in a low-g environment by replacing the role of gravity with the combined effects of surface tension, wetting, and geometry. This technology is under consideration by NASA for further development as a massively parallelized array of open-capillary fluidic channels flowing a CO_2-hungry liquid sorbent.

12.4.3 Plant watering

Growing plants aboard spacecraft is attractive for a variety of space exploration reasons including food production, air revitalization, water purification, and the home-harkening activity it provides for the crew [35]. In fact, without crew involvement, many of the plant growth efforts in space would result in suboptimal results. This is because without gravity, wicking O_2-depleted nutrient-rich water drowns the germinating and sprouting plant, establishing rather hypoxic conditions that can turn acidic as evaporation concentrates nutrients in unearthly ways. Like the gardener on Earth, it is the crew that intervenes to make unanticipated adjustments to water levels, liberate engulfed stems and leaves, prune blight, harvest, etc. Nonetheless, efforts to develop increasingly automated plant growth facilities aboard spacecraft continue [36]. Soilless systems offer potentially cleaner, lower-weight, higher-performance alternatives to traditional substrate (i.e. soil) growth media aboard spacecraft. Hydroponics is one such method for highly autonomous terrestrial growth facilities, but how does one keep the liquid 'down' so that the plants can grow 'up'?

Beginning in 2019, the Plant Water Management (PWM) experiments aboard the ISS have provided technology demonstrations of capillary-based water delivery methods for low-g hydroponic plant growth facilities in space [37]. The methods exploited to date by the PWM experiments are largely those of large open-capillary fluidic channels (figure 12.7), which, as we have seen, provide a passive bubble-separating mechanism despite the partially wetting, partially pinned, highly hysteretic, and highly variable wetting conditions due to nutrient concentration, particulates, biofilms, and other contaminates. Rogue bubbles due to degassing, biochemical processes, bubble ingestion, and aeration are always of concern in practical systems and the open wedge geometry combined with the plant root/stem region provides a passive means of bubble separation in the channel. In the specific case of passive aeration, continuous fresh air bubble streams of a desired size and frequency may be introduced downstream from the channels. These bubbles aerate the solution as they flow along the tubing harness, pass through the pump, and are largely separated by passive inline bubble diverters that send 100% of the liquid to the lower inlet of the channel and a concentrated bubbly two-phase flow to the upper inlet of the channel where the open wedge channel efficiently confines, merges, and coalesces them upstream from the first plant. Again, the role of gravity is replaced by the open wedge channel.

Thus far, the PWM-hydroponic approach has proven successful in providing stable contained flows despite the open-capillary fluidic nature of the channel. From a NASA spacecraft safety perspective, the level of stability achieved is now

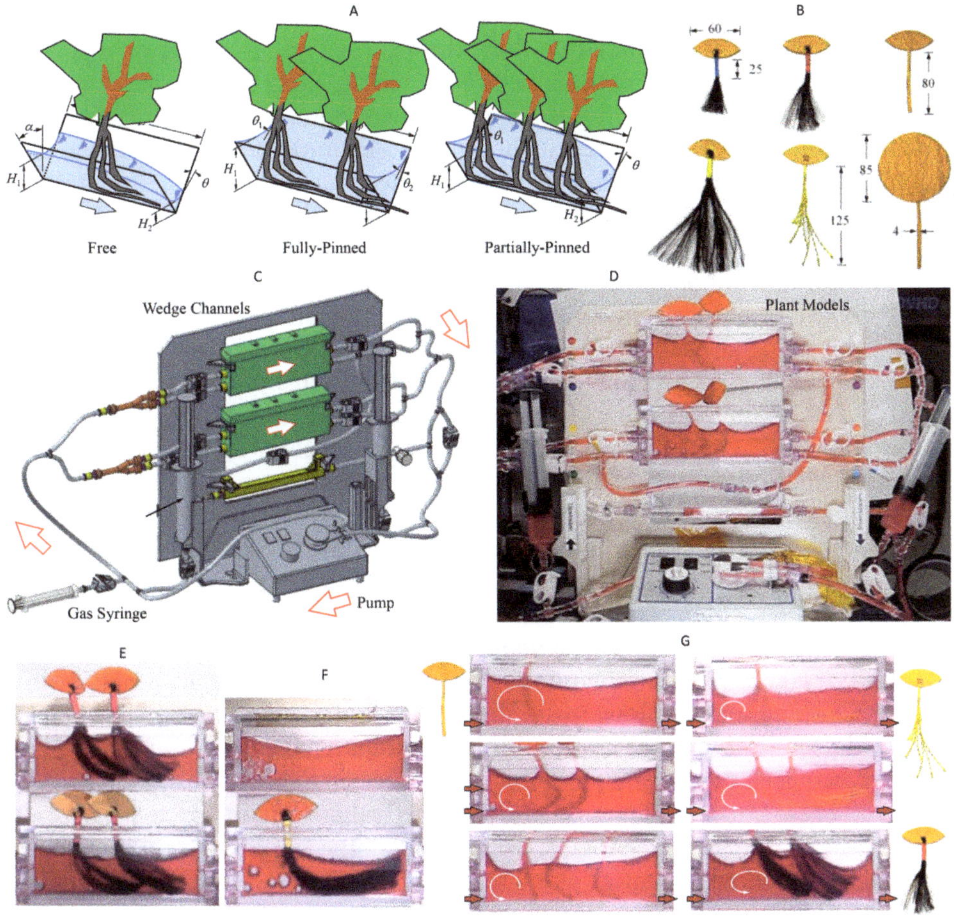

Figure 12.7. (A) Sketches of plant loadings and contact line boundary conditions for large open wedge channels for low-g hydroponics; (B) plant models with dimensions in mm used in the PWM technology demonstration on the ISS; (C) a solid model of the PWM experiment; (D) an image of the PWM experiment in operation on the ISS in two-channel parallel mode with two plants per channel; (E) steady states of the single-phase flow; (F) steady states with passive bubble phase separations for no plants and one plant; (G) steady states for various plant obstructions indicating plant type, active channel inlets and outlets, and strong recirculation zones. Figure courtesy of M M Weislogel.

considered at best a level of containment and at worst a level of control, both of which greatly simplify designs that incorporate them. The bubble separations are passive, and the combination of contact-line pinning, large contact angles, and large contact-angle hysteresis add surprising stability to the system as well as margin to the flow performance. Further demonstrations are planned on the ISS for the germination stage, redundant passive bubble separation devices, and the use of real plants.

The challenges of droplet ejection due to bubble coalescence in the channels are addressed with robust regenerable superhydrophobic lids [38]. The lids are not airtight and, as discussed above, introduce the likelihood of an additional container length scale should the liquid overflow the channel pinning edges. The superhydrophobic substrates have been demonstrated in flight [39], where impacting contaminated aqueous droplets simply rebounded from the substrate back to the bulk liquid where they re-adhered without loss. The long-term performance of this approach is yet to be established, but the ground tests carried out to date suggest that months (i.e. plant timescales) of pristine superhydrophobic function are possible before the substrates need to be either regenerated (washed) or replaced.

Further complications to such a large-length-scale application of open-capillary fluidics aboard spacecraft include changes in channel flow characteristics with changes in root zone geometry. The current channel dimensions are as large as 80 mm deep and 160 mm long and the included wedge angle is 20° ($\approx$ 100 ml capacity). But plant root zone volumes can occupy a significant portion of the channel volume. The fine hydroponic root structure develops along paths of optimal water, nutrient, and oxygen availability. At typical flow rates of <10 ml s^{-1}, ample inertia is present in such flows where $Re \leqslant 1000$, but this is not enough to prevent root development both up- and downstream. The wildcard of plant root response to the low-g environment is plant specific and can be expected to play a major role in the open-capillary channel solution. It could harm or aid channel performance. Other challenges include: plant root geometries that trap merging bubbles that grow 'upwards,' thereby displacing young roots/plants out of the channel; foaming surfactant contaminants; flow instabilities in parallel inertial–capillary channel flows, leading to bubble ingestion in underfilled channels and liquid embolisms in overfilled channels; flow imbalances for parallel systems due to variable biofouling; and more. These concerns must be resolved by further tests in space.

In summary, for single channels, we note that the liquid is stable in all plant models tested to date, that the plant root obstructions to the open-channel flow increase the entrance region recirculation zones, decrease the flow rate limit of the channel, and can cause local depinning wherever roots contact the pinning edge. For parallel flows, a stability limit is observed for increased flow rate and flow resistance imbalance. There is always a local, though stable, distortion of the free surface at the stem.

12.4.4 Spacecraft WetLabs

NASA and its ISS partners, including commercial space partners, are increasingly interested in microelectromechanical systems (MEMS), nanoelectromechanical systems (NEMS), lab-on-a-chip (LOC), organ-on-a-chip (OOC), and other small-scale fluidic systems in space for the development of remote sample collection and processing, point-of-care (POC) diagnostics, biomaterials research for terrestrial applications, and more [40]. Current terrestrial open-capillary fluidic solutions can be directly applied, although some additional considerations are necessary when free microbubbles and droplets are present, as discussed above. New low-g applications of open-capillary fluidics technologies are possible for systems of significantly larger capillary length scales. Such systems are capable of safe and stable operations and offer the promise of dramatic increases in liquid volumes, flow velocities, and flow rates in addition to improvements in test sample throughput, test efficiency, cost, and schedules.

12.5 Conclusions

Open-capillary fluidics solutions are as valued in space as they are on Earth. Numerous applications for open microfluidics solutions need not consider the impacts of microgravity unless free bubbles and droplets are present. In such cases, special considerations must be made to ensure that, say, bubble separations and/or droplet collections proceed as desired. Open-capillary macrofluidics, herein referred to as microgravity or low-g capillary fluidics, or just 'capillary fluidics,' require a change in mindset for the terrestrial open-capillary fluidics designer. One must appreciate anew: (1) the unexpected and potentially ruinous presence of large rogue bubbles and drops, (2) the unearthly presence of significant and lingering inertia in a spontaneous capillary-driven flow, (3) the complexity and metastability of large interfaces serving as reservoirs for smaller-scale open-capillary fluidic phenomena, (4) large-length-scale rogue and planned bubble separation methods for various gas sources (gas producing reactions, degassing, outgassing, phase change heat transfer, etc.), (5) the benefits and pitfalls of perfectly wetting, poorly wetting, and non-wetting systems over large length scales, and (6) the practical size limits of capillary systems in a variable-acceleration environment, say $\pm 10^{-2} g_o$.

It is perhaps strange to note that open-capillary fluidic solutions aboard spacecraft could be so large as to include passive bubble phase separation, CO_2 scrubbing, and plant watering, but it is the mere satisfaction of the condition $Bo \ll 1$ that makes such processes possible at scale. Many more nontraditional open-capillary fluidic solutions are anticipated for spacecraft, extending further into aqueous solution processing, biosample collection and processing, water recycling, and even the domains of propellants, storable fuels, cryogens, coolants, and others.

References

[1] Collicott S H and Weislogel M M 2004 Computation of capillary instabilities using Surface Evolver *AIAA J.* **42** 289–95

[2] Castro-Wallace S L, Chiu C Y, John K K, Stahl S E, Rubins K H, McIntyre A B R and Dworkin J P *et al* 2017 Nanopore DNA sequencing and genome assembly on the International Space Station *Sci. Rep.* **7** 1–12

[3] McIntyre A B R *et al* 2016 Nanopore sequencing in microgravity *Microgravity* **2** 16035

[4] Burton A S, Stahl S E, John K K, Jain M, Juul S, Turner D J and Harrington E D *et al* 2020 Off earth identification of bacterial populations using 16S rDNA nanopore sequencing *Genes* **11** 76

[5] Weislogel M M, Graf J C, Wollman A P, Turner C C, Cardin K J T, Torres L J, Goodman J E and Buchli J C 2022 How advances in low-g plumbing enable space exploration *NPJ Microgravity* **8** 16

[6] Chan K, Arumugam A, Markham C, Jenson R, Wu H-W and Wong S 2023 The Development of a 3D printer-inspired, microgravity-compatible sample preparation device for future use inside the International Space Station *Micromachines* **14** 937

[7] Schneider W, Perry J, Broyan J, Macatangay A, McKinley M, Meyer C, Owens A, Toomarian N and Gatens R 2020 NASA environmental control and life support technology development and maturation for exploration: 2019 to 2020 overview *2020 International Conference on Environmental Systems* (Emmaus, PA: The International Conference on Environmental Systems) https://hdl.handle.net/2346/86395 ICES-2020-200

[8] Nijhuis J, Schmidt S, Nghiep Tran N and Hessel V 2021 Microfluidics and macrofluidics in space: ISS-proven fluidic transport and handling concepts *Front. Space Technol.* **2** 779696

[9] Concus P, Finn R and Weislogel M M 1999 Capillary surfaces in an exotic container: results from space experiments *J. Fluid Mech.* **394** 119–35

[10] Concus P, Finn R and Weislogel R M M 2000 Measurement of critical contact angle in a microgravity experiment *Exp. Fluids* **28** 197–205

[11] Finn R 1999 Capillary surface interfaces *Not. Am. Math. Soc.* **46** 770–81

[12] Weislogel M M, Chen Y, Masica W J, Kohl F J and Green R D 2016 Fifty-plus-year postflight analysis of first fluid experiment aboard a spacecraft *AIAA J.* **11** 4042–52

[13] Brakke K A 1992 The surface evolver *Exp. Math.* **1** 141–65

[14] Chen Y, Schaeffer B M, Weislogel M M and Zimmerli G A 2011 Introducing SEFIT: surface evolver fluid interface tool for studying capillary surfaces *49th AIAA Aerospace Sciences Meeting(4–7 January 2011)* (Reston, VA: American Institute of Aeronautics and Astronautics)

[15] Chen Y, Jenson R, Weislogel M and Collicott S 2008 Capillary wetting analysis of the CFE-Vane Gap Geometry *46th AIAA Aerospace Sci. Meeting and Exhibit (Reno, Nevada, Jan. 7–10 2008)* (Reston, VA: American Institute of Aeronautics and Astronautics) AIAA-2008–0817

[16] Concus P and Finn R 1969 On the behavior of a capillary free surface in a wedge *Proc. Natl Acad. Sci. USA* **63** 292–9

[17] Weislogel M M 2012 Compound capillary rise *J. Fluid Mech.* **709** 622–47

[18] Chen Y, Tavan Y, Weislogel N and M M 2012 A mean curvature model for compound capillary flows in asymmetric containers and conduits *Phys. Fluids* **24** 082111 1–16

[19] Halvorson T 2005 ISS Oxygen Generator Fails for Good, Station Managers (http://space.com/1079-iss-oxygen-generator-fails-good-station-managers.html) [accessed: March 7, 2024]

[20] Conrath M, Canfield P J, Bronowicki P M, Dreyer M E, Weislogel M M and Grah A 2013 Capillary channel flow experiments aboard the International Space Station *Phys. Rev.* E **88** 063009

[21] Jenson R M, Wollman A P, Weislogel M M, Sharp L, Green R, Canfield P J, Klatte J and Dreyer M E 2014 Passive phase separation of microgravity bubbly flows using conduit geometry *Int. J. Multiphase Flow* **65** 68–81

[22] Weislogel M M *et al* 2015 Capillary Channel Flow (CCF) EU2-02 on the International Space Station (ISS): an experimental investigation of passive bubble separations in an open capillary channel *NASA/TM-2015-218720* 20160001341 NASA https://ntrs.nasa.gov/citations/20160001341 68

[23] Metz T, Paust N, Zengerle R and Koltay P 2010 Capillary driven movement of gas bubbles in tapered structures *Microfluid. Nanofluid.* **9** 341–55

[24] Reyssat E 2014 Drops and bubbles in wedges *J. Fluid Mech.* **748** 641–62

[25] Wollman A and Weislogel M M 2013 New investigations in capillary fluidics using a drop tower *Exp. Fluids* **54** 1499

[26] Marchetta J G and Benedetti R H 2010 Simulation of jet-induced geysers in reduced gravity *Microgravity Sci. Technol.* **22** 7–16

[27] Weislogel M M, Chen Y, Collicott S H, Bunnell C T, Green R D and Bohman D Y 2009 *AIAA-2009-0615, 47th AIAA Aerospace Sciences Meeting More handheld fluid interface experiments for the Int. Space Station (CFE-2) (Orlando, January 5–8, 2009)* (Reston, VA: American Institute of Aeronautics and Astronautics)
[28] Weislogel M M, Baker J A and Jenson R M 2011 Quasi-steady capillarity-driven flows in slender containers with interior edges *J. Fluid Mech.* **685** 271–305
[29] McCraney J, Bostwick J, Weislogel M and Steen P 2023 Bubble migration in containers with interior corners under microgravity conditions *Exp. Fluids* **64** 140 (11 pages)
[30] McCraney J, Weislogel M M and Steen P 2022 Capillary flow experiments conducted aboard the international space station: experiments and simulations *Microgravity Sci. Technol.* **34** 63
[31] Viestenz K J, Jenson R M, Weislogel M M and Sargusingh M J 2018 *48th Int. Conf. on Environmental Systems, ICES-2018-241 Capillary Structures for Exploration Life Support ISS Experiment Kit (Albuquerque, New Mexico, 8–12 July 2018)* (Emmaus, PA: The International Conference on Environmental Systems) ICES-2018-241 http://hdl.handle.net/2346/74194
[32] Weislogel M M 2003 Some analytical tools for fluids management in space: isothermal capillary flows along interior corners *Adv. Space Res.* **32** 163–70
[33] Weislogel M M *et al* 2024 Capillary fluidic CO_2 scrubbing aboard spacecraft: the CVS demonstration on ISS: part I overview *54th Int. Conf. on Environ. Systems, ICES-2024-113,* 15 pages (in review)
[34] Weislogel M M *et al* 2024 Capillary fluidic CO_2 scrubbing aboard spacecraft: the CVS demonstration on ISS: part II results *54th Int. Conf. on Environ. Systems, ICES-2024-114,* 15 pages (in review)
[35] Poulet L, Engeling K and Hatch T *et al* 2022 Large-scale crop production for the moon and mars: current gaps and future perspectives *Front. Astronomy Space Sci.* **8** 733944
[36] Anderson M S, Motil B J, Barta D J, Fritsche R F, Massa G D, Quincy C D, Romeyn M W, Wheeler R M and Hanford A J 2017 *Key gaps for enabling plant growth in future missions AIAA SPACE and Astronautics Forum and Exposition (Orlando, FL, 12–14 Sep 2017)* (Reston, VA: American Institute of Aeronautics and Astronautics) AIAA-2017-5142
[37] Wasserman M, Weislogel M, Torres L, Mungin R, Hatch T and McQuillen J 2022 *51st Int. Conf. on Environ. Systems, ICES-2022-012 Plant water management experiments: hydroponics 3 & 4 (St. Paul, MN, 10–14 July 2022)* (Emmaus, PA: The International Conference on Environmental Systems) https://hdl.handle.net/2346/89564
[38] Rasheed R and Weislogel M 2019 *49th Int. Conf. on Environmental Systems (ICES) The unrealized potential of superhydrophobic substrates in advanced life support systems (Boston, 7–11 July)* (Emmaus, PA: The International Conference on Environmental Systems) https://hdl.handle.net/2346/84446 ICES-2019-224
[39] Weislogel M M, Jenson R, Krishcko O, Torres L, Naids A, Graf J and Pettit D 2023 *52nd Int. Conf. on Environ. Systems The collapsible contingency urinal (CCU) for spacecraft (Calgary, 16–20 July 2023)* (Emmaus, PA: The International Conference on Environmental Systems) https://hdl.handle.net/2346/94470 ICES-2023-008
[40] National Research Council 2011 Applied physical sciences *Recapturing a Future for Space Exploration: Life and Physical Sciences Research for a New Era* (Washington, DC: National Academic Press) ch 9, (Decadal Survey)

IOP Publishing

Open-Channel Microfluidics (Second Edition)

Fundamentals and applications

Jean Berthier, Ashleigh B Theberge and Erwin Berthier

Chapter 13

Epilogue

This book was dedicated to the study of open-channel microfluidics and focused on their static and dynamic aspects. We first presented the general condition for spontaneous capillary flow (SCF) in open channels of arbitrary cross-section, showing that a generalized Cassie angle of less than 90° is the condition for SCF. Many different examples corresponding to different channel morphologies were presented. However, it was demonstrated that the dynamics of these capillary flows depends on the average friction length, which has the dimension of length and, combined with the average velocity, determines the shear rate at the wall. In the case of a uniform cross-section, the Washburn dependency of the travel distance on time, i.e. $z \sim t$ 1/2 or $t \sim z^2$, is respected, but the Washburn hydraulic radius must be replaced by the average friction length and the contact angle must be replaced by the generalized Cassie angle (figure 13.1). In the more complicated case of a network, the 'Washburn-like' relation is more complex and the relation between time and travel distance is $t \sim a\, z + b\, z^2$. In fact, the balance between capillary force and wall friction is always respected—after the evanescent inertial regime at the beginning—but the wall friction includes all the different sections of the channel or network. Hence, the Lucas–Washburn–Rideal law must be revisited in order to take account of the whole geometry between the inlet and the advancing meniscus [1].

From an applicative point of view, the configuration of open devices, compared to closed or semi-open microfluidic systems, provides numerous advantages, including: (1) simplification of the manufacturing techniques used (milling, etching, thermal embossing, and 3D printing can be used to fabricate the microcomponent with no need to seal a cover); (2) the easy implementation of surface modifications, such as hydrophilic functionalization; (3) the elimination of failures resulting from bubble formation; (4) and most importantly, the unique ability to access the fluid at any point in the channel in order to input or sample fluid [2–5]. These advantages are schematically shown in figure 13.2[6].

In conclusion, open microfluidics is a burgeoning field on a trajectory of continued growth. The number of publications related to open microfluidics only—not counting

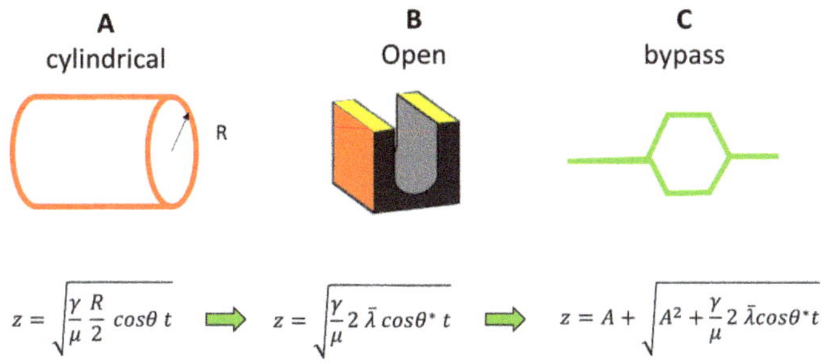

Figure 13.1. Generalizations of the Lucas–Washburn–Rideal law from a cylindrical tube to open channels and networks.

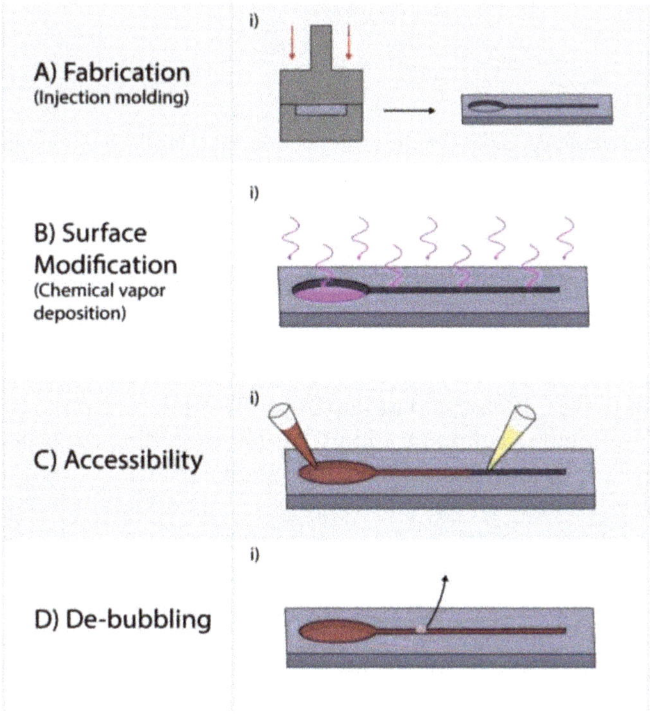

Figure 13.2. (A) In open microfluidics, devices can be ready for use directly after molding or milling, whereas closed microfluidics requires a sensitive bonding step. (B) Surface treatments in open microfluidics can be applied evenly to all surfaces of the channel. (C) In open microfluidic channels, fluids and contents in the channel can be added or sampled at any location along the channel. (D) In open microfluidics, bubbles that form or are added into the channel can be evacuated at any point along the channel length. Reprinted with permission from [6]. Copyright (2019) American Chemical Society.

superhydrophobic and superhydrophobic surfaces—is constantly growing, as shown in figure 13.3.

The ability to rapidly design and fabricate devices using 3D printing and micromilling facilitates prototyping and development in academic research labs. Open

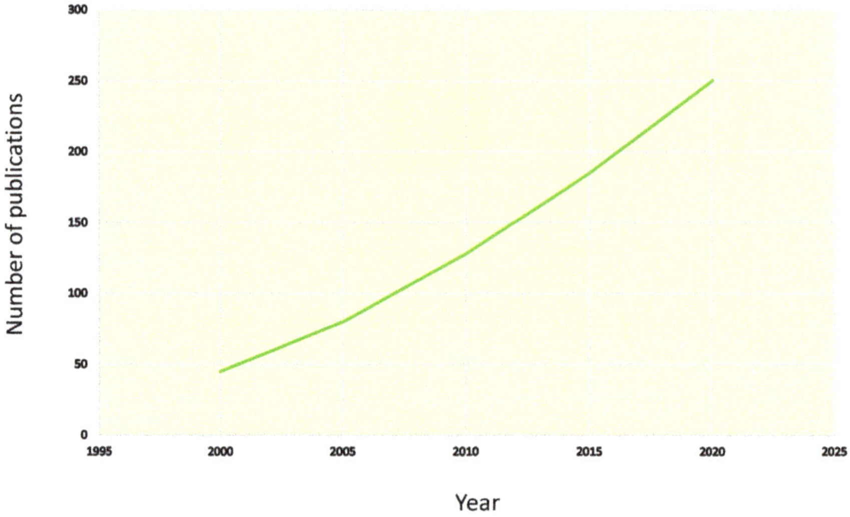

Figure 13.3. The number of publications on the subject of open microfluidics by year. Note that there are still more publications on the topics of superhydrophilic surfaces and superhydrophobic surfaces, which can be considered to be at the boundaries of OM.

microfluidic devices developed using 3D printing and micromilling are then directly injection moldable, allowing them to scale up to larger batch sizes with excellent fidelity across replicates, a key consideration when producing devices for biological and clinical research. The advent of rapid injection molding companies has brought mold costs down to the $1000–5000 range (in contrast to costs in the tens of thousands of dollars for traditional injection molding), further extending the accessibility of open microfluidic device production [5–7]. Simple design principles, ease of prototyping, and the ability to ultimately manufacture open microfluidic devices at scale have resulted in a low barrier to entry for commercial enterprises—both in biomedical start-up companies and large-scale companies and manufacturers.

There are numerous exciting research directions in the field of open microfluidics, and several new publications have come out even in the course of writing this book. Future technological additions to the 'open microfluidic toolbox' include developing more advanced valving methods and directional open capillary flows and enhancing capillary flow through the use of actuation and pumping methods. Biphasic open microfluidic droplet platforms have just recently been conceived, and future work will bring many of the advantages and capabilities of closed-channel droplet-based microfluidics to open systems. Open microfluidic devices with more than one air–liquid interface allow direct access to fluids in multiple directions, offering exciting possibilities to integrate pipetting systems and automation based on robotics. Similarly, the ability to create air–liquid interfaces in multiple directions presents new directions for dynamic and reconfigurable device assemblies fabricated by advanced manufacturing methods such as 3D printing. Finally, advances in our fundamental understanding of the physics of fluid flow in open systems have the potential to spawn new domains and applications in open microfluidics.

References

[1] Berthier J, Gosselin D and Berthier E 2015 A generalization of the Lucas–Washburn–Rideal law to composite microchannels of arbitrary cross section *Microfluid. Nanofluidics* **19** 497–507

[2] Casavant B P *et al* 2013 Suspended microfluidics *Proc. Natl Acad. Sci. USA* **110** 10111–6

[3] Lee J J, Berthier J, Brakke K A, Dostie A M, Theberge A B and Berthier E 2018 Droplet behavior in open biphasic microfluidics *Langmuir* **34** 5358–66

[4] Berry S B, Lee J J, Berthier J, Berthier E and Theberge A B 2019 Droplet incubation and splitting in open microfluidic channels *Anal. Methods* **11** 4528

[5] Oliveira N M, Vilabril S, Oliveira M B, Reis R L and Mano J F 2019 Recent advances on open fluidic systems for biomedical applications: a review *Mater. Sci. Eng.* C **97** 851–63

[6] Berthier E, Dostie A M, Lee U N, Berthier J and Theberge A B 2019 Open microfluidic capillary systems *Anal. Chem.* **91** 8739–50

[7] Lee U N, Berthier J, Yu J, Berthier E and Theberge A B 2019 Stable biphasic interfaces for open microfluidic platforms *Biomed. Microdev.* **21** 16

www.ingramcontent.com/pod-product-compliance
Ingram Content Group UK Ltd.
Pitfield, Milton Keynes, MK11 3LW, UK
UKHW051903100926
8556IPUK00005B/747

9 780750 355056